P9-EDP-933

Pollen and Spores

Proceedings of a Joint International Symposium of the Linnean Society and the Systematics Association held in London, March 1990

The Systematics Association
Special Volume No. 44

Pollen and Spores

Patterns of Diversification

Edited by

S. BLACKMORE
Keeper of Botany

and

S.H. BARNES
Head of Electron Microscopy and Mineral Analysis

The Natural History Museum,
London

Published for the SYSTEMATICS ASSOCIATION by
CLARENDON PRESS · OXFORD
1991

Oxford University Press, Walton Street, Oxford OX2 6DP
Oxford New York Toronto
Delhi Bombay Calcutta Madras Karachi
Petaling Jaya Singapore Hong Kong Tokyo
Nairobi Dar es Salaam Cape Town
Melbourne Auckland
and associated companies in
Berlin Ibadan

Oxford is a trade mark of Oxford University Press

Published in the United States
by Oxford University Press, New York

A catalogue record for this book is available from the British Library

Library of Congress Cataloging in Publication Data
Pollen and spores : patterns of diversification / edited by S. Blackmore and S. H. Barnes.
p. cm. — (The Systematics Association special volume; no. 44)
Based on papers from an international symposium held at the Linnean Society of London and the Natural History Museum, Mar. 28–30, 1990 which was organized by the Palynology Specialist Group of the Linnean Society and the Systematics Association.
1. Pollen — Congresses. 2. Spores (Botany) — Congresses. 3. Pollen, Fossil — Congresses. 4. Spores (Botany), Fossil — Congresses. 5. Palynology — Congresses. 6. Plants — Evolution — Congresses. I. Blackmore, Stephen. II. Barnes, Susan H. III. Systematics Association. IV. Linnean Society of London. Palynology Specialist Group. V. Series.
QK658.P63 1991 582. 13′04463 — dc20 91-15286
ISBN 0-19-857746-X

Set by
Colset Pte Ltd, Singapore
Printed in Great Britain by
St. Edmundsbury Press,
Bury St. Edmunds,
Suffolk

Preface

This volume is based on the papers presented at an international symposium entitled 'Pollen and Spores: Patterns of Diversification' held at the Linnean Society of London and the Natural History Museum on 28–30 March 1990. This was the third international symposium to be organized by the Palynology Specialist Group of the Linnean Society, and the second to be jointly convened with the Systematics Association.

The systematic and palaeontological applications of palynology exploit the complexity and diversity of organization present in pollen grains and spores. The two earlier symposia, 'The Evolutionary Significance of the Exine' (held in 1974) and 'Pollen and Spores: Form and Function' (held in 1985) had demonstrated the exceptional opportunities afforded by pollen grains and spores for bringing together findings from studies of both Recent and fossil material. To capitalize on these opportunities, the meeting focused on the pathways of diversification through which the enormous diversity of pollen grains and spores has arisen. The three principal issues addressed were: the evidence provided by the fossil record, the contribution of ontogenetic data, and methods of systematic analysis.

We thank Professor W.G. Chaloner for his opening remarks and for making everyone welcome to the Linnean Society. We are grateful to all those who presented papers and posters, and the chairmen of the six sessions of the symposium. We are thankful to Dr N.R. Chalmers, Director of the Natural History Museum, for making the facilities of the Museum available, and for welcoming the participants to the Museum on the final day of the symposium.

We are grateful to the many other people who contributed to the success of the symposium. We thank the Executive Secretary of the Linnean Society, Dr J.C. Marsden and his colleagues, especially Ms M. Baird and Miss G.L. Douglas, for their energetic support during the period of preparation for the meeting and the two days it was held at Burlington House. We are also grateful to our colleagues from the Natural History Museum, Mr J. Benfield, Mr K. Jones, Mr P.J. Stafford, Dr R. Scotland, and Mr D.M. Williams who helped with many practicalities, including the catering and the projection of slides. We thank Dr D.J. Galloway and Dr C.E. Jarvis for their valuable assistance with the catering arrangements.

The symposium would not have taken place without the financial backing of the Linnean Society, the Systematics Association, and the Royal Society; we are grateful for their support.

London
October 1990

S. B.
S. H. B.

Contents

Contributors

S.H. BARNES
The Natural History Museum, Cromwell Road, London, SW7 5BD, UK.

S. BLACKMORE
The Natural History Museum, Cromwell Road, London, SW7 5BD, UK.

R.C. BROWN
Department of Biology, University of Southwestern Louisiana, Lafayette LA 70504-2451, USA.

W.G. CHALONER
Biology Department, Huntersdale, Royal Holloway and Bedford New College, Egham Hill, Egham, Surrey, TW20 0EX, UK.

M.E. COLLINSON
Department of Geology, Royal Holloway and Bedford New College, Egham Hill, Egham, Surrey, TW20 0EX, UK.

P.A. COX
Department of Botany and Range Science, Brigham Young University, Provo, Utah 84602, USA.

P.R. CRANE
Field Museum of Natural History, Roosevelt Road at Lake Shore Drive, Chicago, Illinois 60605, USA.

S. CROMAR
Department of Mathematics, Brigham Young University, Provo, Utah 84602, USA.

J.A. DOYLE
Department of Botany, University of California, Davis, CA 95616, USA.

S.C.DUCKER
School of Botany, Univeristy of Melbourne, Parksville, Victoria 3052, Australia.

D. EDWARDS
Department of Geology, University of Wales College of Cardiff, P.O. Box 914, Cardiff, CF1 3YE, UK.

G.A. El-GHAZALY
Naturhistoriska Riksmuseet, Palynologiska Laboratoriet, 104–05 Stockholm, Sweden.

U. FANNING
Department of Geology, University of Wales College of Cardiff, P.O. Box 914, Cardiff, CF1 3YE, UK.

I.K. FERGUSON
Royal Botanic Gardens, Kew, Richmond, Surrey, TW9 3AE, UK.

G.G. FRANCHI
Dipartimento di Farmaco Chimico Tecnologico, Università di Siena, Via P.A. Mattioli 4, 53100 Siena, Italy.

E.M. FRIIS
Department of Palaeobotany, Swedish Museum of Natural History, Box 50007, S-10405 Stockholm, Sweden.

N.I. GABARAYEVA
Komarov Botanical Institute, Professor Popova 2, 157022, Leningrad, USSR.

J. GRAY
Department of Biology, University of Oregon, Eugene, Oregon 97403, USA.

M.M. HARLEY
Royal Botanic Gardens, Kew, Richmond, Surrey, TW9 3AE, UK.

A.R. HEMSLEY
Royal Holloway and Bedford New Colleges, Huntersdale, Callow Hill, Virginia Water, Surrey, GU25 4LN, UK.

J. HESLOP-HARRISON
University College of Wales, Welsh Plant Breeding Station, Plas Gogerddan, Aberystwyth, SY23 3EB, UK.

Y. HESLOP-HARRISON
University College of Wales, Welsh Plant Breeding Station, Plas Gogerddan, Aberystwyth, SY23 3EB, UK.

C.L. HOTTON
Department of Biological Sciences, SUNY, Binghampton, NY 13901, USA.

T. JARVIS
Department of Mathematics, Brigham Young University, Provo, Utah 84602, USA.

R.B. KNOX
School of Botany, University of Melbourne, Parkville, Victoria 3052, Australia.

M.H. KURMANN
Royal Botanic Gardens, Kew, Richmond, Surrey, TW9 3AE, UK.

B.E. LEMMON
Department of Biology, University of Southwestern Louisiana, Lafayette LA 70504-2451, USA.

S. NILSSON
Naturhistoriska Riksmuseet, Palynologiska Laboratoriet, 104–05 Stockholm, Sweden.

E. PACINI
Dipartimento di Biologia Ambientale, Università di Siena, Via P. A. Mattioli 4, 53100 Siena, Italy.

K.R. PEDERSEN
Geological Institute, Århus University, Universitetsparken, DK–8000 Århus C, Denmark.

J.B. RICHARDSON
The Natural History Museum, Cromwell Road, London, SW7 5BD, UK.

R.W. SCOTLAND
Department of Plant Sciences, University of Oxford, South Parks Road, Oxford, OX1 3RB, UK.

J.J. SKVARLA
Department of Botany and Microbiology and Oklahoma Biological Survey, University of Oklahoma, Norman, Oklahoma 73019, USA.

S.S. VANDERPOOL
Department of Biology, University of North Dakota, Grand Forks, North Dakota 58202, USA.

J.H.A. VAN KONIJNENBURG-VAN CITTERT
Laboratory of Palaeobotany and Palynology, State University of Utrecht, Heidelberglaan 23584 C5, Utrecht, The Netherlands.

G.A. VAN UFFELEN
Rijksherbarium/Hortus Botanicus, PO Box 9514, 2300 RA Leiden, The Netherlands.

E. VEZEY
Department of Botany and Microbiology, University of Oklahoma, Norman, Oklahoma 73019, USA.

M.S. ZAVADA
Department of Biology, University of Southwestern Louisiana, Lafayette, LA 70504–2451, USA.

1. Palynological diversity

STEPHEN BLACKMORE
SUSAN H. BARNES
The Natural History Museum, Cromwell Road, London

Abstract

This paper introduces the three main themes of the symposium volume by emphasizing the importance of pollen and spore diversity to plant systematics. These themes are the ontogenetic processes that give rise to morphological diversity, systematic analyses of extant plants, and the fossil record of diversification through geological time.

The first step towards explaining biological diversity is to describe and delimit it, and the second is to make meaningful comparisons. Comparative biology proceeds from the assumption of units of comparison and of characters through a procedure of analysis to the discovery of homologies which define groups. Character analysis leads to the recognition of patterns of diversity and is an essential prerequisite to any attempt at hypothesizing the causal processes underlying these patterns.

Palynological characters, studied in isolation, can generate hypotheses of homology and of the relationships between groups. However, only when integrated into broader character analyses can the systematic potential of palynological data be fully realized. Despite this, the critical assessment of key palynological characters that define major groups is of great importance in plant systematics.

Introduction

Biological diversity demands explanation. It challenges us to chart the limits of natural variation, to question the relationships between the forms we encounter, and to elucidate the processes by which diversity arises and is maintained.

The invention of the microscope revealed a new, and otherwise

Pollen and Spores (ed. S. Blackmore and S. H. Barnes), Systematics Association Special Volume No. 44, pp. 1–8, Clarendon Press, Oxford, 1991.

invisible, realm of biological diversity. This encompasses the enormously diverse groups of microscopic organisms (both prokaryotic and eukaryotic) and the finer details of macroscopic organisms, including the microscopic phases of their life cycles. Pollen grains and spores were soon found to be structurally complex and morphologically diverse, attributes that are now exploited in the science of palynology.

This symposium volume is concerned with patterns of diversification in pollen grains and spores. Three main themes concern the ontogenetic processes that give rise to morphological diversity, systematic analyses of extant plant groups, and the fossil record of diversification through geological time. As an introduction to these topics, we discuss some of the approaches that have been used in interpreting palynological diversity and comment briefly on the status of palynology within systematics.

We begin with a short consideration of biological diversity in a much wider context. Biodiversity has become a familiar word in the vocabularies of scientists, politicians, and the media. The principal reason for this has been growing concern over changes in global ecology and the loss of biological diversity (see, for example, Wilson 1988). As scientists who view the world through microscopes and whose field of vision is often restricted to a few microns, is biodiversity of direct concern to palynologists? We believe it is.

One of the most frequently voiced questions in relation to biodiversity (Wilson 1985; May 1988) is, how many species are there in the world? Recent answers range from 30 million to 80 million species (see, for example, Stork and Gaston 1990). A related question concerns the nature of the processes that give rise to, and maintain, biological diversity. An important component of the answer derives from the microscopic realm, which provides access to the processes of differentiation.

Ontogeny is one of the three principal themes of this symposium volume. Morphogenetic studies provide an opportunity to observe the processes by which diversity arises during the course of development. Indeed, it is only during ontogeny that transformations between one form and another can ever be directly observed in biology (Humphries 1988). Furthermore, developmental observations greatly enhance our ability to compare and interpret recent and fossil forms. Pollen grains and spores are ideal systems in which to study the establishment of such aspects of form generation as cellular polarity and the origin of complex cell walls. Heslop-Harrison (1972) used the highly appropriate expression 'morphogenesis in miniature' to encapsulate this idea. Studies of morphogenesis at the microscopic level are, of course, also relevant to the origin of form at the multicellular level, as Wolpert (1990) has recently argued.

Building on the establishment during the 1960s and 1970s of the basic principles of microsporogenesis, a more recent phase of study has emphasized the systematic and evolutionary implications (see, for example, Lugardon 1980; Guédès 1982; Blackmore and Barnes 1987; Blackmore and Crane 1988; Kurmann 1989, 1990; Blackmore and Knox 1990).

Developmental processes apart, the diversity of pollen grains and spores is merely one aspect of biological diversity in a simpler and much more direct sense. In this respect, palynological diversity has the same potential ability to contribute to the recognition and delimitation of species, and higher taxa, as any other source of comparative data. The second major component of the volume explores the contribution of palynology to systematic analyses.

Pollen grains and spores also constitute a most important source of information concerning the biological diversity of the past as recorded in their extensive fossil record. This was the third theme of the symposium. The contributions published here include discussions of some of the major events in plant evolution, ranging from spore diversification in the earliest land-colonizing plants, to the origins of heterospory and of the major groups of angiosperms.

These themes serve to emphasize the importance of palynological diversity and the considerable contribution that studies of pollen grains and spores can potentially make. This potential is not often achieved because many studies are undertaken without the framework of a valid comparative method. As a result, the value of systematic palynology cannot be articulated as clearly and forcefully as it should be. The remainder of this introduction considers this view further.

Historical approaches to palynological diversity

Comprehensive surveys of the historical development of palynology and plant reproductive biology have been given by Wodehouse (1935) and, more recently, by Ducker and Knox (1985). Consequently, few comments are required here.

Interestingly, the idea that palynological diversity could inform taxonomic decision-making dates from the early nineteenth century (Brown 1811; von Mohl 1835). Palynological characters played a prominent part in the nineteenth century classifications of such flowering plant families as Convolvulaceae (Hallier 1893), Orchidacaeae (Lindley 1852), and Acanthaceae (Lindau 1893). However, the systematic studies of pollen grains on which these classifications were based were unusual and the growth of systematic palynology has largely occurred during the present century. Rather than being systematic, most investigations of pollen grains and spores undertaken during the nineteenth century focused on development, function, or physiology (see Wodehouse 1935).

Systematic and ontogenetic studies have continued in the present century and have benefited enormously from the development of electron microscopy. The most dramatic expansion, however, has been in the use of fossil pollen grains and spores to elucidate subjects as diverse as plant evolution, vegetational history, archaeology, and stratigraphy. Palynology is now a powerful tool that can be applied to issues of global concern, such as the modelling of climatic change (for a recent example see van Campo *et al.* 1990). The demand for increasing precision in discriminating between fossil palynomorphs and determining their affinities has been an important stimulus, providing further impetus for detailed studies of extant taxa.

Recent approaches to palynological diversity

Although a few authors have discussed the potential of palynological data to contribute to higher-level systematics (see, for example, Nowicke and Skvarla 1980), most taxonomic applications have focused on relationships, below the family level, between species or genera.

A wide variety of methods have been adopted in applying palynological evidence below the familial level, and characterizing these is complex. Nevertheless, some generalizations concerning typical approaches in systematic palynology are possible. Frequently, palynological characters are regarded as subsidiary in importance to other characters, such as gross morphology. Consequently, in any particular instance, palynological data may be found either to agree with an existing classification, and so lend support to it, or not. Generally, when there is disagreement between the palynological evidence and that of other characters, palynology either gives way to the other evidence, or is considered to indicate the need for a macromorphological revision. This viewpoint apparently stems from the notion that palynological evidence is of a special or distinct nature, and should be compared with the totality of other evidence, in the form of an accepted classification of the group in question. An unfortunate consequence of this view is that palynology must inevitably be considered secondary to other data in taxonomy.

What is missing in this approach is a method for evaluating palynological characters together with other kinds of evidence (Blackmore *et al.* 1991). Ideally such a method would enable us to ask, what relationships are suggested by the palynological evidence, by the non-palynological evidence, and by a combination of both sets of data?

It is only constructive to undertake specialized palynological studies of taxa if such methods are available. Otherwise there is no mechanism for integrating the products of palynological investigations into the mainstream of plant taxonomy. Palynological information must, furthermore, be capable both of evaluation against groups and classifications proposed on the basis of other evidence, and of integration and analysis in a combined data set.

Fortunately, methods are available that can meet these requirements. In recent years, systematics has undergone a period of rapid methodological advance with the introduction of new techniques for comparing taxa and assessing the significance of individual characters. In particular, cladistic methods have led to the publication of explicit data sets and the use of repeatable analytical procedures. Although cladistic methods were initially slow to be adopted by botanists, and are still considered highly contentious by some, we believe that their merits are clear. Among zoologists such arguments are now largely over, although there is still much to be discussed, not least because of the existence of several distinct schools of cladistic practice. Two recent international symposia of the Systematics Association, one zoological, the other botanical, provide an interesting contrast in this context. In the proceedings of a meeting on tetrapod phylogeny (Benton 1988) every contributor adopted cladistic methods whereas, in contrast, only 35 per cent of contributors to a meeting on the Hamamelidae (Crane and Blackmore 1989) did so.

Explicit methodology, in which the characters present in the taxa studied are clearly defined and tabulated, and the methods of character analysis are specified, is now essential in systematics. By enabling the significance of individual characters or groups of characters to be assessed, in terms of their potential to define groups, explicit methods of character analysis have made it possible to compare the merits of competing classifications.

These advances have had a profound impact on, for example, our understanding of the relationships of angiosperms to other seed plants (Hill and Crane 1982; Crane 1985; Doyle and Donoghue 1986). Not only have such studies greatly clarified seed plant relationships, they have also highlighted the importance of a number of critical palynological characters. Doyle (1988) and Crane (1990) have provided detailed discussions of the significance of a number of these characters. These discussions clearly demonstrate that palynological characters are significant at the higher taxonomic level as well as below the family level.

Conclusions

By adopting systematic concepts and procedures now widely used in biology, palynology can make a much greater contribution to comparative biology. If this is not achieved, then the results of studies in systematic palynology are merely descriptive data. Once published, this valuable data is available for interpretation by other biologists but, without analysis, much of it will never be used. If we palynologists do not interpret our data, we must appreciate that it is unlikely that anyone else will. If the descriptive data generated by palynologists is not used in broader systematic studies, there is every likelihood that palynology will become a peripheral subject, considered to lie outside mainstream biology. This should not be so: as this symposium volume demonstrates, palynology is an exciting area of fundamental science that produces results of great systematic value.

References

Benton, M.J. (1988). *The phylogeny and classification of the tetrapods*, Vols 1 and 2. Oxford University Press, Oxford.

Blackmore, S. and Barnes, S.H. (1987). Pollen wall morphogenesis in *Tragopogon porrifolius* (Compositae: Lactuceae) and its taxonomic significance. *Review of Palaeobotany and Palynology* **52**, 233–46.

Blackmore, S. and Crane, P.R. (1988). Systematic implications of pollen and spore ontogeny. In *Ontogeny and systematics*, (ed. C.J. Humphries), pp. 83–115. Columbia University Press. New York.

Blackmore, S. and Knox, R.B. (1990). *Microspores: evolution and ontogeny*. Academic Press, London.

Blackmore, S., Scotland, R.W., and Stafford, P.J. (1991). The comparative method in palynology, in press.

Brown, R. (1811). On the Proteaceae of Jussieu. *Transactions of the Linnean Society of London* **10**, 15–226.

Crane, P.R. (1985). Phylogenetic analysis of seed plants and the origin of angiosperms. *Annals of the Missouri Botanical Garden* **72**, 716–93.

Crane, P.R. (1990). The phylogenetic context of microsporogenesis. In *Microspores: evolution and ontogeny*, (ed. S. Blackmore, and R.B. Knox), pp. 11–42. Academic Press, London.

Crane, P.R. and Blackmore, S. (1989). *Evolution, systematics and fossil history of the Hamamelidae*, Vols 1 and 2. Oxford University Press, Oxford.

Doyle, J.A. (1988). Pollen evolution in seed plants: a cladistic perspective. *Journal of Palynology* **23–4**, 7–18.

Doyle, J.A. and Donoghue, M.J. (1986). Seed plant phylogeny and the origin of angiosperms: an experimental cladistic approach. *Botanical Review* **52**, 321–431.

Ducker, S.C. and R.B. Knox, (1985). Pollen and pollination: a historical review. *Taxon* **34**, 401–9.

Guédès, M. (1982). Exine stratification, ectexine structure and angiosperm evolution. *Grana* **21**, 161–70.

Hallier, H. (1893). Versuch einer natürlichen Gliederung der Convolvulaceen auf morphologischer und anatomische Grundlage. *Botanischer Jahrbucher* **16**, 453–591.

Heslop-Harrison, J. (1972). Pattern in plant cell walls: morphogenesis in miniature. *Proceedings of the Royal Institution of Great Britain* **45**, 335–51.

Hill, C.R. and P.R. Crane. (1982). Evolutionary cladistics and the origin of angiosperms. In *Problems of phylogenetic reconstruction*, (ed. K.A. Joysey and A.E. Friday), pp. 269–361. Academic Press, London.

Humphries, C.J. (ed.) (1988). *Ontogeny and systematics*. Columbia University Press, New York.

Kurmann, M.H. (1989). Pollen wall formation in *Abies concolor* and a discussion on wall layer homologies. *Canadian Journal of Botany* **67**, 2489–504.

Kurmann, M. (1990). Exine ontogeny in conifers. In *Microspores: evolution and ontogeny*, (ed. S. Blackmore and R.B. Knox), pp. 157–72. Academic Press, London.

Lindau, G. (1893). Beiträge zur Systematik der Acanthaceen. *Botanischer Jahrbucher* **18**, 36–64.

Lindley, J. (1852). *Folia orchidacea. An enumeration of the known species of orchids*. London.

Lugardon, B. (1980). Comparison between pollen and pteridophyte spore walls. *Proceedings at the 4th International Palynological Conference, Lucknow* **1**, 199–206.

May, R.M. (1988). How many species are there on earth? *Science* **241**, 1441–9.

Nowicke, J. and Skvarla, J.J. (1980). Pollen morphology: the potential influence in higher order systematics. *Annals of the Missouri Botanical Garden* **66**, 633–700.

Stork, N. and Gaston, K. (1990). Counting species one by one. *New Scientist* **1729**, 43–7.

van Campo, E., Duplessy, J.C., Prell, W.L., Barratt, N., and Sabatier, R. (1990). Comparison of terrestrial and marine temperature estimates for the past 135 kyr of southeast Africa: a test for GCM simulations of palaeoclimate. *Nature* **348**, 209–12.

von Mohl, H. (1835). Sur la structure et les formes des grains de pollen. *Annales des Sciences Naturelles* **23**, 148–80.

Wilson, E.O. (1985). The biological diversity crisis: a challenge to science. *Issues in Science and Technology* **2**, 20–9.

Wilson, E. O. (1988). *Biodiversity*. National Academy Press, Washington, DC.

Wodehouse, R. P. (1935). *Pollen grains. Their structure, identification and significance in science and medicine*. McGraw-Hill, London.

Wolpert, L. (1990). The evolution of development. *Biological Journal of the Linnean Society* **39**, 109–24.

2. Sporogenesis in simple land plants

R.C. BROWN
B.E. LEMMON
Department of Biology, University of Southwestern Louisiana, Lafayette, USA

Abstract

Quadripartitioning of the sporocyte into a tetrad of spores during meiosis is one of the most distinct examples of precise geometrical division in the plant kingdom. The cytoskeleton is of fundamental importance in the establishment and maintenance of quadripolarity. The microtubule systems associated with establishment of polarity in sporocytes are quite different from those in mitosis of vegetative tissue where the future division plane is marked by a component of the cytoskeleton known as the preprophase band of microtubules (PPB). PPBs do not occur in meiocytes. Instead, quadripolarity in simple land plants is marked in a variety of ways. Cytoplasmic lobing of the spore mother cells of bryophytes dramatically predicts future cytokinetic planes. Quadripolarity is most elegantly marked in monoplastidic meiosis of mosses, hornworts, and the vascular cryptogams *Isoetes* and *Selaginella* by plastid polarity manifested in plastid migration and the development of a unique quadripolar microtubule system (QMS). The QMS, which develops in meiotic prophase, contributes to predictive quadripositioning of plastids and participates directly in spindle ontogeny. Plastid arrangement prior to meiosis determines future spore domains in monoplastidic sporocytes, whereas in polyplastidic sporocytes the spore nuclei play a major role in claiming cytoplasmic domains. In simple land plants, quadripolarity established early in meiotic prophase is reflected in subsequent events of sporogenesis: organelle migration and apportionment, chromosome movement, cytokinesis, initiation and pattern of the spore wall, and placement of apertures.

Pollen and Spores (ed. S. Blackmore and S. H. Barnes), Systematics Association Special Volume No. 44, pp. 9–24, Clarendon Press, Oxford, 1991.

Introduction

The production of tetrads of spores with complex protective walls is a characteristic feature of all land plants. The fundamental components of sporogenesis are the same in free-sporing plants, where the spores serve to disperse new plants, and in seed plants, where the walled microspores have evolved into pollen grains capable of dispersing male gametes. In all cases

(1) the nuclear division is meiotic and results in four nuclei containing the reduced sets of chromosomes;

(2) the cytoplasm is quadripartitioned and distributed equally to the spores/microspores of the tetrad; and

(3) a distinctively patterned spore wall containing sporopollenin (with few exceptions) develops around each spore/microspore.

The control of division plane is a key component in the developmental pathway leading to the mature spore or pollen grain. In many simple plants, the division sites for quadripartitioning are established in the early sporocyte. The control of division plane is a function of the cytokinetic apparatus. As defined by Gunning (1982), the cytokinetic apparatus is a complex cytoskeletal-based organelle system that functions both in determination of the plane of division and completion of the process of cytokinesis. Whereas the cytokinetic apparatus in vegetative growth is remarkably uniform, that of meiotic cells shows considerable variation.

In vegetative cells, the cytokinetic apparatus includes a preprophase band of microtubules (PPB) that marks the division site prior to mitosis, and a phragmoplast that is instrumental in new cell wall formation after mitosis. The PPB predicts the division site in both symmetrical and asymmetrical divisions and serves as convincing evidence that the cytoplasm is committed to divide in a specific plane before the nucleus enters prophase. Development of the PPB in the G_2 stage of interphase (Mineyuki *et al.* 1988) is the first morphological signal that a cell is preparing for mitosis.

The cytokinetic apparatus of meiotic cells is different from that of vegetative cells in that PPBs are lacking (e.g. Brown and Lemmon 1989*a*). Nevertheless, sporocytes do cleave along precisely determined planes, resulting in spore arrangements that have long been recognized as reliable taxonomic characters. Selection of the division plane in sporocytes appears to be endogenously controlled and is independent of the influence of neighbouring cells. Sporocytes are isolated by a sporocyte wall which is composed of callose in angiosperms but is quite variable in the simple land plants. This isolation of sporocytes apparently plays a key role in establishing the conditions for sporogenesis (Heslop-Harrison 1971) and sporocytes typically undergo meiosis as free cells released into the mucilaginous fluid of the sporangium.

While the cytokinetic apparatus is highly variable in sporogenesis of the major taxa of bryophytes and vascular cryptogams, the following important generalization can be made. Premeiotic establishment of division quadripolarity is typical of the simple land plants and is probably primitive; postmeiotic determination of division plane is a character typical of higher plants. In the hornworts, mosses, and hepatics (except the Marchantiidae), and in the vascular cryptogams *Isoetes* and *Selaginella*, the future division planes are marked by one or more of the following: precocious lobing of the cytoplasm, morphogenetic plastid migration, and organization of microtubule systems associated with the establishment of polarity and selection of cleavage planes.

Cytoplasmic lobing

The premeiotic sporocytes of mosses, hornworts, and hepatics (except the Marchantiidae) become clearly lobed into four spore domains. The furrows predict the future planes of spore cleavage that will occur following the second nuclear division. In *Sphagnum* (Brown *et al.* 1982) cytoplasmic lobing is less pronounced but the division sites are precisely marked by thickened infurrowings of the sporocyte wall. While the mechanism of sporocyte lobing is unknown, it appears that continued wall ingrowths at the division sites results in lobing of the cytoplasm. Microtubules have not been directly implicated in the initiation of cytoplasmic lobing in bryophytes (Brown and Lemmon 1987*a*, *b*). Likewise, in mitosis, peg-like wall ingrowth may develop in the division site after microtubules of the PPB have disappeared (Gunning 1982). Thus, in both cases the deposition of wall materials is evidence for a division site independent of microtubules.

Lobing of the cytoplasm in meiotic prophase is most extreme in hepatics of the Jungermanniidae and in mosses of the Polytrichidae, in which the cleavage furrows reach nearly to the nucleus lying in the small central portion of the sporocyte. Lobing, which begins early in prophase and results in pronounced definition of the future spores, appears to be independent of nuclear position. The sporocyte nucleus regularly undergoes a dramatic two-way migration during meiotic prophase (Brown and Lemmon 1982, 1987*a*, 1989*a*; Busby and Gunning 1988*a*). In early prophase, the nucleus moves from the centre of the sporocyte to the periphery and then in mid-prophase it returns to the centre. Although this unusual nuclear migration is seldom mentioned in meiotic studies, it is so frequently illustrated that we assume it to be a characteristic feature of sporogenesis in plants. Certainly, it is a regular component of meiosis in bryophytes and fern allies. Our examinations of large numbers of sporocytes of the mosses *Atrichum* and *Pogonatum* showed no correlation between the pattern of cytoplasmic lobing and the position of the nucleus. In its acentric position the nucleus was as often seen to lie between lobes as in one of the lobes. In the hepatic *Pellia*, Davis (1901) illustrated a nucleus migrating from an early prophasic position in one of the cytoplasmic lobes to the central portion where the spindle develops. However, as discussed later in this paper, in microsporocytes of *Selaginella* the acentric nucleus and overlying plastid establish an axis that is predictive of the future spindle which will lie perpendicular to it.

Unlike all other bryophytes, members of the Marchantiidae undergo meiosis in sporocytes that are not lobed or marked in any way that would predict the eventual cleavage planes (Farmer 1895; Schuster 1966; Brown and Lemmon 1988*b*). Rather, the spore domains appear to be determined by interaction of nuclear-based radial arrays of microtubules that are developed after the second nuclear division of meiosis. The Marchantiidae are the only bryophytes known to exhibit post-meiotic determination of cleavage plane, a character that appears to be typical of simultaneous cytokinesis in angiosperm microsporogenesis.

Monoplastidic cell division

Monoplastidic meiosis, which occurs in bryophytes (except hepatics) and the heterosporous vascular cryptogams *Isoetes* and *Selaginella*, has provided a wealth of information on the reorganization of the cytoplasm in preparation for division. Monoplastidic sporocytes exhibit the phenomenon of plastid polarity in which

(1) division and migration of the plastid in prophase establishes the cytoplasmic domains of the future spore tetrad; and
(2) microtubules organized at the plastids contribute to both spindle and cytokinetic apparatus (reviewed by Brown and Lemmon 1990).

The behaviour of the plastids in monoplastidic cell division mimics the behaviour of centrosomes in animal cells and provides a useful model for probing the little understood nature of the microtubule organizing centre (MTOC) in plant cells. Whereas the material responsible for nucleation of microtubules in cells of higher plants is thought to be associated with the nuclear envelope (reviewed by Wick and Duniec 1983; Clayton *et al.* 1985; Wick 1985*a*, *b*), microtubules in monoplastidic cells clearly emanate from the plastid envelope

(Brown and Lemmon 1987*a*, 1988*a*, 1989*a*, 1990; Busby and Gunning 1988*b*). The observation that microtubules reappear at plastids in monoplastidic sporocytes recovering from drug-induced microtubule depolymerization strengthens the concept of the plastid MTOC (Busby and Gunning 1989).

Plastids serving as MTOCs result in unique microtubule systems that make up both the cytokinetic apparatus and spindle apparatus. A quadripolar microtubule system (QMS) emanating from four plastids (or the four tips of two dividing plastids) is associated with predictive positioning of the plastids and organization of the meiotic spindle in hornworts (Brown and Lemmon, in manuscript), true mosses (Brown and Lemmon 1987*a*, *b*; Busby and Gunning 1988*a*, *b*), sphagnum mosses (Brown and Lemmon 1986), megasporocytes of *Isoetes* (Brown and Lemmon 1989*a*), and microsporocytes of *Isoetes* (this paper). A similar QMS occurs in lobed hepatic sporocytes (Farmer 1895), but it is not plastid-based; and a modified plastid-based microtubule system occurs in microsporocytes of *Selaginella* (Brown and Lemmon 1985*a*).

We have chosen sporogenesis in hornworts as a model to demonstrate development and organization of the QMS, its role in plastid division and migration, its relationship to establishment of division planes, and its transformation into the meiotic spindle. The hornworts are an interesting group of primitive land plants, having many characteristics in common with the higher green algae such as *Coleochaete*, and are the only group of land plants that exhibit monoplastidic cell division throughout their life history. Recent studies has shown that plastid migration and unique microtubule systems serve to mark the progress of prophasic reorganization of the cytoplasm in preparation for division, and are involved in the establishment of division polarity (Brown and Lemmon 1985*b*, 1988*a*). It is informative to compare plastid polarity in mitosis and meiosis. In both, the single plastid divides and migrates in response to division polarity and serves to nucleate microtubules involved in nuclear division and cytokinesis. In preparation for mitosis, the plastid migrates to a position parallel to the future spindle axis. A unique axial microtubule system (AMS) develops parallel to the isthmus of the dividing plastid and precisely intersects the division site, which is girdled by a PPB (Fig. 2.1a). Since plastid migration and AMS development occur prior to PPB formation, both serve as reliable markers of polarity. As the plastid divides, the AMS elongates into two opposing sets of microtubules focused on the tips of the daughter plastids. These microtubules contribute directly to the development of the mitotic spindle.

A similar AMS is associated with successive plastid divisions that result in predictive positioning of a plastid at each of the future tetrad

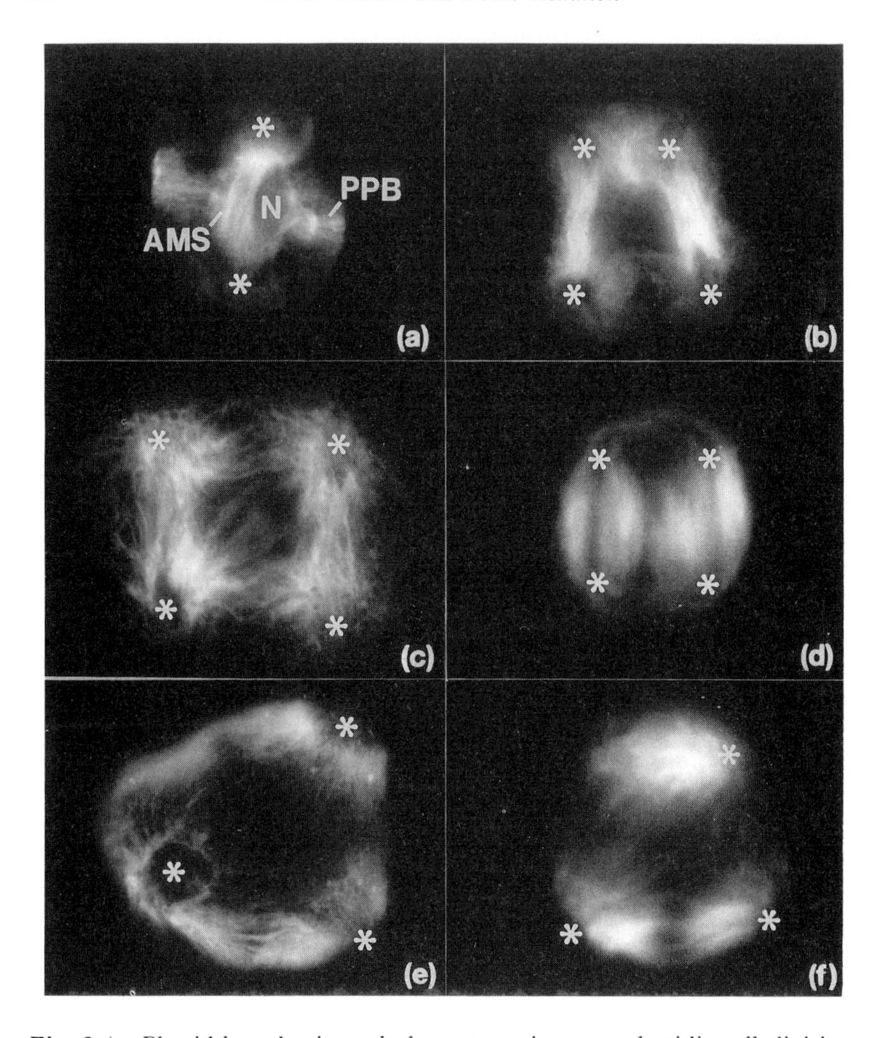

Fig. 2.1 Plastid-based microtubule systems in monoplastidic cell division. Plastids/plastid tips are indicated by asterisks. (a) In mitotic prophase of hornworts, an AMS in the isthmus region of the dividing plastid lies at right angles to the PPB which girdles the cell in the future plane of division. The AMS will elongate and give rise to the spindle extending between plastid tips at opposite poles. (b) In meiotic prophase of hornworts, an AMS is associated with each of the two plastids undergoing second plastid division. The two AMSs will contribute to the development of the QMS, which predicts spore domains and ultimately contributes to spindle ontogeny. (c) Typical meiotic quadripolarity in *Isoetes* megasporogenesis. The QMS in late meiotic prophase interconnects the four plastids which have precociously divided and migrated to tetrad poles. All four plastids can be seen in this flattened cell; normally the plastids are in tetrahedral arrangement. (d) Sequen-

poles in prophase of meiosis. An AMS arises in association with the first plastid division that gives rise to two plastids which elongate on either side of the nucleus. An AMS reappears at the midregion of each plastid as they simultaneously prepare for a second division (Fig. 2.1b). The two AMSs develop into the primary arrays which connect sister plastids after second plastid division. Additional microtubules radiate from the four plastid MTOCs to produce secondary arrays which interconnect the non-sister plastids. Together the two systems comprise the mature QMS which interconnects the four tetrahedrally arranged plastids and encages the nucleus. In *Isoetes* and mosses, the plastids divide without an AMS associated with the isthmus. However, microtubule arrays connecting pairs of sister plastids are often recognizable as primary arrays before the secondary arrays develop. This is especially obvious in *Funaria*, where the second plastid division is delayed until after meiosis I and plastid tips, which are positioned at the tetrad poles, serve to organize microtubules of the QMS (Busby and Gunning 1988*a*, *b*).

It is clear from the study of hornworts that plastid division and AMS formation is essentially the same in meiosis and mitosis. In meiosis, spindle development is delayed until after the four plastids are positioned. Thus, in meiotic cells the plastids mark the four poles of second division and not those of first division (Fig. 2.1b,c). The development of the first division spindle involves a partial merger of the four opposing arrays of the QMS and establishment of a division axis that terminates between non-sister plastids. The prophase II nuclei are each positioned between plastids which will serve as poles of the second meiotic division (Fig. 2.1e,f).

A summary diagram of variations in the organization of the QMS is shown in Fig. 2.2. The complex division quadripolarity established in prophase ensures that each spore receives a plastid in addition to a haploid nucleus. In the bryophyte type (Fig. 2.2a,b), the plastids remain in tetrahedral arrangement throughout meiosis and the foci of microtubules migrate toward the spindle axis during development

---

tial establishment of division polarity in *Selaginella* microsporogenesis. Spindle-shaped components of the procytokinetic plate (PCP) flank the two plastids lying on either side of the nucleus in late prophase. The PCP will contribute directly to development of the meiotic spindle. (e) Plastid-based microtubule arrays at second meiotic division in *Isoetes* megasporogenesis resemble the QMS of meiotic prophase. The two spindles of second meiotic division extend between plastids at the tetrad poles. (f) After first meiotic division of microsporogenesis in *Selaginella*, the plastids migrate to a tetrahedral position where they serve as poles of the second division spindles.

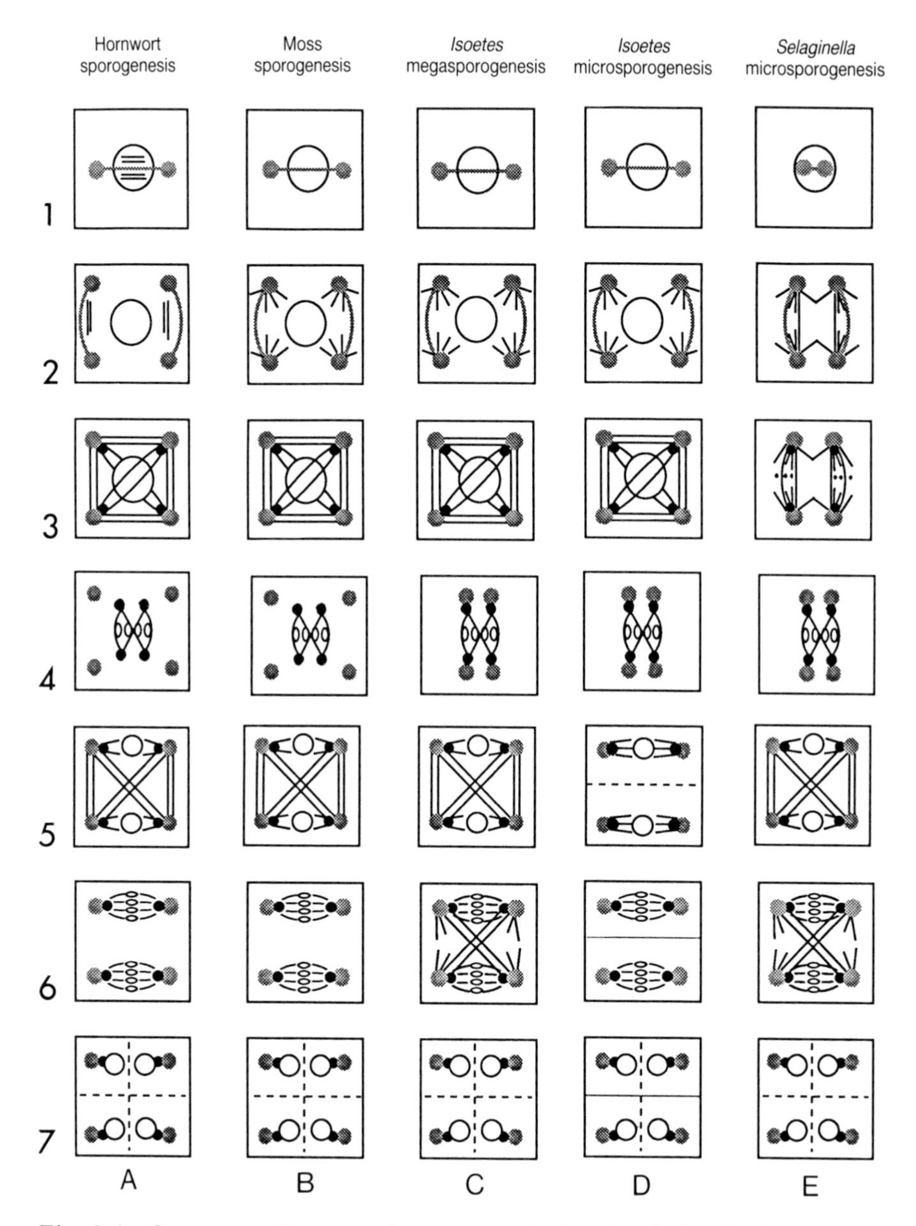

Fig. 2.2 Summary diagram of sporogenesis in simple land plants. For simplicity, all examples are shown in two dimensions (as though flattened) and numbers of microtubules are much reduced. Hornworts (column A) and mosses (column B) constitute the bryopsid type; *Isoetes* (columns C and D) and *Selaginella* (column E) constitute the lycopsid type. Plastids are shaded; MTOCs are black; nuclei/chromosomes are outlined; lines repre-

of the first division spindle. In the lycopsid type (Fig. 2.2c,d,e), the plastids are located at poles of both the first and second divisions. In *Isoetes* (Fig. 2.2c,d) the plastids with associated microtubules leave their prophasic positions to converge in pairs upon the first division spindle axis. In *Selaginella* (Fig. 2.2e) the plastids do not move to the tetrahedral position until after the first division.

Transitions of microtubules from one configuration to another during sporogenesis provide convincing evidence of how microtubules can be reoriented or 'sorted' in response to changes in polarity. We assume that this reorganization of microtubules into stage-specific arrays is a function of MTOC migration and replacement of microtubules. Although dynamic instability of microtubules as shown in animal cells (Cassimeris *et al.* 1987) has yet to be documented for plant microtubules, there is reason to believe that plant microtubules can be cycled and that their reorganization into different arrays results

---

sent orientation of microtubules. Row 1, Early prophase. The acentric nucleus appears to be centrally located when seen in top view. An AMS occurs in association with the isthmus of the dividing plastid only in hornworts. In *Selaginella*, the two plastids remain in association with the nucleus; in all others the plastids assume a more peripheral position in the cytoplasm. Row 2, Mid-prophase. An AMS develops in association with each of the two dividing plastids in hornworts. The late-dividing plastids of *Selaginella* remain attached to the nuclear envelope. Row 3, Late prophase. Quadripolarity is established in all except *Selaginella*, where bipolarity is established. The QMS in A–D interconnects the four plastids and encages the central nucleus. A typical QMS is not formed in *Selaginella* (E). Rather, two components of the procytokinetic plate (PCP) flank the nucleus and terminate in plastids located close to the division axis. Row 4, Metaphase I. The spindle lies between non-sister plastids. In the bryopsid type (A–B), the spindle poles leave the plastids and converge on the division axis. In the lycopsid type (C–E) the plastids MTOCs are located at poles. This requires a second migration from tetrahedral position in *Isoetes* (C–D). In *Selaginella* (E) the plastids are closely associated with the nucleus from earliest prophase. Row 5, Prophase II reflects the position of telophase I nuclei between plastids. The QMS is re-established (A–C) or established for the first time in *Selaginella* (E), but not in *Isoetes* microsporogenesis (D) which undergoes cytokinesis after first division. Row 6, Metaphase II. In the bryopsid type (A–B), microtubules are concentrated in the second division spindles exclusively, but in the lycopsid type (C and E) microtubules radiating from the plastids in QMS pattern interconnect all four plastids as well as form spindles between non-sister plastids. Row 7, In all cases of monoplastidic meiosis, the four spores of the tetrad each contain a plastid as well as a nucleus. Cytokinesis typically occurs simultaneously after meiosis is completed, with successive cytokinesis in *Isoetes* microsporogenesis being the exception.

from movement of MTOCs in response to changes in polarity (Brown and Lemmon 1989*a*). In the case of cells with plastid polarity, the material responsible for nucleating microtubules is associated with the plastids, except for the brief period in bryophyte meiosis when the foci of microtubules leave the plastids and converge on the poles of the first division spindle axis.

A typical QMS is not formed during microsporogenesis of *Selaginella* (Figs 2.1d, 2.2e). This major departure from the usual pattern in monoplastidic meiosis is an interesting variation of plastid polarity that is particularly informative of the evolution of meiotic division polarity (Brown and Lemmon, in manuscript). *Selaginella* is the only example of direct establishment of the first division spindle without prior establishment of second division poles. The sequential determination of meiotic spindle axes is a character that *Selaginella* shares with the higher plants. Additionally, the orientation of the meiotic spindle can be related to polarity of the very early sporocyte in which the acentric nucleus is capped by a single plastid. Because the second plastid division does not occur until just before metaphase, the spindle axis can be seen to lie at right angles to the plastid–nuclear axis of the early sporocyte. In all other examples of monoplastidic meiosis, where a QMS interconnects four plastids in tetrahedral arrangement, the sporocyte is symmetrical and all ability to relate to the plastid–nuclear axis of the early sporocyte is lost.

In *Selaginella* microsporocytes, the two plastids resulting from first plastid division rotate apart and come to lie parallel to each other on opposite sides of the nucleus (Fig. 2.2e). A complex system of microtubules, the procytokinetic plate (PCP), develops parallel to the long axes of the plastids and defines the polarity of the first nuclear division (Figs 2.1d, 2.2e). The spindle axis will lie between the plastids, the tips of which define the poles, and the equator is marked by a plate of vesicles midway across the two spindle-shaped microtubule arrays associated with the plastids. The second plastid division occurs just before metaphase. The meiotic spindle develops from merger of the two spindle-shaped components of the PCP and from additional microtubules emanating from the pair of plastids at each pole (Fig. 2.2e). Although the plastids were not in tetrahedral arrangement during prophase (as is typical of other monoplastidic sporocytes), the plastid pairs at each pole tend to tilt away from each other, resulting in some twisting of the spindle at metaphase. It is not until after completion of the first nuclear division that the four plastids migrate to a tetrahedral arrangement, where they serve as poles of the second division (Figs 2.1f, 2.2e).

Meiotic cytokinesis

In sporogenesis, cytokinesis may be successive with wall formation after the first nuclear division, resulting in a dyad (Fig. 2.2d), or it may be simultaneous, in which case development of the intersporal walls is delayed until after the second nuclear division (Fig. 2.2a–c,e). In simultaneous cytokinesis, the processes of nuclear and cytoplasmic division are uncoupled and the undivided sporocyte is coenocytic during the second nuclear division. This variation provides additional insight into the nature of the cytokinetic apparatus.

Simultaneous cytokinesis is typical of the simple land plants and, among seed plants, occurs in some gymnosperms, some monocots, such as the Orchidaceae, and most dicots. Although a cell plate is not produced after the first nuclear division, it is typical for the first division site to be occupied by a band of organelles that effectively divides the cytoplasm into two domains (Rodkiewicz *et al.* 1984; Brown and Lemmon 1987*b*, 1988*c*). In monoplastidic sporocytes, where the plastids are located at tetrad poles, the organelle band consists of mitochondria, vesicles, and lipid bodies, whereas in polyplastidic sporocytes the plastids are also included in the organelle band.

Cellularization during simultaneous cytokinesis involves formation of microtubules between non-sister nuclei as well as the usual phragmoplasts between sister nuclei. Cytokinetic planes are defined by interaction of these nuclear-based radial arrays of microtubules, which have been termed 'secondary spindles' in earlier literature (Heslop-Harrison 1971). In simultaneous cleavage in microsporocytes of flowering plants, the position of telophase II nuclei defines the pattern of cytokinesis which determines the arrangement of the tetrad (Heslop-Harrison 1971; Huynh 1976) and influences the location of germinal pores (Heslop-Harrison 1971; Dover 1972; Sheldon and Dickinson 1983, 1986). Division planes appear to be selected after the completion of meiosis, when microtubule systems radiating from the spore nuclei claim cytoplasmic domains and new walls are laid down wherever opposing systems of microtubules interact (Brown and Lemmon 1988*c*). When lagging chromosomes result in supernumerary nuclei, these small nuclei claim a proportionately smaller cytoplasmic domain and supernumerary microspores are produced in addition to the usual tetrad members (Brown and Lemmon 1989*b*). Cytokinesis controlled by postmitotic nuclear-based arrays is also known in cellularization of coenocytic endosperm (Gunning 1982), but is atypical of cell division where cytokinesis immediately follows mitosis.

The cytokinetic apparatus in microsporogenesis of flowering plants differs from monoplastidic meiosis, where prophasic positioning of the plastids and development of the QMS signal predetermination of division planes. The QMS is analogous to the PPB in that both are early expressions of the cytokinetic apparatus. In all cases, the cytokinetic apparatus is microtubule-based, but further analogies are unclear.

Development of intersporal septa, which appear to be similar in composition to the surrounding sporocyte wall, is different from the centrifugal expansion of the cell plate of vegetative cells. Whereas premeiotic deposition of wall material in the future cleavage planes, leading to quadrilobing, occurs only in the simple land plants, the microsporocytes of many angiosperms exhibit some postmeiotic lobing of the cytoplasm at the beginning of cytokinesis (Gifford and Foster 1989). Cytoplasmic infurrowing in the cleavage planes proceeds centripetally, but the septa are completed by the coalescence of vesicles that occurs more or less evenly along the entire planes of cytokinesis. This form of cytokinesis is plant-like and is unlike the process of cytoplasmic infurrowing that accomplishes cytokinesis in animal cells.

Spore wall development

Initiation of the spore wall in bryophytes is clearly related to quadripolarity established early in meiotic prophase. Immediately after cleavage each spore has a rounded distal surface and a trilete proximal surface shaped by contact with the other spores. In hepatics with deeply lobed sporocytes, the only portion of the spore circumference not covered by the original sporocyte wall is the proximal pole. In certain hepatics of the Jungermanniidae, the mature sculptured exine is prepatterned by exine precursors produced while the sporocytes are still in meiotic prophase (Brown *et al.* 1986; Brown and Lemmon 1987*c*). The exine on the proximal face is usually thinner and less ornate than elsewhere along the perimeter of the spore and appears to function as a germinal aperture. Thus, both cleavage planes and mature exine are prepatterned in the early sporocyte. In moss spores with specialized proximal apertures, highly polar microtubule systems are associated first with the initiation of exine on the distal surface and then with development of the aperture on the proximal surface (Brown and Lemmon 1983). Comparative aspects of bryophyte spore wall development have been treated elsewhere (Neidhart 1979; Brown and Lemmon 1988*d*).

Conclusions

Comparative studies of sporogenesis in simple land plants have demonstrated that meiosis, cytokinesis, and spore wall development comprise an integrated morphogenetic pathway leading to the tetrad of distinctive spores. Quadripolarity is established early in meiotic prophase and is reflected in all subsequent events: organelle migration and apportionment, chromosome movement, cytokinesis, initiation and pattern of the spore wall, and placement of apertures. Complex cytoskeletal systems involved in the establishment and maintenance of division quadripolarity are early expressions of the cytokinetic apparatus peculiar to the sporocytes of lower plants. In higher plants, there is no structural evidence of predivision determination of cleavage planes, and spore domains appear to be claimed after nuclear division by nuclear-based microtubule arrays. Thus, the cleavage plane reflects the position of nuclei, which results from the orientation of spindle axes. Whereas early expression of the cytokinetic apparatus orients the meiotic spindles in bryophytes and lycopsids, the mechanisms that result in typical patterns of cleavage in angiosperms remain unknown. Since the control of division plane is of critical importance in the morphogenesis leading to taxonomically significant characters of mature spores and pollen, it will be of fundamental importance to decipher the role of microtubules in the transduction chain involved in the determination of division polarity.

Acknowledgements

We thank Monica Miller for technical assistance. This study was supported in part by NSF grant BSR 8610594 and by grants from the Louisiana Quality Education Trust Fund, LaSER 86USL-1-126-07 and RD-A-16.

References

Brown, R.C. and Lemmon, B.E. (1982). Ultrastructural aspects of moss meiosis: Review of nuclear and cytoplasmic events during prophase. *Journal of the Hattori Botanical Laboratory* **53**, 29–39.

Brown, R.C. and Lemmon, B.E. (1983). Microtubule organization and morphogenesis in young spores of the moss *Tetraphis pellucida* Hedw. *Protoplasma* **116**, 115–24.

Brown, R.C. and Lemmon, B.E. (1985*a*). A cytoskeletal system predicts division plane in meiosis of *Selaginella*. *Protoplasma* **127**, 101–9.

Brown, R.C. and Lemmon, B.E. (1985*b*). Preprophasic establishment of division polarity in monoplastidic mitosis of hornworts. *Protoplasma* **124**, 175–83.

Brown, R.C. and Lemmon, B.E. (1986). Three-dimensional relationship of the meiotic spindle to predivision microtubules in *Sphagnum*. *Journal of Cell Biology* **103** (5,2), 55a.

Brown, R.C. and Lemmon, B.E. (1987*a*). Division polarity, development and configuration of microtubule arrays in bryophyte meiosis I. Meiotic prophase to metaphase I. *Protoplasma* **137**, 84–99.

Brown, R. C. and Lemmon, B. E. (1987*b*). Division polarity, development and configuration of microtubule arrays in bryophyte meiosis II. Anaphase I to the tetrad. *Protoplasma* **138**, 1–10.

Brown, R. C. and Lemmon, B. E. (1987*c*). Involvement of callose in determination of exine patterning in three hepatics of the subclass Jungermanniidae. *Memoirs of the New York Botanical Garden* **45**, 111–21.

Brown, R. C. and Lemmon, B. E. (1988*a*). Preprophasic microtubule systems and development of the mitotic spindle in hornworts (Bryophyta). *Protoplasma* **143**, 11–21.

Brown, R. C. and Lemmon, B. E. (1988*b*). Cytokinesis occurs at boundaries of domains delimited by nuclear-based microtubules in sporocytes of *Conocephalum conicum* (Bryophyta). *Cell Motility and the Cytoskeleton* **11**, 139–46.

Brown, R.C. and Lemmon, B.E. (1988*c*). Microtubules associated with simultaneous cytokinesis of coenocytic microsporocytes. *American Journal of Botany* **75**, 1848–56.

Brown, R.C. and Lemmon, B.E. (1988*d*). Sporogenesis in bryophytes. *Advances in Bryology* **3**, 159–223.

Brown, R. C. and Lemmon, B. E. (1989*a*). Morphogenetic plastid migration and microtubule organization during megasporogenesis in *Isoetes*. *Protoplasma* **152**, 136–47.

Brown, R.C. and Lemmon, B.E. (1989*b*). Minispindles and cytoplasmic domains in microsporogenesis in orchids. *Protoplasma* **148**, 26–32.

Brown, R. C. and Lemmon, B. E. (1990). Monoplastidic cell division in lower land plants. *American Journal of Botany* **77**, 559–71.

Brown, R.C., Lemmon, B.E., and Carothers, Z.B. (1982). Spore wall development in *Sphagnum lescurii*. *Canadian Journal of Botany* **60**, 2394–409.

Brown, R.C. and Lemmon, B.E., and Renzaglia, K.S. (1986). Sporocytic control of spore wall pattern in liverworts. *American Journal of Botany* **73**, 593–6.

Busby, C.H. and Gunning, B.E.S. (1988*a*). Establishment of plastid-based quadripolarity in spore mother cells of the moss *Funaria hygrometrica*. *Journal of Cell Science* **91**, 117–26.

Busby, C.H. and Gunning, B.E.S. (1988*b*). Development of the quadripolar meiotic cytoskeleton in spore mother cells of the moss *Funaria hygrometrica*. *Journal of Cell Science* **91**, 127–37.

Busby, C.H. and Gunning, B.E.S. (1989). Development of the quadripolar meiotic apparatus in *Funaria* spore mother cells: analysis by means of anti-microtubule drug treatments. *Journal of Cell Science* **93**, 267–77.

Cassimeris, L., Walker, R.A., Pryer, N.K., and Salmon, E.D. (1987). Dynamic instability of microtubules. *BioEssays* **7**, 149–54.

Clayton, L., Black, C. M., and Lloyd, C. W. (1985). Microtubule nucleating sites in higher plant cells identified by an auto-antibody against pericentriolar material. *Journal of Cell Biology* **101**, 319–24.

Davis, B. M. (1901). Nuclear studies on *Pellia*. *Annals of Botany* **40**, 148–80.

Dover, G. A. (1972). The organization and polarity of pollen mother cells of *Triticum aestivum*. *Journal of Cell Biology* **11**, 699–711.

Dunlop, D. (1949). Notes on the cytology of some lycopsids. *Bulletin of the Torrey Botanical Club* **76**, 266–77.

Farmer, J. B. (1895). On spore-formation and nuclear division in the Hepaticae. *Annals of Botany* **9**, 469–529.

Gifford, E. M. and Foster, A. S. (1989). *Morphology and evolution of vascular plants*, (3rd edn). Freeman, New York.

Gunning, B. E. S. (1982). The cytokinetic apparatus: Its development and spatial regulation. In *Cytoskeleton in plant growth and development*, (ed. C. W. Lloyd), pp. 229–92. Academic Press, London.

Heslop-Harrison, J. (1971). Wall pattern formation in angiosperm microsporogenesis. *Symposia of the Society for Experimental Biology* **25**, 277–300.

Huynh, K. L. (1976). Arrangement of some monosulcate, disulcate, trisulcate, dicolpate, and tricolpate pollen types in the tetrads and some aspects of evolution in the angiosperms. In *The evolutionary significance of the exine*, (ed. I. K. Ferguson and J. Muller), pp. 101–24. Academic Press, London.

Mineyuki, Y., Wick, S. M., and Gunning, B. E. S. (1988). Preprophase bands of microtubules and the cell cycle: Kinetics and experimental uncoupling of their formation from the nuclear cycle in onion root-tip cells. *Planta* **174**, 518–26.

Neidhart, H. V. (1979). Comparative studies of sporogenesis in bryophytes. In *Bryophyte systematics*, (ed. G. C. S. Clark and J. G. Duckett), pp. 251–80. Academic Press, New York.

Rodkiewicz, B., Kudlicka, K., and Stobiecka, H. (1984). Patterns of amyloplast distribution during microsporogenesis in *Tradescantia*, *Impatiens* and *Larix*. *Acta Societatis Botanicorum Poloniae* **53**, 437–41.

Schuster, R. M. (1966). *The Hepaticae and Anthocerotae of North America*, Vol. 1. Columbia University Press, New York.

Sheldon, J. M. and Dickinson, H. G. (1983). Determination of patterning in the pollen wall of *Lilium henryi*. *Journal of Cell Science* **63**, 191–208.

Sheldon, J. M. and Dickinson, H. G. (1986). Pollen wall formation in *Lilium*: the effect of chaotropic agents, and the organization of the microtubular cytoskeleton during pattern development. *Planta* **168**, 11–23.

Wick, S. M. (1985*a*). The higher plant mitotic apparatus: redistribution of microtubules, calmodulin and microtubule initiation material during its establishment. *Cytobios* **43**, 285–94.

Wick, S. M. (1985*b*). Immunofluorescence microscopy of tubulin and microtubule arrays in plant cells. III. Transition between mitotic/cytokinetic and interphase microtubule arrays. *Cell Biology International Reports* **9**, 357–71.

Wick, S. M. and Duniec, J. (1983). Immunofluorescence microscopy of tubulin and microtubule arrays in plant cells. I. Pre-prophase band development and concomitant appearance of nuclear envelope-associated tubulin. *Journal of Cell Biology* **97**, 235–43.

3. A review of *in situ* spores in Silurian land plants

U. FANNING

Department of Geology, University of Wales College of Cardiff, Cardiff, UK

J.B. RICHARDSON

The Natural History Museum, Cromwell Road, London, UK

D. EDWARDS

Department of Geology, University of Wales College of Cardiff, Cardiff, UK

Abstract

Spores are described from the sporangia of six rhyniophytoid taxa from early Pridoli (late Silurian) strata at Perton Lane, Hereford and Worcester, England. Those from species of *Cooksonia* are crassitate: *C. pertoni* subsp. *pertoni* contains *Ambitisporites* sp.; *C. pertoni* subsp. *synorispora*, *Synorisporites verrucatus*; and *C. cambrensis* has a crassitate spore with micrograhulate sculpture on both proximal and distal surfaces, which has not yet been recorded in dispersed spore assemblages and will probably require a new genus. Retusoid spores of diverse aspect have been isolated from *Salopella* sp. and two new taxa. The *Salopella* spores are exceptionally large and provisionally assigned to *Retusotriletes* sp., the genus also used for spores of *Caia langii* which has elongate spinous sporangia. Spores of *Pertonella dactylethra* (characterized by sporangia of *Cooksonia* shape, but with spines) have a distinctive sculpture following the curvaturae perfectae, and are close to *Retusotriletes coronadus* from the Silurian of Spain. Elongate sporangia, superficially similar to *Salopella* but generally longer, have yielded laevigate hilate cryptospores occurring in loose dyads. Discussion includes documentation and significance of geographical and stratigraphical distribution of the *in situ* spores in the dispersed record, and the value of *in situ* spores on taxon delimitation and assessment of affinity.

Pollen and Spores (ed. S. Blackmore and S. H. Barnes), Systematics Association Special Volume No. 44, pp. 25–47, Clarendon Press, Oxford, 1991.

Introduction

Records of assemblages of palynomorphs far outnumber those of megafossils in Silurian sediments (Richardson and Edwards 1989) and consequently Silurian palynology has become of major importance in biostratigraphy (Richardson and Ioannides 1973; Richardson 1988), and more recently in hypotheses concerning early colonization of the land by plants (Gray 1985). The isolation of such spores from sporangia would not only give a more complete picture of the plant and provide further characters for recognition and taxon delimitation, but would also allow a more sensible reading of the dispersed spore record in terms of coeval vegetation.

Prior to this study little was known of spores in Silurian plants, although it is quite remarkable that Lang (1937) in the initial description of the best-known genus, *Cooksonia*, isolated spores on film-pulls from the type species *C. pertoni*, describing them as trilete and smooth-walled. The film-pull itself obscures detail, but later workers have related the spores to *Ambitisporites* or *Leiotriletes* (Allen 1980). The only other illustrated records are even less satisfactory and are again based on film-pulls from coalified sporangia, in this case from Freshwater East, Dyfed. The spores could not be placed in any dispersed taxa (Edwards 1979).

Here we bring together the results of a collaborative project between a palynologist, John Richardson (NHM) and palaeobotanists, Dianne Edwards and Una Fanning (UWCC). We propose to describe spores isolated from sporangia terminating smooth, isotomously branching axes belonging to plants termed rhyniophytoids (Edwards and Edwards 1986). This informal grouping is used because we have no anatomical evidence that would indicate affinity with the broadly morphologically similar plants of the Rhyniophytina (Banks 1975).

Materials and methods

All new information comes from coalified compressions at Perton Lane. The exposure, in the east side of the lane near Stoke Edith, Herefordshire (SO 5971 4035) is less than half a metre above the base of the Rushall Beds (the local equivalent of the Downton Castle Sandstone Formation) in the Downton Group in the Woolhope Inlier (Squirrell and Tucker 1960). The plants occur above a bone bed thought equivalent to the Ludlow Bone Bed of Shropshire in sediments of early Pridoli age (late Silurian).

After examination under a dissecting microscope, specimens thought to contain spores were photographed and measured. Coalified material was then loosened using tungsten needles, picked off on a fine brush and mounted on SEM stubs, or cleared in Schultze's solution and mounted for light microscopy. In some cases it was treated with hydrofluoric acid to remove sediment. In a few cases spores were recovered from stubs, cleared, and mounted for light microscopy. Scanning electron microscopes used were the Cambridge S180 (NHM), Jeol JSM-T100 (National Museum of Wales), and Jeol JSM-35CF (UWCC).

Taxa with *in situ* spores: description and discussion

Cooksonia pertoni Lang (1937)

Basis for identification. Our new specimens (Fig. 3.1a) encompass and extend the variability in sporangial outline illustrated by Lang (1937) in compressions from Perton Lane, and their detailed analysis, particularly relating to relative orientations of sporangium and axis during burial, has permitted a three-dimensional reconstruction of fertile regions (Fanning, Richardson and Edwards 1988). Limited anatomical evidence for the sporangial wall shows no evidence for a specialized dehiscence mechanism.

Spores. Spores were isolated from *c.* 13 per cent of sporangia ($n = 311$) and are of two distinct types. Both have an equatorial crassitude, but in one case the exine is laevigate and the other verrucate, with verrucae occasionally fused into small groups.

The laevigate spores, placed in *Ambitisporites* sp. (Fig. 3.1b), are the larger (maximum diameters range between 22 and 43 μm, $\bar{x} = 31$ μm) and occur in the majority of sporangia (56 per cent). They closely resemble the *in situ* spores described by Lang, which have been re-examined, although it has not been possible to study them at ×1000 magnification. The ornamented spores (Fig. 3.1c), assigned to *Synorisporites verrucatus* (maximum diameters 15–30 μm, $\bar{x} = 23$) have been isolated from only 44 per cent of sporangia. Their recovery from so comparatively few sporangia is in marked contrast to their common occurrence in Lower Downton rocks.

Comparison with dispersed spores. The earliest trilete miospores to date belong to the genus *Ambitisporites* Hoffmeister. Two species (*A. avitus* and *A. dilutus*) are found in rocks of early Silurian age (late *sedgwicki* graptolite Biozone–Upper Aeronian Stage). They are absent or rare in the Upper Silurian of the Welsh Borderland but *A. dilutus* extends into the early Devonian, although this may partly be due to reworking.

(a)
(b)
(c)
(d)
(e)
(f)
(g)
(h)

A. avitus and *A. dilutus* are present in Welsh Borderland assemblages but the latter are more common. The size range of *A. avitus* (35–87 μm) overlaps that of the spores from sporangia, but the crassitude is much narrower in the *in situ* spores than in the typical *A. avitus*. These differ from the dispersed specimens of *Ambitisporites* in having a laevigate or scabrate exine; a narrower, ill-defined crassitude; and narrow lips.

Smaller miospores with an indistinct crassitude and a thinner wall than either *A. avitus* or *A. dilutus* were recorded from the Wenlock shales (Richardson and Lister 1969) and are known to have a range from the Sheinwoodian to the Gedinnian of the Welsh Borderlands, being widespread but not common in the Pridoli (JBR, work in progress). Richardson and Lister originally placed these spores in the genus *Retusotriletes* but they are now regarded as belonging to *Ambitisporites* sp. Spores of this type resemble those of *C. pertoni* subsp. *pertoni*.

Discussion. Thus we have recorded the earliest known miospore genus in sporangia. The earliest megafossils resembling *Cooksonia* have been described from upper Wenlock sediments (*ludensis* graptolite Biozone) from Ireland, but are completely lacking in anatomy (Edwards *et al.* 1983). The specimens exhibit variation in sporangial outline and it seems likely that at least three different species are present, including *C. pertoni*. This megafossil occurrence follows the next major event in the evolution of trilete spores as seen in the dispersed spore record, viz. the first appearance of sculpture and, more important to this study, that of verrucate sculpture, in spores called *S.* cf. *verrucatus* from late Wenlock (Homerian) strata. Thus the two subspecies of *C. pertoni* similar to those in the Pridoli probably occurred in the Wenlock, connecting elements of a macroplant/spore lineage from the Wenlock (or even Llandovery) to the Downton (Pridoli). Spores isolated from Lower Devonian *C. pertoni* with similar sporangial anatomy belong to the dispersed spore genera *Streelispora*/*Aneurospora* (Fanning *et al.* 1988).

Fig. 3.1 All figured material comes from Pridoli strata at Perton Lane, near Hereford. (a) *Cooksonia pertoni* Lang. Coalified compression showing typical morphology, but from which spores have not been isolated. NMW 90. 41G.1, × 8. (b) SEM of *Ambitisporites* sp. isolated from *C. pertoni* subsp. *pertoni*. V.62776 (stub), × 1600. (c) SEM of *Synorisporites verrucatus* isolated from *C. pertoni* subsp. *synorispora*. V.62777 (stub), × 1500. (d) *C. cambrensis* Edwards. Coalified compression. NMW 89. 49G.3, × 23 (e) SEM of crassitate spores isolated from *C. cambrensis*. NMW 89. 49G.4 (stub), × 2000. (f) Close-up of wall of spore in (e), showing microgranular sculpture, × 4800. (g) Disc-like structure comprising the spores, *Synorisporites tripapillatus*. V.62775 (stub), × 60. (h) *Synorisporites tripapillatus* from disc in (g), × 1100.

These are structurally similar to the equatorially crassitate spores isolated from Silurian representatives but with apiculate ornament. We discussed the implications of these observations in the naming of *C. pertoni* in 1988 (and see p. 39) together with some comments on evolutionary issues.

Isolated disc with Synorisporites tripapillatus

Spores have been isolated from a number of circular discs (Fig. 3.1g) with dimensions and shape similar to those of sporangia of *C. pertoni* when viewed from above. In all except one case they are assignable to *Ambitisporites* sp. or *Synorisporites verrucatus*. The exception (0.97 mm × 0.75 mm; Fig. 3.1g) contains spores similar in size (maximum diameter ranges between 14 and 17 μm, $\bar{x} = 16$ μm) and structure to *S. verrucatus*, but with distal sculpture more murinate and with three prominent inter-radial papillae (on the proximal surface). They are referable to the dispersed taxon *Synorisporites tripapillatus* Richardson and Lister (1969). The shape of the spore mass suggests that the parent plant had discoidal sporangia of similar form to those of *C. pertoni*. The lack of such megafossils is most puzzling in that the spore is present in most lower Downton samples in the Anglo-Welsh regions and is very abundant in some areas (e.g. the Downton at Ludlow). The earliest known occurrence of *S. tripapillatus* is in the basal part of the Ludlow Bone Bed Member of the Downton Group (Pridoli) along with *S. verrucatus*. No proximally tripapillate miospores are known in older rocks, and *tripapillatus* is one of the nominal species of the *tripapillatus–spicula* miospore Biozone. Present records indicate that this tripapillate species is confined to the Lower Pridoli. *Cooksonia pertoni* sporangia containing proximally tripapillate but distally apiculate spores occur in the Lower Devonian (Fanning *et al.* 1988).

Cooksonia cambrensis Edwards (1979)

Basis for identification. As circumscribed on Pridoli specimens from Freshwater East, Dyfed, by Edwards (1979), this species encompasses two forms of sporangia, both of which terminate parallel-sided axes showing little change in diameter at the sporangium/stalk junction. In forma α the sporangial outline is circular, and in forma β, it is elliptical. The seven specimens recorded at Perton Lane are ± circular in outline (Fig. 3.1d) with sporangial width/height ratios ranging between 0.81 and 1.07 ($\bar{x} = 0.97$). The largest is 1.01 mm high and 0.94 mm wide and the smallest 0.47 mm by 0.52 mm. Axis diameter ranges between 0.13 and 0.49 mm ($\bar{x} = 0.27$ mm). One specimen exhibits branching. In the single, three-dimensionally preserved, almost spherical sporangium, only the outermost layer of the wall is anatomically preserved. It consists of cells of one kind, with no indication of tissue specialized for dehiscence. In surface view the cells are longitudinally elongate (20 by 35 μm) and have thickened outer periclinal and anticlinal walls. This was the only specimen to yield spores.

Spores. The spores have been examined only by scanning microscopy (SEM). Ambs are ?subcircular to subtriangular and the spores are trilete with narrow crassitude (Fig. 3.1e). The maximum diameter is 9–22 μm ($\bar{x} = 13$ μm). The exine is characterized by regular microgranulate sculpture on both proximal and distal surfaces (Fig. 3.1f). Such sculpture makes these spores highly distinctive under SEM, but would not be detectable in light microscopy. The spores would thus be recorded as a species of *Ambitisporites* if found in dispersed spore assemblages, which had been examined by conventional means, but those in the sporangia either belong to a new genus for crassitate spores, or to an existing genus that will have to be emended to accommodate them.

Discussion. The Dyfed specimens were assigned to *Cooksonia*, because the terminal sporangia are short and wide. In this study such morphological evidence for generic placement is reinforced by anatomical evidence, in that there are no specialized cells for dehiscence and the spores are crassitate. The distinguishing specific characters of *C. cambrensis* relate to sporangium/axis junction and now to spore sculpture.

Pertonella dactylethra Fanning, Edwards, and Richardson (1991)

The discoidal sporangia, which terminate smooth isotomously branching axes, are very similar in form to those of *C. pertoni*, but possess more or less evenly distributed emergences (Fig. 3.2a). Widest at their bases, the latter are parallel for most of their length and have rounded apices.

Spores. Isospores isolated from six sporangia, are usually preserved in polar compression. They are 34–45 μm in diameter ($\bar{x} = 39$ μm, $n = 20$) and have subcircular to subtriangular ambs (Fig. 3.2b). The trilete mark consists of prominent folds (tectum) which increase in height towards the equator. The exine is 2–4 μm thick and slightly thickened equatorially. Contact areas are well defined. The proximal sculpture consists of a band (3–4 μm wide) of variable sculpture, following the curvature perfectae and confluent with a broad band of grana and microconi flanking the trilete mark. The curvatural band consists of sculpture which varies radially on an individual spore and also between specimens. As seen on the SEM, from the equator polewards, the sculptural type changes from low ridges separated by canaliculae, to crowded ridges surmounted by sharply pointed coni decreasing in size to grana and microconi (> 1 μm); this innermost band is confluent with the broad sculptural band (2–6 μm wide) enclosing the trilete folds. The tectum (trilete fold) also bears sparse grana and microconi. The distal surface is laevigate, 'roughened' and punctate, or sparsely granulate. The distal exine is *c*. 2 μm thick.

Comparison with the dispersed spore record. The spores closely resemble one of those included in the dispersed miospore species *Retusotriletes coronadus* Rodriguez Gonzalez (1983, Pl. 8, Fig. 8) from the Silurian of Spain but not the holotype or other figured specimens. It has been recorded in palynological assemblages at Perton Lane (Fanning *et al.* 1991).

Caia langii Fanning, Edwards, and Richardson (1990)

We erected this new taxon for smooth, isotomously branching axes with elongate, sometimes bifurcating, terminal sporangia that are parallel-sided with obtuse tips and bearing up to 10 conical emergences, mostly concentrated near the apex (Fig. 3.2c), and containing retusoid spores (Fig. 3.2f).

Spores. These have more or less circular to irregular ambs and were originally elliptical in cross-section. They are laevigate, with a thin-

---

Fig. 3.2 All figured material comes from Pridoli strata at Perton Lane, near Hereford. Light micrographs (LM) were taken using Nomarski interference contrast. (a) *Pertonella dactylethra* Fanning, Edwards and Richardson. Coalified compression. NMW 84. 14G.1, × 10. (b) SEM of spore from *P. dactylethra*. NMW 84. 14G.7 (stub), × 1700. (c) *Caia langii* Fanning, Edwards and Richardson. Coalified compression. NMW 89. 38G.1, × 15. (d) New taxon containing cryptospores. NMW 90. 41G.3, × 8 (e) *Salopella* sp. coalified compression, specimen lost, × 8. (f) SEM of spores (*Retusotriletes* sp.) isolated from *Caia langii*. NMW 90. 41G.4 (stub), × 2200. (g) Light micrograph of spore isolated from *Salopella* sp. NMW 90. 41G.5, (slide). England Finder (EF) no. K56/2/4, × 350. (h) Part of (g) magnified to show curvatural ridge (i) LM of *Ambitisporites* isolated from *C. pertoni* subsp. *pertoni*, NMW 90. 41G.6 (slide). EF no. 048/P48, × 750. (j) LM of 'loose' dyad isolated from new genus with elongate sporangia. NMW 90. 41G.7 (slide). EF no. P39, × 500. (k) LM of single cryptospore from new taxon. NMW 90. 41G.8 (slide). EF no. H41, × 600.

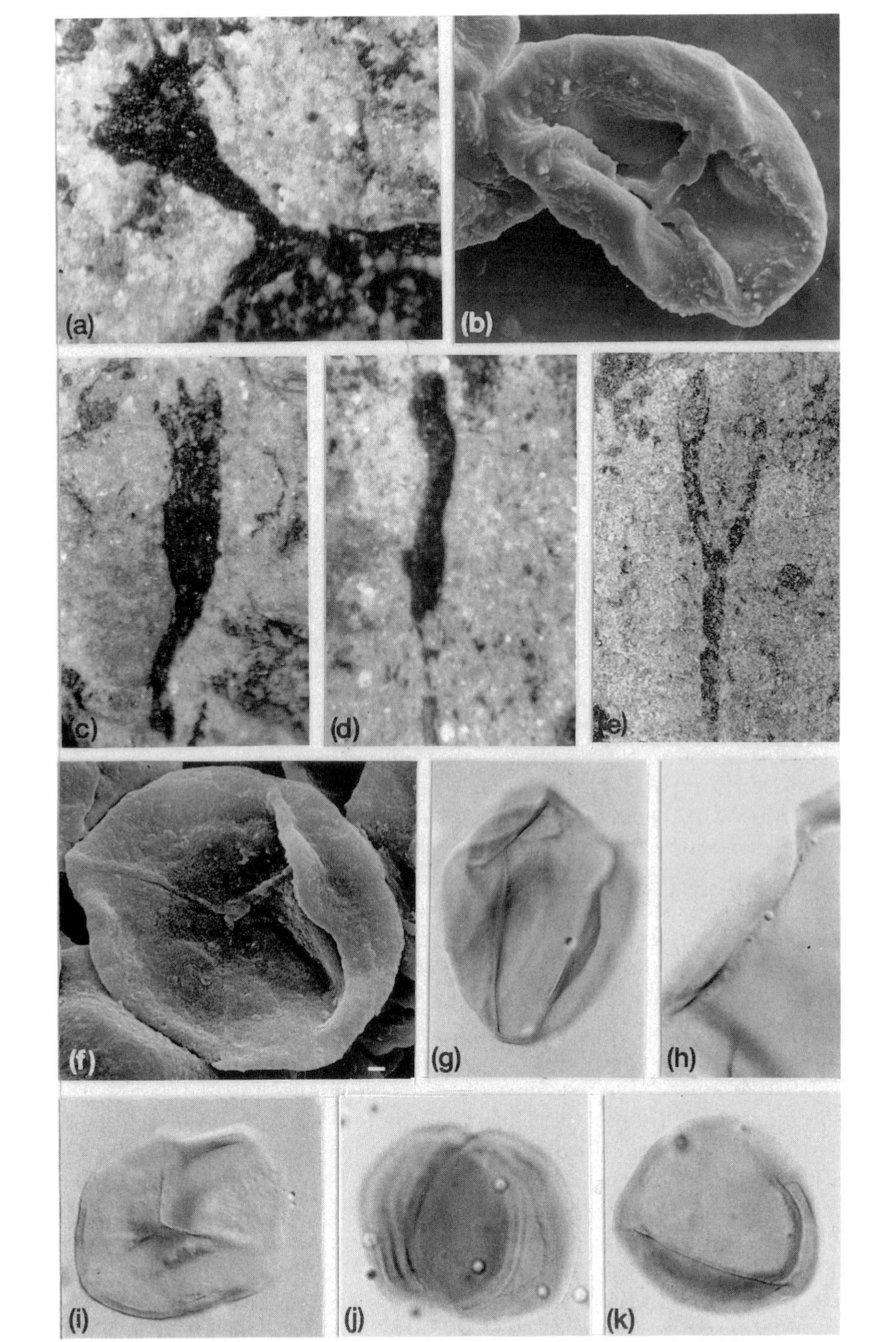
(a)
(b)
(c)
(d)
(e)
(f)
(g)
(h)
(i)
(j)
(k)

walled exine (1 μm thick) and are often folded. Their triradiate marks are frequently indistinct and have low, even lips, 0.05 μm wide, which are confluent with a narrow curvatural fold of the same width. They range from 23 μm to 32 μm in diameter. Occasionally scanning electron micrographs show a granular spore surface, but this may be a degradational feature. Most of the spore masses recovered from sporangia appear quite 'clean', but some are covered by a thin microgranulate 'sac'. These retusoid spores belong to the dispersed spore genus *Retusotriletes* (Naumova) Richardson, and are discussed further on p. 39.

Comparison with the dispersed spore record. Gedinnian specimens of *R.* cf. *triangulatus* (Streel) in Richardson and Lister (1969) have a more rigid wall than the *in situ* Silurian spores, and show a darkened/thickened triangular area at the proximal pole. Traces of such a proximal feature are occasionally seen in *Caia* spores. Rarely, the Lower Devonian dispersed spores are individually enclosed by a loose, sculptured, possibly perisporal layer (Richardson and Lister 1969).

Salopella sp.

Three specimens with fusiform terminal sporangia are assigned to the genus *Salopella* erected by Edwards and Richardson (1974) for compression fossils with this sporangial morphology. All the Silurian specimens are of similar size, with height ranging between 2.6 and 2.99 mm ($\bar{x} = 2.82$ mm) and maximum widths 0.55–0.78 mm ($\bar{x} = 0.65$ mm): height to width ratios vary between 3.67 and 5.44 ($\bar{x} = 4.43$). The sporangia are more or less parallel-sided in the lower two-thirds and taper to an acute point (Fig. 3.2e). One specimen appears to have a sterile tip, in that coalified material typical of sporangia does not extend to the apex, which is defined by fragments of coalified material and limonite tightly adhering to the rock surface. As it is possible that this is a preservational feature and the sample is small, we are reluctant to separate the specimen from the other two. The sporangia are subtended by parallel-sided axes which range from more or less the same width as the sporangium to half its width (0.29–0.65 mm). In the former, the exact proximal limit of the sporangium is difficult to determine. Isotomous branching occurs in two of the specimens, with daughter branches placed symmetrically above the branch point at angles of 40° and 50°. The longest specimen measures 8.16 mm. In one axis, only peripheral tissues are preserved and comprise elongate tapering cells 26 μm wide and of unknown length.

Of the three better known species of *Salopella*, the Perton Lane material differs from the Australian *S. australis*, described from Ludlow and Lower Devonian sediments, in size and possibly position of sporogenous tissue, and from *S. caespitosa* in size and sporangial shape (Timms and Chambers 1984). There is at least a threefold difference in size when compared with the type *S. allenii* and no evidence for branching close to sporangia. *In situ* spores will be compared below, but confirm affinity with *Salopella*. However, because of the small sample and its fragmentary nature, we leave the Pridoli specimens as *Salopella* sp.

Spores. The trilete spores, with circular to subcircular ambs, are thin-walled (*c.* 1 μm) and show taper-pointed folds. The exine is minutely scabrate to 'infrascabrate' (i.e. with minute irregular wrinkles). The narrow lips of the trilete ($<$ 1 μm near the apex and expanding to 2 μm near the equator) are more or less equivalent to the radius, such that curvaturae perfectae are coincident with the equator for most of their length, and thus not clearly seen. One specimen in a mass of spores is invaginated at the radial apex in polar compression (Fig. 3.2g). Curvaturae are sometimes slightly darker (possibly thicker) than the rest of the exine and *c.* 1 μm wide. In the 17 specimens measured, the maximum diameter has a range of 80–97 μm. The spores are tentatively referred to the dispersed genus *Retusotriletes*.

Comparison with the dispersed spore record. Spores of this size have not yet been recorded from the matrix at Perton Lane. Indeed, azonate spores with good retusoid contact faces are rare in the Silurian, exceptions include *Apiculiretusispora spicula* and *Retusotriletes* cf. *minor* in the lower Pridoli, but become increasingly common in the Lower Devonian, especially in Siegenian and Emsian rocks (this may well reflect an increase in numbers of zosterophylls, which are known to contain spores of this type, e.g. Allen 1980; Gensel 1980; Edwards *et al.*, in press).

Discussion. Of the three taxa possessing retusoid spores, only those of *Pertonella* are typical. Spores of *Caia* and *Salopella* sp. have curvaturae more or less coincident with the equator for much of its length. In the case of *Caia* the curvaturae are 'folds' and in *Salopella* may be thickened. Indeed, one can easily imagine that in these two taxa, the curvaturae are minor modifications of a crassitate structure, as seen in certain *Cooksonia pertoni* subsp. *pertoni*. The fragmentary enclosing microgranulate layer in *Caia* (Fanning *et al.* 1990) is also unusual (but this could be equivalent to the layers seen in certain zosterophyllums, e.g. *Z. fertile*, Edwards 1969*a*; *Z. llanoveranum*, Edwards 1969*b*). Thus, although spores from the three taxa are united in their possession of curvaturae, it seems unlikely that they are closely related. *Salopella* spores are exceptionally large for spores of this age; *Pertonella* spores have a distinctive proximal sculpture closely similar to a new dispersed *Apiculiretusispora* sp. from the Gedinnian; *Caia* spores, in contrast, lack distinctive characters.

The appearance of retusoid miospores is much later (Wenlock and early Ludlow–late Homerian to early Gorstian) than those with an equatorial crassitude (upper Llandovery–uppermost Aeronian) and at least some retusoid spores probably had equatorially crassitate ancestors, as in both cases curvaturate thickenings are developed. In other retusoid spores the curvaturae are marked not by a band of thickening but by either an abrupt change of exine thickening over the proximal face, or absence or change in sculpture. In some cases, as in the spores of *Pertonella*, a distinct type of sculpture marks the curvature. Thus curvaturae perfectae are heterogeneous in structure and sculpture, probably reflecting diverse origins.

New genus—new species

All the sporangia described above are morphologically distinct from their subtending axes. They can thus be identified as putative sporangia even in the absence of *in situ* spores. In contrast, certain specimens superficially similar to axes are identified as sporangia only because they contain spores. Three sporangia are attached to axes. They are parallel-sided, of approximately the same to twice the diameter as their naked subtending axes, and have bluntly rounded apices (Fig. 3.2d). Their most striking distinguishing features are their great length and the nature of the *in situ* spores. Considering the former, sporangial height/width ratios range between 5.45 and 6.80 ($\bar{x} = 6.01$). The attached sporangia measure 5.03 × 0.87 mm, 4.36 × 0.80 mm, and 5.03 × 0.74 mm. Axes range from 0.34 to 0.74 mm. The axes are unbranched but of relatively short length (< 4.7 mm). The sporogenous region is easily distinguished from its subtending axis because it consists of coalified material that readily detaches from the matrix, either as a single sheet or in large fragments. Such separation occurs immediately on splitting the rock and, surprisingly, the compression leaves no impression on the underlying matrix. In contrast, the axes consist of very fragmentary material which adheres tightly to the rock. Spores were extracted from the sporangial material, but we found no evidence for a layer identifiable as a sporangial wall.

Spores. These are laevigate, hilate cryptospores with circular amb (Fig. 3.2k), some occurring as loose dyads (Fig. 3.2j) in which individuals are of unequal size (anisomorphic). Some specimens show a narrow equatorial crassitude around the hilum. The maximum diameter ranges between 39 and 68 μm. Such spores are similar to the dispersed spore '*Archaeozonotriletes*'? cf. *divellomedium*, which is currently being renamed as a cryptospore species (Richardson, work in progress) and are of similar size (39–60 μm; $\bar{x} = 48$ μm).

Comparison with dispersed spore record. Hilate cryptospores are common in Silurian and lower Devonian rocks. Loose dyads of such spores occur in the Wenlock but the spores are usually found as separate monads in later Silurian and Devonian sediments. However, dyads of '*A*'? cf. *divellomedium* occur rarely in the Ditton Group, and a single dyad of this species has been found covered by a thin fragmentary reticulate envelope, assumed to be sporopollenin impregnated. Envelope-enclosed dyads and pseudodyads are common in the late Ordovician and early Silurian (Gray *et al.* 1982; Richardson 1988). There is thus a general sequence of morphological phases, from 'permanent' dyads to 'loose' dyads to hilate monads (separated from dyads). However, 'permanent' dyads of other species commonly occur with '*A*'? cf. *divellomedium* in the Pridoli and early Gedinnian.

General discussion

In situ dyads

The discovery of hilate spores produced in dyads within sporangia marks one of the most significant advances of our studies on early land plants. We can now be certain that at least some dyads were formed in sporangia of rhyniophytoid aspect, closest in shape to the form genus, *Salopella*, but which contain trilete spores. The apparent absence of branching in the subtending axes may indicate bryophyte (?hepatic) affinity, but we need anatomy and less fragmentary specimens. Similarly, we have as yet no means of assessing the relationship with the producers of permanent dyads, or indeed tetrads, although current ultrastructural studies of the spores concerned may have some relevance here. However, Lang (1937) recovered elongate spore masses from several Downtonian localities, one (Tin Mill Race) containing smooth dyads he described as 'bicellular'. Other examples were thought to be 'invested' by interlacing tubes, similar to those described for *Nematothallus*.

Dyads and life cycles

Direct evidence for the life cycles of Silurian plants is lacking. Sporangium-bearing fragments are interpreted as parts of the sporophyte generation of a homosporous plant because the contained isospores are produced in tetrads, assumed to be the haploid products of meiotic divisions of diploid spore mother cells. Such spores are then considered to develop exosporically into free-living gametophytes. Thus, assuming the new megafossil taxa are indeed diploid sporophytes, then the products of meiosis are two dyads rather than a single tetrad. This would involve separation after the first division of meiosis and subsequent separate deposition of exine on the products of the second division. This seems to be the most parsimonious explanation. Incidentally, if indeed the plants producing 'membrane'-bound obligate groups of spores were ancestral to those producing dyads and tetrads, the 'membrane' too would have to be deposited after the separation following the first division in the case of dyads! We also considered the possibility that the dyads represent a cessation of meiosis after the first meiotic division, such that the cells had the diploid amount of DNA but were essentially haploid, and the sporangia are thus homologous to those containing 'normal' tetrads. The 'normal' haploid condition of the gametophyte could have been restored by completion of meiosis (i.e. the second meiotic division) during the first division of the spore nucleus, although this would have resulted in a genetic mosaic in the gametophyte, crossing-over having occurred at the first meiotic division.

Finally, we considered the possibility that the dyads were of the same ploidy level as the parent plant. Thus, haploid dyads could have been produced mitotically on a haploid sporophyte which developed apogamously under conditions unsuitable for fusion of gametes, as is recorded in certain ferns today (Hayes 1924). On the other hand, in certain angiosperms diploid dyads are produced following meiosis in megaspore mother cells because there is no cytokinesis following the first division of meiosis and each restitution nucleus so formed produces two unreduced cells after the second division (Moore 1976). In a homosporous pteridophyte, the resultant diploid gametophyte could again produce a sporophyte apogamously under dry conditions.

The possibilities illustrated above are just a few of the wide range of deviations from 'normal' life cycles of extant plants, and clearly it is impossible to do more than speculate on the nature of dyads and the life cycle of the dyad-producer described here. However, we can be certain that dyads did not result from haphazard failure of meiosis and that, in at least one instance, the plants that produced the cryptospores possessed terminal sporangia morphologically similar to those of rhyniophytes. It thus becomes increasingly important to discover more of the anatomy of the group and, in particular, the nature of conducting tissues.

Spores and delimitation of taxa

Rhyniophytoids are distinguished on sporangial morphology. Thus, species of *Cooksonia* are based on sporangial shape, but all are short and wide. Given the dearth of morphological characters, the use of anatomical features, including those on *in situ* spores, would clearly provide a more realistic indication of diversity and perhaps of relationships (or, alternatively, of convergence). However, as will be shown below, information on such characters leads to further uncertainty in deciding on their hierarchical status and to more practical problems relating to the recognition and naming of the taxa when preserved as compressions.

These dilemmas are nicely illustrated in our recent researches on *C. pertoni*, where we have demonstrated four different spore types in sporangia of similar morphology. Two have been described above, the others occur in Lower Devonian representatives (Fanning *et al.* 1988). The crassitate spores are structurally similar but differ in ornament, and we decided to place the plants in three subspecies, thus emphasizing spore structure as at least a specific character, and sculpture a subspecific character. Our decision has the added advantage that we can continue to use *C. pertoni* for sporangia of appropriate shape, but which lack information on spores. The alternative possibility, to regard spore sculpture as a specific character would have restricted the use of *C. pertoni* to sporangia containing *Ambitisporites*, as illustrated by Lang (1937). That spore structure may be a generic character in *Cooksonia* is suggested by our demonstration of crassitate spores in *C. cambrensis*, and this, in turn, partly influenced our decision to erect the genus, *Pertonella*. In the latter, the sporangia are discoidal and of identical shape to those in *C. pertoni*, but may be distinguished morphologically by prominent spinose emergences. On axes of Lower Devonian plants spines are usually considered specific characters (e.g. Gensel 1979 for *Psilophyton*; although whether or not these remarkable sporangial outgrowths should be similarly treated is debatable), therefore on morphological grounds the new plant might well be considered a new species of *Cooksonia*. However, the distinctive nature of the spores left us in no doubt that this was a distinct genus. The spores of *Pertonella dactylethra* are retusoid and thus broadly similar to those in *Caia* and *Salopella* (see p. 31). Here, although retusoid spores may characterize a genus—the best known example is *Zosterophyllum* (Edwards *et al.* 1991—and may suggest that *Caia* is congeneric with *Salopella*, we placed it in a new genus because of differences in sporangium shape, particularly at the apex. Our comments on sporangial emergences in *Pertonella* are equally relevant for *Caia*, and indeed, combined with possession of retusoid spores, may suggest a close relationship between the two, viz. that *Caia langii* should be a species of *Pertonella*. Again we decided that evidence from sporangium shape should have the greater weighting. To sum up, we are of the opinion that the possession of retusoid spores does not necessarily indicate a close relationship, especially when they are as distinctive, and unusual, as those in *Caia* and the Pridoli *Salopella* sp. (see p. 34).

Dispersed spores and the habit, habitat, and diversity of early plants

Relative abundance. In the Lower Downton (Ludlow Bone Bed and Platyschisma Members) there are 28 taxa (Richardson and Lister 1969, and unpublished data) and at least 16 cryptospore taxa (unpublished provisional data). The species list (Table 3.1) for Perton Lane appears quite extensive, but in a spore count of 250, 17 out of a total of 28 known miospore taxa are absent. They were thus produced by only a small proportion of the local flora (which thus statistically escaped preservation in the megafossil record) or by plants that grew in some remote part of the sediment 'catchment area'. Most provisional cryptospore 'taxa' figured in the count (Table 3.1).

At Perton Lane, eight of these sporomorph taxa (total 44, Table 3.1) are recorded in sporangia: one is a cryptospore and seven are miospores. Although, generally, the number of cryptospore taxa described so far is small, their abundance (based on counts of 250) in the matrix from Perton Lane and the type Ludlow area is greater than that of miospores (Table 3.2). This statistic is not matched in the megafossil record, where sporangia containing cryptospores are few (Table 3.3). Only two species have been found *in situ* and both are hilate: one is laevigate (the new taxon) and the other apiculate/granulate. The latter occur in isolated discs at another locality. None of the 'permanent' tetrads or dyads, nor any non-hilate cryptospore species, have been found in sporangia and, although the 'permanent' tetrads may be reworked, most of the spores are undoubtedly coeval. The discrepancy between numbers of dispersed spores and megafossils may perhaps result from non-recognition of the latter, poor preservation (e.g. from lack of a substantial or resilient sporangium wall), or growth well away from depositional environments. The evidence of the nature of the cryptospore-bearing plants is new and more limited than that for rhyniophytoid sporangia with miospores. However, at this stage, our data show nothing to indicate that either the numbers of cryptospores per sporangia, or the number of sporangia per plant, were greater than in plants producing trilete spores. Again, from limited evidence, the method of sporangia production is similar to the traditional rhyniophytoids, i.e. both are terminal. No branching has been

Table 3.1. Sporomorph taxa in the Lower Downton Group of the Welsh Borders, *in situ* spores and spores present in a count from Perton Lane (*)

	In situ	Count
Miospores (Total taxa 28)		
Azonate		
Laevigate		*
Retusotriletes coronadus	x	
R. dubius		
R. cf. *minor*		
R. sp. A		
R. sp.	x	*
Apiculate		
Apiculiretusispora spicula		
A. synorea		*
A. sp. A		
A. sp. B		
A. sp. C		
Murinate		
?*Dictyotriletes* sp. B		
Emphanisporites cf. *micrornatus*		
E. cf. *neglectus*		
E. sp.		
Zonate		
Equatorially crassitate		
Laevigate		
Ambitisporites dilutus	x	*
A. cf. *warringtonii*	x	*
A. warringtonii		*
Amicosporites miserabilis		
Apiculate-scabrate		
?*Aneurospora* sp.	x	
?*Streelispora* cf. *granulata*		*
Verrucate-murinate		
Synorisporites downtonensis		
S. tripapillatus	x	*
S. verrucatus	x	*
Patinate		
Laevigate		
Archaeozonotriletes chulus		
A. chulus var. *chulus*		
A. chulus var. *nanus*		*

Table 3.1 *(con't)*

	In situ	Count
Distally murinate		
nov. gen.		
var. *inframurinatus*		
Apiculate		
Cymbosporites echinatus		*
Verrucate		
Cymbosporites verrucosus		
?Perinate		
Laevigate		
Perotrilites sp. A		
Cryptospores (Total taxa 16)		
Monads (non-hilate)		
Laevigate sp. 1		*
Laevigate sp. 2		*
Apiculate sp. 1		*
Verrucate sp. 1		*
Hilate		
Laevigate		
?*Archaeozono.* cf. *divellomedium*	x	*
Laevigate sp. 3		*
Apiculate/granulate		
Apiculate sp. 2	+	*
Apic. prox. murinate 1		*
Murinate-verrucate		
?*Archaeozonotriletes dubius*		*
Verrucate sp. 1		*
'Permanent' dyads		
Laevigate		
Dyadospora sp. 1		*
Dyadospora sp. 2		*
Apiculate/granulate		
'*Dyadospora*' sp. 3		*
Murinate		
'*Dyadospora*' sp. 4		*
'Permanent' tetrads		
Laevigate		
'*Tetraletes*' sp. 1		*
Apiculate (spinose)		
'*Tetraletes*' sp. 2		*

Most of the cryptospores are not formally described and therefore the total number of taxa is provisional. Rare, undescribed miospores are not included. + : sporangium from another locality.

Table 3.2. Relative proportion of sculptural types in a count of 250 at Perton Lane

	Miospores (%)	Cryptospores (%)
Laevigate	9.7	36.6
Verrucate/murinate	21.6	3.5
Apiculate	7.0	21.6

Percentages have been recalculated excluding indeterminate forms.

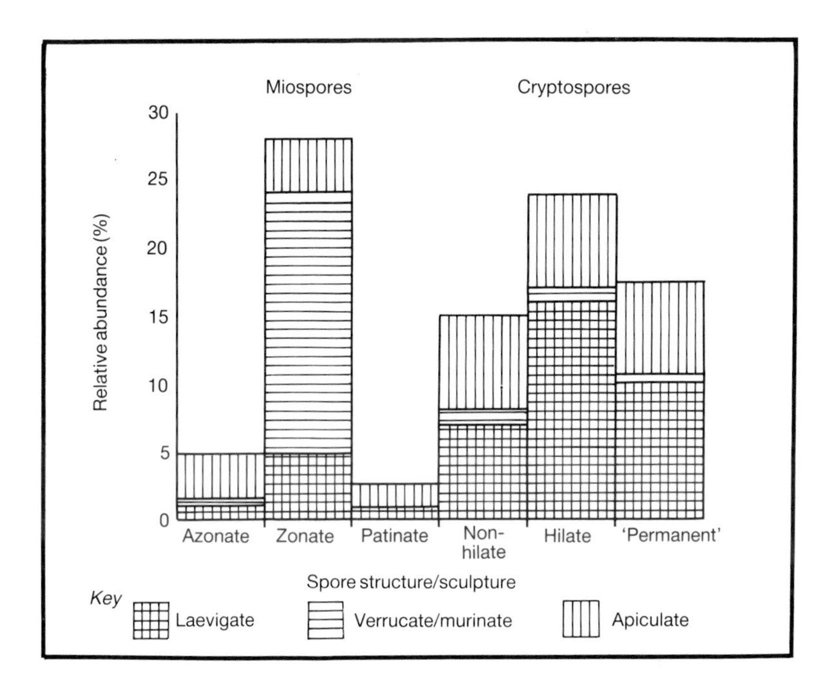

Table 3.3 Macroplants identified at Perton Lane

	Absolute number	%
Cooksonia pertoni	311	82.5
C. cambrensis	7	1.8
Pertonella	29	7.7
Caia	24	6.4
Salopella	3	0.8
New taxon	3	0.8

discovered yet and, consequently, given a unit area of 'soil' it seems likely that the parent plants of both these spore types would be producing similar numbers of sporangia. The above explanations, non-recognition and preservation, may well be applicable to the new taxon (see p. 36) but, so far, their relative contributions are incalculable.

Considering the miospores in the matrix, the most important structural form is zonate (equatorially crassitate), and the most abundant species *Synorisporites verrucatus* (16.4 per cent). This fits in with the fact that *C. pertoni* is the commonest fossil at Perton Lane (*c.* 83 per cent: Table 3.3), although the low numbers of *Ambitisporites* in the dispersed spore record, and conversely the single record of a plant with *S. tripapillatus*, remain perplexing (see pp. 29–30). The spores of *Pertonella* are not common but figure in the counts; those of *C. cambrensis* are small and are not easy to identify in strew mounts under the light microscope; the other miospore-bearing plants, *Caia* and *Salopella*, both have retusoid spores and both figure in the counts, but their lack of diagnostic characters make them less easily identifiable and therefore their numbers may be distorted. Hilate cryptospores of the new taxon figure regularly in the count (3.6 per cent) but this contrasts with the abundance of *Cooksonia* spores.

Range of sculpture and structure in dispersed spores. The spore morphologies found in sporangia include: trilete azonate retusoid laevigate, equatorially crassitate laevigate, microgranulate and verrucate forms, and proximally alete hilate, laevigate, and apiculate spores derived from dyads. Missing when compared with the dispersed miospore record are patinate spores with a wide variety of distal sculpture, laevigate, verrucate, radially murinate and spinose: azonate retusoid distally apiculate spores, and proximally murinate azonate forms with laevigate or granulate distal sculpture, and other species of structural and sculptural types represented. Amongst the cryptospores the major omissions are alete monads and 'permanent' tetrads and dyads.

Now that we have recovered spores from all but a few Silurian rhyniophytoid megafossils from the Anglo-Welsh region (exceptions are *Steganotheca* and *Torticaulis*), it would seem that most of these spore producers are not represented in the megafossil record. A good example is *Apiculiretusispora*, several species of which occur in Lower Pridoli rocks but which has not yet been found *in situ*. However, since all sporangia sampled did not contain spores, since we already have some evidence that sporangia of similar morphology produce different spores, and since there is the possibility that we have not recognized some megafossils as sporangia (as in the case of the new taxon), we reached this assumption with caution.

In deliberations on the possible adaptive significance of exine sculpture and on temporal variations in the latter, we speculated that the sculpture change in our proposed *Cooksonia pertoni* lineage (from distally verrucate to a distal sculpture of grana or coni) could have been related to environmental pressures as the early land colonizers radiated into drier habitats—possibly ensuring more reliable germination (Fanning *et al.* 1988, pp. 13, 22). According to such a hypothesis, the producers of the spores with apiculate distal sculpture may have lived in drier areas and, in the Silurian, perhaps more remote from the depositional area, with few if any of their parent plant remains reaching the depositional site at Perton Lane. Indeed, there is a marked change in facies between the Lower Downton Group (Ludlow Bone Bed and Platyschisma Shale Formations) (fluvio-marine environments) and the Lower Ditton Group (floodplain–alluvial environment remote from any marine influence). A similar explanation may hold for the producers of rare spores, e.g. *Perotrilites* sp.A, *Amicosporites miserabilis*, and patinate forms showing a range of sculpture. Clearly, more records from Silurian continental sediments are required to test this hypothesis.

Finally, comparing the sculpture in cryptospores and miospores, the data (Table 3.2) show that a relatively high percentage of the cryptospores are smooth and, when ornamented, usually apiculate. Table 3.2 is based on a sample from Perton Lane, and there may have been local floral variations within the Anglo-Welsh Basin, but it is interesting that the percentage of major sculptural types differ between miospores and cryptospores. Laevigate cryptospores are dominant, whereas in miospores the distally verrucate/murinate spores are most abundant. Apiculate/granulate cryptospores are second in abundance and are over three times more abundants than similarly sculptured miospores. The three main categories of apiculate/granulate cryptospore, non-hilate, hilate, and 'permanent' dyads and tetrads had identical representation in the counts, of around 7 per cent. In contrast, the total for the three main structural types of miospore came to the same figure. Such statistics show that our proposal, where in crassitate miospores the producers with verrucate/murinate sculpture lived in coastal habitats and those with apiculate sculpture perhaps lived in drier alluvial habitats farther from the Perton Lane site, is clearly highly speculative and over-generalized, in that the same scenario does not fit the cryptospore evidence. The fact that similar dispersal mechanisms (similar plant and sporangial morphologiles) apparently applied to three miospore groups of different structure on the one hand, and three cryptospore groups on the other, may imply that, in spite of the morphological similarities, there were two quite distinct groups of plant, possibly with different ecophysiology, and, from the dominance of the cryptospores, that the greater part of the local plant 'population' consisted of cryptospore-bearing plants (?bryophyte-like rhyniophytoids). On the other hand, the commonest recognizable plant species in the flora is *Cooksonia pertoni*, which on dispersed spore evidence usually bore verrucate miospores. The only two plant megafossils collected that contained cryptospores differ morphologically from each other, but not much from other rhyniophytoids, and contain spores with exines of the two commonest sculptural groups in the cryptospores, i.e. smooth and apiculate. The major morphological differences between these two sporangia thus illustrate the enigma that known variations in sporangial morphology do not allow the distinction of cryptospore- and miospore-bearing plants. Evolution was therefore cryptic, and at this exciting stage of land colonization is revealed to a major extent by differences in spores.

References

Allen, K.C. (1980). A review of *in situ* late Silurian and Devonian spores. *Review of Palaeobotany and Palynology* **29**, 253–70.

Banks, H.P. (1975). Reclassification of Psilophyta. *Taxon* **14**, 401–13.

Edwards, D. (1969*a*). *Zosterophyllum* from the Lower Old Red Sandstone of South Wales. *New Phytologist* **68**, 923–31.

Edwards, D. (1969*b*). Further observations on *Zosterophyllum llanoveranum* from the Lower Devonian of South Wales. *American Journal of Botany* **56**, 201–10.

Edwards, D. (1979). A late Silurian flora from the Lower Old Red Sandstone of south-west Dyfed. *Palaeontology* **22**, 23–52.

Edwards, D. and Edwards, D.S. (1986). A reconsideration of the Rhyniophytina Banks. In *Systematic and taxonomic approaches in palaeobotany*, (ed. R.A. Spicer and B.A. Thomas), pp. 199–220. Oxford University Press, Oxford.

Edwards, D. and Richardson, J.B. (1974). Lower Devonian (Dittonian) plants from the Welsh Borderland. *Palaeontology* **17**, 311–24.

Edwards, D., Feehan, J., and Smith, D.G. (1983). A late Wenlock flora from Co. Tipperary, Ireland. *Botanical Journal of the Linnean Society* **86**, 19–36.

Edwards, D., Fanning, U., and Richardson, J.B. (1991). *In situ* spores in early land plants from Wales and the Welsh Borderland: a review. *Proceedings of the Geologists Association*, in press.

Fanning, U., Richardson, J.B., and Edwards, D. (1988). Cryptic evolution in an early land plant. *Evolutionary Trends in Plants* **2**, 13–24.

Fanning, U., Edwards, D., and Richardson, J.B. (1990). Further evidence for diversity in late Silurian land vegetation. *Journal of the Geological Society of London* **147**, 725–8.

Fanning, U., Edwards, D., and Richardson, J.B. (1991). A new rhyniophytoid from the late Silurian of the Welsh Borderland. *Neues Jahrbuch für Geologie und Paläontologie*, in press.

Gensel, P.G. (1979). Two *Psilophyton* species from the Lower Devonian of eastern Canada with a discussion of morphological variation within the genus. *Palaeontographica B* **168**, 81–99.

Gensel, P.G. (1980). Devonian *in situ* spores: a survey and discussion. *Review of Palaeobotany and Palynology* **30**, 101–32.

Gray, J. (1985). The microfossil record of early land plants: advances in understanding of early terrestrialization, 1970–1984. *Philosophical Transactions of the Royal Society of London* **B309**, 167–95.

Gray, J., Massa, D., and Boucot, A.J. (1982). Caradocian land plant microfossils from Libya. *Geology* **10**, 197–201.

Hayes, D.W. (1924). Some studies on apogamy of *Pellaea atropurpurea* (C). *Transactions of the American Microscopical Society* **43**, 119–35.

Lang, W.H. (1937). On the plant-remains from the Downtonian of England and Wales. *Philosophical Transactions of the Royal Society of London* **B227**, 245–91.

Moore, D.M. (1976). *Plant cytogenetics*. Chapman and Hall, London.

Richardson, J.B. (1988). Late Ordovician and early Silurian cryptospores and miospores from Northeast Libya. In *Subsurface palynostratigraphy of north-east Libya*, (ed. A. El-Arnauti *et al.*), pp. 89–109.

Richardson, J.B. and Edwards, D. (1989). Sporomorphs and plant megafossils. In *A global standard for the Silurian system*, (ed. C.H. Holland and M.G. Bassett), pp. 216–26. National Museum of Wales Geological Series Number 9, Cardiff.

Richardson, J.B. and Ioannides, N. (1973). Silurian palynomorphs from the Tanezzuft and Acacus Formations, Tripolitania, North Africa. *Micropaleontology* **19**, 257–307.

Richardson, J.B. and Lister, T.R. (1969). Upper Silurian and Lower Devonian spore assemblages from the Welsh Borderland and South Wales. *Palaeontology* **12**, 201–52.

Rodriguez Gonzalez, R.M. (1983). *Palynologia de Las foramaciones del Silurico-Superior-Devonico-Inferior de la Cordillera Cantabrica*. Publicaciones Universidad de Leon (Spania).

Squirrell, H.C. and Tucker, E.U. (1960). The geology of the Woolhope inlier (Herefordshire). *Quarterly Journal of the Geological Society of London* **116**, 139–85.

Timms, J.D. and Chambers, T.C. (1984). Rhyniophytina and Trimerophytina from the early land floras of Victoria, Australia. *Palaeontology* **27**, 265–79.

4. *Tetrahedraletes*, *Nodospora*, and the 'cross' tetrad: an accretion of myth

JANE GRAY
Department of Biology, University of Oregon, Eugene, Oregon, USA

Abstract

An important group of pre-Late Silurian land plants produced spores that first appear in the mid-Ordovician. This benchmark may mark their first adaptive radiation on land or merely the acquisition of a preservable structure.

Until they essentially disappeared in the late Early Silurian, these spores were united in obligate tetrahedral tetrads. Two Early Silurian genera: *Nodospora* and *Tetrahedraletes* are founded on misinterpretation of tetrad geometry. In addition, distinct tetrad types may not be taxonomically unique at the genus/species level. Accordingly, tetrad geometry provides no basis for assigning *Tetrahedraletes* and *Nodospora* to different taxa.

Attribution of Ordovician/Silurian spore tetrads to embryophytes is considered controversial in the apparent absence of morphological analogues among extant embryophytes. Morphological analogues cannot be found among cyanobacteria or algae, but are found among a few bryophytes, although not among vascular cryptogams. Tetrahedral tetrads are the dominant tetrad type in embryophyte spore/pollen morphogenesis. Obligate tetrahedral tetrads are here regarded as the ancestral embryophyte diaspore.

Principal evolutionary trends in pre-Late Silurian spore tetrads are an increase in pre-late Llandovery tetrad size and rapid replacement of tetrads by single, trilete spores in the late Llandovery (late Early Silurian).

Pollen and Spores (ed. S. Blackmore and S. H. Barnes), Systematics Association Special Volume No. 44, pp. 49–87, Clarendon Press, Oxford, 1991.

Introduction

The earliest land plants for which preserved structures are known dispersed reproductive spores in obligate (permanent) tetrahedral tetrads (Gray and Boucot 1971, 1972, 1975; Boucot and Gray, 1982; Gray *et al.* 1982, 1985, 1986; Gray 1985*a*, *b*, 1988, 1989). Spore tetrads provide the only fossil record for these early Palaeozoic plants. Some authors continue to suggest an 'algal cyst' origin (cf. Meyen 1987, p. 291) for the tetrads, although no morphological counterparts among fossil or living algae are suggested. Structures with the morphology and durability of even the oldest spore tetrads are unknown among fossil or extant algae or cyanobacteria. The nearest morphological analogues are found among extant bryophyte spores, particularly among some hepatic spores. Obligate spore tetrads are scarce among vascular cryptogams. Pre-late Silurian spore-producing plants are interpreted to have been at a post-algal but pre-tracheophyte level of organization. They would therefore be the earliest embryophytes, potentially with non-vascular, possibly bryophytic organization (Gray 1985*a*, *b*). Their embryophytic status rests on the correlated characters of the spore type, just as the embryophytic status of late Silurian trilete spore producers does, since archegonia or gametophytes are seldom preserved in the fossil record.

The evolutionary tetrad stage, referred to Microfossil Assemblage (MA) Zone I (Gray 1985*b*), has been documented in North America from Tennessee to Nova Scotia, in South America (Brazil and Paraguay), in Europe (Sweden, Norway, Czechoslovakia, Belgium, UK), Africa (Libya, Egypt, Ghana, South Africa), Arabia, and Australia. This extensive suite of fossil specimens provides the basis for remarks in this paper.

MA Zone I spore tetrads are small, commonly less than 50–60 μm in diameter. Individual spores are frequently so compactly adherent that the tetrad may appear as a single globular sphere. The tetrads are smooth-walled, and may bear an ornamented or psilate 'supraexinous' envelope surrounding the tetrad. Spores rarely break from their tetrad; those that do invariably display a triradiate scar on the proximal face. From their morphology, tetrads are interpreted as meiotic tetrahedral tetrads, and individual spores as meiospores.

MA Zone I lasted about 40 Ma from the mid-Ordovician (Llanvirn–Llandeilo: Gray *et al.* 1982; Gray, unpublished) until the late Llandovery (Gray *et al.* 1982, 1986; Gray 1985*b*, 1988, 1989), when tetrads were 'replaced' abruptly by radially symmetrical trilete spores near the beginning of Zone C_5 of the late Llandovery (Gray 1989). The trilete spore is as common a morphology among vascular cryptogams as the obligate tetrad is rare. Radially symmetrical trilete spores are also found among hepatics and hornworts. Most moss spores lack a triradiate scar. This may be due to a heterochronic decrease in the time, following meiosis, that individual spores of the tetrad remain in contact before exine development. In *Fissidens*, for example, exine formation only begins during spore enlargement, after the spores separate following meiosis (Mueller 1974). This timing is neotenic in the sense that spore release from the tetrad occurs before exine is formed on which a triradiate scar can be imprinted. Nevertheless, in mosses and other bryophytes, in vascular cryptogams, and in seed plants the tetrahedral tetrad remains the prevailing tetrad type in spore morphogenesis. This is interpreted to mean that the fossil spore tetrad is the primitive, or ancestral, embryophyte diaspore, and the tetrahedral tetrad is the ancestral meiotic tetrad type in embryophyte sporogenesis. Late Llandovery replacement of spore tetrads by trilete spores is interpreted as the adaptive radiation of vascular plants, or their immediate predecessors, plus possible bryophyte lineages.

In this paper, morphologies of embryophytic spore tetrads are examined as a prelude to considering morphologies and affinities of fossil tetrads. Named fossil tetrads, attributed to distinctive tetrad types, are examined in the light of this information and redefined elsewhere (Gray, submitted). Evidence is presented for the explosive late Llandovery change-over to trilete spores.

Obligate tetrads in spores and microspores

Modern embryophytes and tetrads

Tetrad morphogenesis. Spores of bryophytes and spore-producing vascular plants, and microspores of seed plants, are produced from spore mother cells by meiosis. Meiosis takes place by two correlated successive or simultaneous divisions of the spore mother cell. Four reproductive spores, or meiospores, result, each with a haploid number of chromosomes. Meiospores are variously aligned in the tetrad depending on the orientation of the meiotic spindle axes (parallel, at right angles, obliquely, etc.) and the timing of cytokinesis and permanent wall formation. Characteristically, simultaneous cytokinesis gives rise to a tetrahedral tetrad. Other tetrad arrangements are due to successive cytokinesis. The reasons for differences in orientation of the spindle axes and timing of the divisions are unknown.

Melville (1981) has postulated that some tetrad types are determined by 'surface tension and increasing pressure' on cells (within sporangia) that are mobile before rigid walls are formed. Under 'pressure', related to their increasing size within a confined space, tetrads may be transformed from tetrahedral into different types if held in such a position until rigid exine forms. Interestingly, the occasional tetrahedral tetrads reported for cyanobacterial cells that result from successive cell division within a gelatinous sheath, are also seen as a pressure-related packing phenomenon of already separated cells within a confining space (Knoll and Golubic 1979), but from some other tetrad configuration *to* the tetrahedral, the opposite direction from that postulated by Melville (1981).

The tetrad as a dispersal unit. A tetrad as a dispersal unit is unusual among extant higher cryptogams but is more common among seed plants, where cohesion in tetrads and other permanent units has been examined in detail for a comparatively small number of taxa (reviewed by Knox 1984; Knox and McConchie 1986; Blackmore and Crane 1988). The wide spectrum of unrelated angiosperm families having obligate tetrads appears to be the result of convergent, neotenous evolution. Among spore-producers, the obligate tetrad appears to be the ancestral condition, judged by the fossil record, the incidence of obligate tetrads among hepatics that may provide the nearest modern analogues to ancestral embryophytes (Gray 1985*a*, *b*; Mishler and Churchill 1984, 1985), and the early occurrence of trilete spores in late Silurian/early Devonian tracheophytes and tracheophyte-like plants.

In permanently adherent pollen, cohesion is due to at least three, not mutually exclusive, possibilities:

(1) fusion of exine or part of the exine (tectum) to form an uninterrupted envelope over the outer surface of the tetrad;
(2) fusion of exine or part of the exine on proximal faces; and
(3) exine links or bridges on proximal faces between individual microspores without direct exine contact between inner walls of adjacent microspores.

(See, for example, Van Campo and Guinet 1961; Skvarla and Larson 1963; Roland 1965, 1971; Ayensu and Skvarla 1974; Skvarla *et al.* 1975, 1976; McGlone 1978*a*; Takahashi and Sohma 1980, 1984; Knox 1984, table 5, 2; Barnes and Blackmore 1986; Knox and McConchie 1986, table 2; Le Thomas *et al.* 1986; Blackmore and Crane 1988.)

In *Sphaerocarpos*, one of the few modern higher cryptogams with obligate spore tetrads, tetrads are held together by fusion of a continuous enveloping outer exine (an 'external' exine which does not intrude between the spores) and fusion of other exine ('middle' and 'internal' exine) on the intrasporal walls (Denizot 1976). The external exine of *Sphaerocarpos* is not homologous with the tectum of pollen (tectum has no counterpart in spores of higher cryptogams). The fusion of continuous tectum in some obligate pollen tetrads and 'external' exine in some *Sphaerocarpos* tetrads is therefore a remarkable example of morphological convergence.

In some cases, pollen/spore dispersal in tetrads or singly has been shown to have a genetic basis, sometimes being controlled by a single gene. In *Sphaerocarpos*, where a tetrad is the regular dispersal unit, *S. donnellii* has two genetic races. In one, spores remain in obligate tetrahedral tetrads; in the other they are singly dispersed (Allen 1923, 1925; Schuster 1966, p. 181; Doyle 1975). Doyle (1975) has demonstrated significant correlated differences in how 'external' exine, forming ornamentation, is fused in these two races. Complete fusion across the junction of the individual spores holds tetrads together in one race, a possibility suggested for fossil tetrads with supraexinous ornamentation. Spores that separate have only incompletely fused exine. Among angiosperms, a recessive gene governs the occurrence of pollen tetrads in *Petunia*, which normally disperses single grains (Levan 1942).

Tetrad configurations. Varied tetrad morphologies found during the development of meiospores that separate at maturity, as well as in mature adherent tetrads. Principal types are the tetrahedral and tetragonal tetrads. In seed plants, especially angiosperms, other tetrad configurations are known, but these types are infrequent. Wodehouse (1935, p. 159) suggested, in specific reference to angiosperm pollen, that 'all possible arrangements of four cells in contact are found', consequently transitional types are also recognized, including 'asymmetric' arrangements (cf. Skvarla and Larson 1963). In contrast, tetrad configurations other than tetrahedral and tetragonal are basically unknown among spores of higher cryptogams, although the decussate tetrad, recognized in sporogenesis of *Dryopteris bangii* (Huynh 1973) may play a part in ontogeny of other similarly bilateral, monolete spores normally formed in tetragonal tetrads.

However, some hepatic spores show considerable variation in meiotic tetrad types. Udar *et al.* (1974, cited in Gupta and Udar 1986, p. 9) and Gupta and Udar (1986) indicate that most of the spectrum of angiosperm tetrad types occur among some hepatics, e.g. some *Riccia* species, thus documenting what appears to be a remarkable example of convergent evolution between hepatics and angiosperms. Additionally, Gupta and Udar (1986) describe a wide variety of scar types in hepatic spores, most of which they do not relate to spore orientation in the tetrad. Hassel de Menendez (1986) illustrates spores of the hornwort *Leiosporoceros* that originate in what she refers to as a 'tetragonal bilateral alterno-opposite' tetrad with transition to an isobilateral tetragonal (i.e. normal 'square' or 'isobilateral') tetrad. A spore from such a tetrad has a monolete scar, which ramifies into two short, divergent branches at one pole, to give a pseudotrilete scar. Spores of most hornwort genera, as denoted by their radial symmetry and triradiate scars, originate in tetrahedral tetrads (Hassel de Menendez 1986).

The tetrahedral tetrad: ancestral type. Spores of higher cryptogams are characteristically oriented in tetrahedral or tetragonal tetrads during sporogenesis, with tetrahedral prevailing.

The tetrahedral tetrad can be defined simply as spores 'arranged in two planes' (Erdtman 1943, p. 52), or as spores aligned as if they were located at the apices of a tetrahedron. Tetrahedral tetrads are isodiametric, the individual spores meet at the centre of the tetrad and each has three equally spaced contact points with its neighbours. The spores themselves are also characteristically isodiametric, of uniform diameter and radial symmetry (at least in polar view); heteropolar spores have no horizontal plane of symmetry.

The tetrahedral tetrad is generally illustrated with one spore uppermost, lying over the other three, or some minor variation of this orientation (Fig. 4.1). Seldom is a tetrahedral tetrad illustrated in a 'true' lateral position, those of Erdtman (1943), for example, are all in the conventional orientation.

That tetrahedral tetrads are ancestral or 'primitive' in embryophyte sporogenesis is strongly supported by fossil tetrahedral tetrads as the oldest spore type and by the prevalence of trilete spores, whose ontogeny includes a tetrahedral tetrad, in the late early and late Silurian and Devonian. The tetrahedral tetrad prevailed in ontogeny as the only tetrad type for 60 or 70 million years. All other tetrad types in meiospore morphogenesis are derived and appeared later in time. Maheshwari's (1950, p. 45) suggestion that simultaneous cytokinesis is ancient, and successive cytokinesis derived, would appear to be confirmed by the fossil record.

The tetragonal tetrad. In the tetragonal tetrad (also called 'square', 'isobilateral', or 'cruciate'), spores are aligned parallel to one another in a single plane, each being in contact with its adjacent neighbour along two sides, and all four meeting along a line parallel to the long axis of the spore. Previous alignment of a mature spore in a tetragonal tetrad is denoted by its bilateral symmetry and the single, monolete

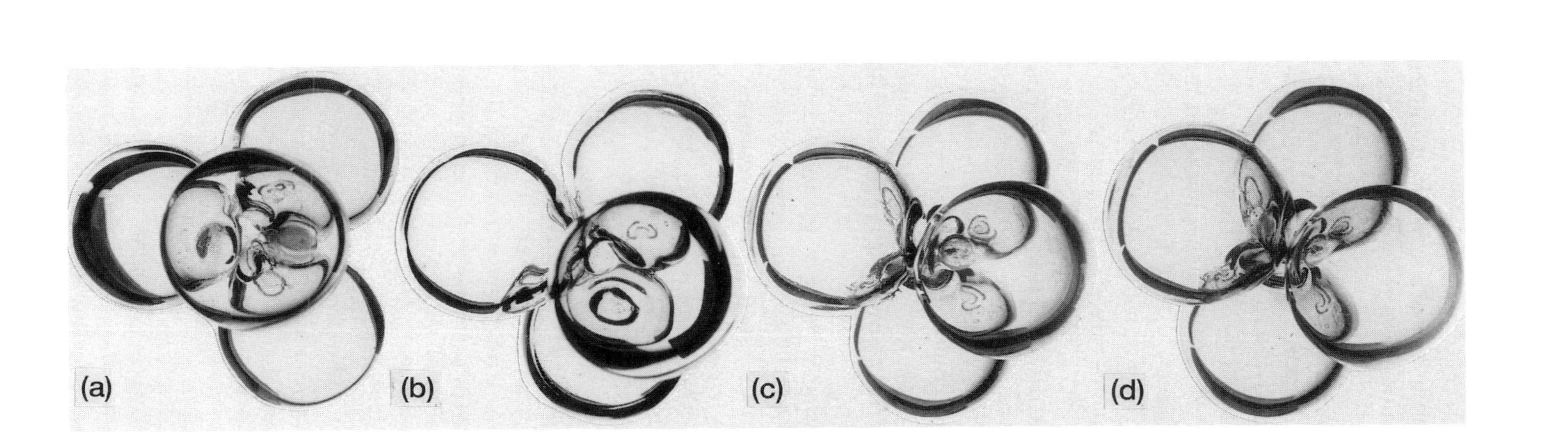

Fig. 4.1 A glass model of a tetrahedral tetrad rotated from the conventional one-spore-uppermost position (a), to the 'cross' tetrad position (d), which is the fully lateral position. These configurations have been used as the basis for recognizing distinct tetrad types and as the basis for Palaeozoic spore tetrad genera (see text).

scar which marks the boundaries of the two contact faces. The decussate tetrad may provide the same symmetry and scar pattern (Huynh 1973) in a separated meiospore.

The first spores indicating a tetragonal (or possibly decussate) tetrad ontogeny occur in the Devonian. The tetragonal tetrad would thus appear from the fossil record to be the first of the derived tetrad types.

The cross tetrad. The cross tetrad has a checkered history. Erdtman's (1945) figure for the cross tetrad, the prototype for this configuration, was unaccompanied by a definition. It is principally reported in pollen grains, especially among mature dicotyledon pollen (Erdtman 1945; Selling 1947; Cranwell 1953; Walker 1971; Walker and Doyle 1975; Melville 1981). Gupta and Udar (1986, p. 57, pl. XX) state that cross tetrads (called 'tetragonal' in their paper) also occur during sporogenesis in the hepatic *Riccia cruciata*, although the meiospores separate at maturity. Their's is the only report of cross tetrads in higher cryptogam spores. A cross tetrad and a so-called 'contiguous cross' tetrad have also figured in the literature of Precambrian algal cells (Oehler *et al.* 1976).

Despite such references, it is easy to demonstrate that the cross tetrad is merely a tetrahedral tetrad in fully lateral view. Oehler *et al.* (1976, Fig. 3C,D) recognized, but did not comment on, the fact that Erdtman's (1945) cross tetrad is merely an alternate orientation for the tetrahedral tetrad. Their contiguous cross tetrad (their fig. 3B) falls into the same category in my opinion. As conceived by some, the cross tetrad may be due to minor malformation of a tetrahedral tetrad in fully lateral view or conceivably may be a 'pressure'-related phenomenon (*sensu* Melville 1981).

Cross tetrads have also been referred to as 'decussate', a term not used by Erdtman in 1945. They have surfaced in the bryophyte literature with regard to spores of the hepatic *Riccia cruciata*, as a 'tetragonal' tetrad (Gupta and Udar 1986, pl. XX).

An early user of 'decussate' was Walker (1971). Walker and Doyle (1975, p. 686, Fig. 12) and Melville (1981) also referred to Erdtman's cross tetrads as decussate. The decussate microspore tetrads of embryologists have quite different morphology (Maheshwari 1950; Huynh 1973, 1976). Tetrads with the same morphology were illustrated and labelled 'hexahedral' by Erdtman (1943, p. 52). In 1969, Erdtman defined and illustrated tetrahedral, rhomboidal, tetragonal, and decussate (but not cross) tetrads. His definition of decussate conformed to the usage of embryologists, '*Decussate* tetrads with the same more or less square outline as tetragonal tetrads but with the (elongated) pollen grains pairwise at right angles to each other . . .' (p. 38). This distinction was not made by Walker (1971), Walker and Doyle (1975), or Melville (1981).

The above literature provides the principal evidence for the existence of the cross tetrad (and its morphological equivalents; decussate of Walker 1971; Walker and Doyle 1975; Melville 1981; and tetragonal of Gupta and Udar 1986). The cross tetrad is always indicated to be rare compared with other tetrad types (Walker 1971) and is seldom recognized as occurring independently of other tetrad morphologies. For example, in a partial list of angiosperm pollen in permanent tetrads (Erdtman 1945), only *Chilopsis linearis* (Bignoniaceae) and *Prepusa connata* (Gentianaceae) are identified as having exclusively cross tetrads. Taxa in the same families, including another species of *Chilopsis*, are indicated to have pollen in tetrahedral tetrads sometimes in combination with other tetrad types. The cross tetrad thus occurs jointly with one or more other tetrad types, most commonly with tetrahedral tetrads. Selling (1947) mentions cross tetrads in *Vaccinium* and *Gardenia* but only in conjunction with more regular occurrence of tetrahedral tetrads. Cranwell's (1953) cross and tetrahedral tetrads are both present in *Pentachronda pumila*, which belongs to a family (Epacridaceae) almost all of whose pollen is released regularly in tetrahedral tetrads, except in tribe Styphelieae where the tetrad condition is modified in some taxa by early abortion of some of the grains (Erdtman 1952; McGlone 1978*a*, *b*). Walker's (1971) decussate tetrad was mentioned only in conjunction with other tetrad types in the same taxa. Melville (1981) mentions decussate tetrads in *Calluna vulgaris* and other Ericaceae, where the prevailing configuration is a tetrahedral tetrad. Gupta and Udar's (1986) tetragonal tetrad occurs only in *Riccia cruciata*, where the tetrahedral tetrad is most common, as denoted by the general occurrence of trilete spores. Conceivably, Melville's suggestion of a pressure- or physical space-related sequence, tetrahedral tetrad–decussate–rhomboidal–linear, might have merit.

Further confusion was promulgated by Kremp (1965, pp. 34, 167, pl. 1, 2) who reproduced originally uncaptioned tetrad figures from Iversen and Troels-Smith (1950, pl. XI) and Faegri and Iversen (1950, pl. V) and captioned them incorrectly. Iversen and Troels-Smith clearly intended their illustration as three separate sets of plan *and* side (or lateral) views of three tetrad types: tetrahedral (row 1); square (or isobilateral or tetragonal) (row 2); and rhomboidal (row 3). Kremp identified two of the three tetrads (row 1) as tetrahedral tetrads; the third, the fully lateral view of the tetrahedral tetrad, Kremp identified as a cross tetrad.

This misunderstanding results from failure, with one exception (Iversen and Troels-Smith 1950), to illustrate the tetrahedral tetrad in genuinely lateral orientation. A glass model tetrahedral tetrad illustrates the point (Fig. 4.1) that a cross tetrad is merely a laterally oriented tetrahedral tetrad.

Apparently, some have construed a departure from radial symmetry of individual spores of a tetrahedral tetrad in genuinely lateral orientation as a cross or decussate tetrad (*sensu* Erdtman 1945; Melville 1981), i.e. as an authentic change in configuration. Because pollen and spores are prepared chemically, and mounted beneath a coverslip, any such departure is best explained as an artefact due to coverslip compression. Similar distortion may be due to pressure exerted within the sporangium by developing microspores (Melville 1981). In fossil material the potential for flattening or distortion is ever present, especially where an extensive sedimentary overburden may have led to regional metamorphism (Gray and Boucot 1975).

Is tetrad variation taxonomically significant? Diverse tetrad types may be associated with the tetrahedral tetrad, regardless of mechanisms related to spindle axes, timing of cytokinesis, and wall deposition. Angiosperm genera with pollen customarily in permanent tetrahedral tetrads occasionally have tetragonal or rhomboidal tetrads; genera with tetragonal tetrads frequently have rhomboidal tetrads (Walker 1971, p. 10). Monocot taxa often have square (isobilateral) and decussate tetrads (Huynh 1976). Pollen of *Hedycara angustifolia* (Monimiaceae) may be arranged in more or less tetragonal tetrads, decussate tetrads, or intermediates (Sampson 1977, p. 62). Takahashi and Sohma (1982) note that pollen of *Drosera pygmaea* and *D. glanduligera* (Droseraceae) are united in tetragonal, rhomboidal, and tetrahedral tetrads. Tetrad variation in the monocot *Typha latifolia* (Typhaceae) is well known (Wodehouse 1935; Skvarla and Larson 1963). The same is true among some hepatics whose spores remain permanently adherent. Thus the tetragonal tetrad (= 'cruciate') is predominant in *Riccia perssonii*, but occasional rhomboidal, tetrahedral, or decussate tetrads occur (Gupta and Udar 1986, p. 55, Pl. XX). Spore tetrads of *R. curtisii* (Gupta and Udar 1986, p. 57, Pl. XX) may be tetrahedral or rhomboidal.

Distinctions based on different tetrad configurations have no necessary taxonomic significance, any more than variations in the number of pores have taxonomic significance in pollen of any one species of *Carya* or *Alnus*.

Although coherent postmeiotic tetrads are not involved, tetrahedral and tetragonal tetrads may both occur in sporogenesis of genera or species of vascular cryptogams, though one condition is usually fixed at the generic level. Selling (1944, 1946) lists more than a dozen genera of spore-producing vascular cryptogams with tetrahedral, trilete *and* monolete, bilateral spores, many in the same species. He records similar diversity even within the same sporangium (Selling 1946, p. 22) and suggests that many examples of co-occurring trilete and monolete spores from his 1944 paper (pp. 7–8) are also from single sporangia (Selling 1946, p. 22). Nayar (1963) reports both trilete and monolete spores in the tropical fern *Loxogramme*. Within a single sporangium of *L. lanceolata* there may be tetrahedral trilete spores, bilateral monolete spores, and spores intermediate in shape and length and angle of divergence of the radii. Tryon and Tryon (1982, p. 10) and Tryon (1986) note other examples of the same phenomenon at the generic level. Gupta and Udar (1986) report many similar instances among hepatics, noting wide variation in spore shapes and aperture types within generic and specific level taxa, including the absence of scars or indistinct scars in taxa whose spores also have conspicuous scars.

Fossil embryophytes and tetrads

The relevance to the fossil record of the above discussion is that Erdtman's misconception of tetrad geometry has been extended into the Palaeozoic by Strother and Traverse (1979) who erected two Silurian (Llandovery) genera for obligate tetrads. They used Erdtman's distinction between tetrahedral and cross tetrads as the basis for two genera: '*Tetrahedraletes* is constructed to include spore tetrads which have a tetrahedral configuration; the genus *Nodospora* includes tetrads with a cross configuration (*sensu* Erdtman 1943) [*sic* Erdtman first illustrated the cross tetrad in 1945] . . . These consistent configurations combine with other morphological features including size, shape, sculpture, and the abovementioned thickenings, to provide us with a strong basis for differentiating taxa.' (Strother and Traverse 1979, p. 8.)

Their formal diagnoses for each taxon make clear that tetrahedral or 'cross' tetrad configuration is the *only* consistent difference used to distinguish *Tetrahedraletes* (one species) from *Nodospora* (three species). The issue has been additionally complicated by the fact that tetrads of both taxa are smooth-walled, but sometimes, in *Nodospora*, have a detachable, ornamented supraexinous envelope. Since '*Nodospora*' is merely a lateral tetrahedral tetrad, it is clear that all pre-Devonian spore tetrads are obligate tetrahedral tetrads and that all originally 'smooth-walled *Nodospora*', as well as those that have lost their supraexinous ornamentation, could be assigned to *Tetrahedraletes* were both 'genera' not reduced to *nomina dubia* (Gray, submitted). Distinctions between these two 'genera' are merely an accretion of myth.

The fossil tetrad as a dispersal unit. Obligate tetrads as dispersal units, especially those with supraexinous envelopes, have occasioned much discussion with regard to affinities of the fossil tetrads. As discussed above, adherence in tetrads is far from a novelty in embryophyte diaspores and would appear to be neotonous. In contrast, meiotic tetrads are unknown among those algae considered to be most closely related to embryophytes, and mature obligate meiotic tetrads are unknown among any algae (see p. 63). Nor are envelope-bearing tetrads a novelty among embryophytes, since such 'envelopes' (as fused exine or fused tectum) provide one of the possible mechanisms explaining physical cohesion in embryophyte tetrads.

By analogy with various possibilities for cohesion in pollen and *Sphaerocarpos* tetrads, fossil tetrads must share the absence, or limited duration of, intrasporal polysaccharide walls during tetrad sporogenesis. In fossil tetrads with adherent, ornamented, supraexinous envelopes, tetrads may be held together by fusion of exine ornamentation from spore to spore, as in *Sphaerocarpos*. Coherence of non-envelope-bearing tetrads, or tetrads with loosely adherent envelopes, suggests that proximal faces of adjacent spores also share fused exine, as in many mature pollen tetrads. When MA Zone I tetrads break apart, damage to the proximal spore face suggests exine fusion (and a thinner exine) between adjacent spores, similar to adherent pollen tetrads whose proximal exine-sharing faces are thin compared with distal exine (cf. Ford 1971; Roland 1971: Sampson 1977; McGlone 1978*a*; Le Thomas *et al.* 1986). Cohesion is also correlated with incomplete cytokinesis during meiosis, which leaves links or bridges between adjacent spores. John Rowley (personal communication, 1988) suggests that wall links are a 'definite possibility' for the non-envelope-bearing spore tetrad illustrated in Gray (1985*b*, Pl. 2, Fig. 18), and examples of wall-linked tetrads are common in MA Zone I.

Cohesion mechanism(s) will not be clear until fossil tetrads have been sectioned and studied with TEM and the reasons for dissociation of tetrads into trilete spores in the late Llandovery also remain unclear. A simple mutational change may have led to various physical changes correlated with tetrad dissociation and dispersal of single spores and pollen grains.

Envelope- and non-envelope-bearing spore tetrads

Morphology. With the optical microscope, Ordovician/early Silurian spore tetrads seem to fall cleanly into two categories: smooth-walled tetrads with an outer, ornamented or loosely fitting psilate supraexinous layer that encompasses the tetrad and obscures the sutures between individual spores, here referred to as envelope-bearing; and smooth-walled tetrads, here referred to as non-envelope-bearing. *Tetrad envelope* is a neutral term for the supraexinous layer in the absence of information about its potential developmental homologues among extant embryophyte spores. The several types of ornamentation in MA Zone I spore tetrads are always confined to the tetrad envelope, and may be stripped from the tetrads much as perispore in fern spores can be stripped to leave smooth-walled spores. The timing in differentiation of the exine walls of the spore tetrad in conjunction with differentiation of the tetrad-envelope is unknown. Some 'psilate' envelopes, particularly those that are loosely fitting (cf. Gray 1985*b*, Pl. 1, Figs. 2, 3, Pl. 2, Figs 21, 22), may be the result of diagenetic degradation of ornamentation; others cannot be accounted for by this means (cf. Gray 1988, Fig. 3), especially in the Malvinokaffric Realm, where tetrads with reticulate and rugulate envelopes are unknown.

With the optical microscope, what appear to be smooth-walled non-envelope-bearing tetrads comprise the majority of Ordovician/early Silurian tetrads at most localities on a global basis (cf. Gray *et al.* 1982, Figs 2–4, 8, 9; Miller and Eames 1982, Pl. 6. Fig 1, 3; Duffield 1985, Fig. 3(2, 4–6, 9); Johnson 1985, Pl. XI, Figs 1, 3, 4; Gray *et al.* 1986, Fig. 6). Scanning electron microscopy (SEM) examination makes it clear that many of these tetrads are genuinely smooth-walled, and non-envelope-bearing. Not only do they lack evidence of ornamentation, but a suture between tetrad spores is evident (Miller and Eames 1982, Pl. 5, Figs 1, 5; Gray 1985*b*, Pl. 2, Fig. 18; Gray 1988, Figs 1, 5), and exine bridges can often be discriminated within the suture, indicating the absence of a supraexinous envelope. Other tetrads, however, that appear 'optically' to belong to the same category, can be seen with SEM to be covered by a thin, closely adherent smooth-walled tetrad envelope concealing suture lines between the tetrad spores (Gray, submitted). In other words, many spore tetrads believed 'optically' to be non-envelope-bearing do, in fact, have tetrad-envelopes. This is analogous to the recognition of 'optically' inconspicuous perispores in fern spores. With the optical microscope, the only perispore-bearing fern spores were those with obvious, prominantly sculptured, and/or loosely adherent and often readily detachable perispores. Thin, tightly adherent, lightly sculptured, or unsculptured perispores of many fern spores were not readily discriminated until the wide use of transmission electron microscopy (TEM) and SEM in examining fern spores (cf. Lugardon 1972, 1974; Lloyd 1981; Tryon and Tryon 1982).

In Ordovician/early Silurian tetrads, recognition of closely adherent, unornamented tetrad envelopes is exacerbated because poor preservation and coalification make the contact area between individual tetrad spores, and consequently the presence of a tetrad-envelope, impossible to see *except* with SEM. Consequently, extensive SEM work is necessary to discriminate what appear to be genuinely smooth-walled, non-envelope-bearing tetrads from those that bear tightly adherent psilate envelopes. This also means that, in the absence of adequate SEM work, Strother and Traverse's 'smooth-walled *Tetrahedraletes*' is so inadequately described that published taxa attributed to it cannot be confidently assigned, although *tentative* assignments are possible on the basis of apparent biostratigraphic/phytogeographic restrictions. Within the Malvinokaffric Realm, Ordovician/Silurian spore tetrads with reticulate and rugulate tetrad envelopes (cf. Gray 1985*b*, Pl. 1, Figs 4–16; Pl. 2, Figs 17, 19, 20; Gray 1988, Figs 2, 4, 6) are unknown, only tetrads with closely adherent psilate envelopes and non-envelope-bearing, smooth-walled tetrads are recognized. In the North Silurian Realm non-envelope-bearing tetrads and tetrads with closely adherent, psilate envelopes are recognized in the post-early C_5 Llandoverian of North America (MA Zone II), in the late Wenlockian of Norway (MA Zone II), and in the mid-Ludlovian of Sweden (MA Zone III) (Gray and Boucot, in preparation) following extinction of tetrads with ornamented tetrad envelopes. Both are rare in MA Zone III. Prior to the late Llandovery extinction event in the North Silurian Realm, tetrads with psilate envelopes appear to be genuinely rare. This being the case, one can infer that most extra-Malvinokaffric Realm tetrads that appear optically to be non-envelope-bearing in pre-C_5 beds are probably truly non-envelope-bearing.

Moreover, SEM examination reveals that *all* tetrad envelopes, including psilate, firmly adherent envelopes, can be stripped from the tetrad (Gray, submitted), not just those that appear optically to be only loosely attached or in the process of being shed (for example, *Nodospora retimembrana*, Miller and Eames 1982, Pl. 6, Figs 4–6; Gray 1985*b*, Pl. 1, Figs 2–6, 14–16; *Nodospora burnhamensis*, Strother and Traverse 1979, Pl. 2, Fig. 5; Duffield 1985, Fig. 3(8)).

Consequently, non-envelope-bearing, smooth-walled tetrads may have different origins. Some may *never* have had tetrad envelopes. Others may naturally have shed envelopes that were only loosely adherent, or may have lost them through inadequate preservation or as the result of harsh extraction. This concept parallels circumstances in the Quaternary and Holocene, where spores assignable to various Polypodiaceous taxa have a readily detached ornamented perispore. Smooth-walled spores that result from perispore loss are generally indistinguishable (see Selling 1946, p. 62).

In sum, SEM work is essential to discriminate genuinely smooth-walled tetrads and tetrads with thin, closely adherent psilate envelopes, including previously defined genera (*Tetrahedraletes* and *Nodospora* based on the illustrated 'smooth-walled' type specimen) that potentially fall into one or both categories. Consequently, neither taxon is usable, and both are *nomina dubia* (Gray, submitted).

Affinities. The affinities of fossil spore tetrads with and without envelopes have been, and continue to be, the subject of speculation because of the near absence of morphological analogues among modern embryophytes. Attention has often centred on obligate tetrads as 'an indication of a semi-aquatic algal origin', and potential affinities with red algae have been singled out (Hemsley 1991) because they are the only algae that produce tetrahedral tetrads of meiospores that could potentially carry triradiate scars like those of the trilete spores of higher cryptogams. Although meiotic tetrahedral tetrads are known within Ceramiales: Florideophyceae, as Hemsley noted it bears stressing that red algal tetrads and spores arising from them are not resistant to drying or to chemical treatment and do not have a rigid wall on which a triradiate scar could be imprinted (see Gray and Boucot 1977). Fritsch (1945, p. 604) pointed out that 'liberated tetraspores are always naked and gliding movements . . . are reported . . .' together with 'amoeboid changes of shape'. Chamberlain and Evans (1973) studied spore production in the red alga *Ceramium* and reported the following:

(1) tetraspores squeeze at maturity one by one through a small aperture in the tetrasporangium prior to rounding up;
(2) each liberated spore is surrounded by a sheath of hyaline mucilage;
(3) the mucilage seems to play a protective role, preventing immediate bacterial attack on the 'naked spore'; and
(4) the spore is naked in the sense that it is enclosed only by plasma membrane.

Taylor and Taylor (1987) found no visible wall in tetraspores of the red alga *Polysiphonia* and, contrary to Hemsley's assertions of what Taylor and Taylor reported, specifically state that spores of neither the brown alga *Dictyota* (whose spore tetrads were square or isobilateral) nor *Polysiphonia* were chemically resistant.

The occurrence of meiotically produced, obligate tetrahedral tetrads or dissociated trilete spores with durable walls, is unknown among any extant algae. Attempts to attribute Ordovician/Silurian spore tetrads, with or without envelopes, to algal meiospores simply obfuscates attempts to understand their affinities and phylogenetic significance.

Even more novel affinities have been sought for envelope-bearing tetrads, quite apart from smooth-walled tetrads, among unicellular green algae (see below). Attributions to green algae (Strother 1982; Johnson 1985; Satterthwait and Playford 1986) necessarily divorce the tetrad envelope from the tetrad in seeking to explain its origin for, as discussed below, green algae have no meiotic tetrads; some have mitotically produced tetrads but these are not obligate.

In the fossils, since the tetrad envelope encompasses a spore tetrad, potential analogues for it cannot be divorced from structures that are not also spore tetrads. Moreover, the fact that the envelope is detachable means that affinities cannot be suggested for envelope-bearing tetrads that do not include tetrads without envelopes. In seeking ontogenetic and phylogenetic analogues among modern organisms, the complete unit, tetrad with tetrad envelope, must be considered, as well as the possibility that smooth-walled tetrads are merely tetrads from which the envelope has been dislodged. Potential morphological analogues, and by implication phylogenetic relationships, among any groups of fossil or living algae cannot be demonstrated (see below).

Envelope-bearing spore tetrads. Attempts have been made to explain envelope-bearing tetrads in terms of persistent algal mother cell walls or spore mother cell walls (Strother 1982; Johnson 1985). Potential ontogenetic sources for tetrad envelopes among modern embryophytes spores, such as perispore and exine, have been mostly ignored (see Gray 1985*b*). These three possibilities are discussed in detail because of their bearing on the embryophytic relationships of envelope-bearing tetrads.

Mother cell wall-, spore mother cell wall-, and sheath-enclosed tetrads. Could tetrahedral tetrads like those produced mitotically among some green algae, or even among cyanobacteria by cellular shifting of vegetative cells, be a potential source of fossil envelope-bearing spore tetrads as some have implied? In some modern unicellular Chlorophyceae, particularly dry-soil Chlorosarcinales, vegetative cell division may form dyads, isobilaterial or tetrahedral tetrads, as well as complexes that include several tetrad generations. Tetrahedral tetrads are particularly common in the life cycle of *Borodinella, Borodinellopsis, Friedmannia, Pseudotetracystis*, and *Tetracystis* (Fritsch 1935; Chantanachat and Bold 1962; Brown and Bold 1964; Dykstra 1971; Arneson 1973). Chlorosarcinalean daughter tetrads may be temporarily retained within the parent cell wall, but tetrads are transitory, and dissociate with age due to deterioration and rupture of the parent wall; moreover, daughter cells, whether single, in tetrads, or tetrad complexes, commonly divide to produce zoosporangia, that rupture to release zoospores or aplanospores (Brown and Bold 1964, text figs 1 and 4). Vegetative tetrads also may be found among a few aquatic/soil Chlorococcales such as some *Chlorella*. In *Chlorella vulgaris* the four autospores are released singly from the mother cell before they reach maturity (Griffiths and Griffiths 1969). It is the presence of sporopollenin in 'mother cells walls' in some Chlorophyceae, such as *Chlorella* (Atkinson *et al.* 1972; Honneger and Brunner 1981), that led to the suggestion that envelope-bearing spore tetrads of the early Silurian might be related to green algae (Strother 1982; Johnson 1985).

Only rarely has a *potential* fossil example been reported. Oehler *et al.* (1976, pp. 266–7) describe a possible mitotically produced tetrahedral tetrad from the Amelia Dolomite (*c.* 1500 Ma). They conclude that the 'Amelia Dolomite tetrahedral tetrad most closely resembles a mitotically derived green algal autospore tetrad enclosed within the wall of the parent cell', but they remain uncertain whether the tetrahedral tetrad geometry is original or resulted from 'cellular shifting'. There are no associated unicells with 'triradiate marks'.

Tentative attribution to green and red algae (Schopf and Blacic 1971) of 'tetrahedral tetrads' of spores with 'sheaths', or what could be interpreted as spore mother cell wall, in late Precambrian deposits may have encouraged some to consider extinct groups within Chlorophyceae and/or Rhodophyceae, to which the Ordovican/Silurian spore tetrads might belong.

Eotetrahedrion princeps is the principal evidence for this morphology. One specimen comprised of 'four sporelike unicells' encompassed in a hyaline membrane (Schopf and Blacic 1971, p. 939) comes from the Bitter Springs Formation (*c.* 900–1000 Ma). Solitary unicells, associated with and morphologically resembling *E. princeps*, some showing 'triradiate scars', are interpreted as freed meiotically produced algal spores. Others (Knoll and Golubic 1979) interpret *E. princeps* as fortuitously aligned cells of procaryotic cyanobacteria, having nothing to do with meiosis. In comparison with modern cyanobacteria, Knoll and Golubic (1979, Fig. 9) interpret *E. princeps* as due to a secondary shift in the original four cells from a square tetrad to a tetrahedral tetrad as the cells increase in size and separate from one another within the confinement of the 'parental envelope' in which daughter cells divide. The 'tetrahedral tetrad' arrangement is believed to be the 'most stable packing arrangement' within the parental envelope, which bursts when it reaches the limits of its elasticity, releasing the separate daughter cells. The tetrahedral tetrad among cyanobacterial cells is thus a temporary and fortuitous occurrence. Four daughter cells in tetrad arrangement within a common gelatinous, extracellular envelope is also fortuitous and does not provide a counterpart for the envelope-bound tetrahedral tetrads described here. Knoll and Golubic (1979) regard triradiate scars that Schopf and Blacic (1971) observed on dispersed spores as folds that resulted from post-mortem diagenetic shrinkage. Some specimens include more than one triradiate scar (Knoll and Golubic 1979).

No basis exists for suggesting a phylogenetic relationship between Ordovician/Silurian spore tetrads and meiotic or mitotic tetrahedral tetrads produced by any extinct or living algae. Visual or potential preservational counterparts of only a part of a structure (such as a mother cell wall with sporopollenin) in a modern organism provide no justification for suggesting a relationship to envelope-bearing spore tetrads. In searching for likely phylogenetic analogues, durability *and* morphological similarity are equally important. Ultimately, the homologous nature of tetrad envelopes to other structures must be demonstrated, as well as entire morphological continuity, in suggesting a phylogenetic relationship.

The presence of sporopollenin in some algal mother cell walls may assure their durability, but sporopollenin provides no phylogenetically significant information once its presence is divorced from a structure with morphology analogous to that of the fossil spore tetrads. Sporopollenin has been demonstrated, for example, in cell walls of myxobacteria (Strohl *et al.* 1977), in phycomycete and ascomycete sexual spore walls (Gooday *et al.* 1973, 1974; Furch and Gooday 1978; Furch and Pambor 1978), and in algal cell walls (Honneger and Brunner 1981, Table 1), as well as in embryophyte spores.

It might be noted that algal zygotes, or zygospores, in some freshwater and terrestrial green algae are often variously ornamented and might resemble spores of higher cryptogams. These walls are ruptured when the zygote germinates and divides meiotically to produce zoospores (meiospores). The durable zygote wall does not surround a tetrahedral tetrad of spores.

Perispore and tetrads Seeking morphological analogues among modern cryptogams for ornamented and loosely adherent psilate tetrad envelopes of Ordovician/early Silurian spore tetrads, I used the term perispore (Gray 1985*b*). By describing detachable spore tetrad ornamentation as perispore, I used the term as Bower (1959) had for classically defined perispore of fern spores: descriptively and topographically for a conspicuous, loosely fitting, readily detached, ornamented or often more or less unornamented or membranous layer, external to exine in single fern spores (see Denizot 1974, who reviews wall terminology applied to pteridophyte and bryophyte spores). Thus defined, perispore is typically or classically regarded as enveloping single spores of dryopteroid, blechnoid, and asplenoid ferns (Wagner 1974). In some fern spores, e.g. *Asplenium* studied in ultra-thin sections, a space between perispore and spore exine may account for the easy detachment of perispore (Pettitt 1966, p. 242, Pl. 12, Fig. 2). Conversely, perispore may remain firmly adherent to the spore, as the tetrad envelope does in some fossil tetrads. The only difficulty in the morphological analogue between tetrad envelope and perispore is that perispore is differentiated around individual spores rather than around the tetrad unit from which spores dissociate.

Few cryptogam spores remain in adherent tetrads in postmeiotic phases of sporogenesis, compared with pollen where obligate tetrads and other multicellular units are common. Regular separation of spores must mean either that there is limited opportunity for exine fusion, or that wall links are small and transitory. If spores regularly remained in tetrads, it seems reasonable to assume that tetrads would share perispore, just as closely appressed pollen tetrads may share a fused exine (tectum) across the juncture of the individual microspores. The morphological analogy between envelope-bearing fossil tetrads and perispore-bearing modern spores seems a poor one only because no ferns or other vascular cryptogams and few non-vascular cryptogams, have mature spores normally remaining in tetrads.

Several examples support this view. Perispore will differentiate around more than individual spores in circumstances potentially analogous to adherence of fossil spores in a tightly appressed tetrad, when meiosis or cytokinesis fails. Among vascular cryptogams (and bryophytes), dyads and occasional triads may occur under circumstances of meiotic irregularity (Allen 1945; Hickok and Klekowski 1973; Klekowski 1973*a*, *b*; Hickok 1979; Pettitt 1979). These results of meiotic failure may come to lie within a common wall. Erdtman illustrated a dyad of *Selaginella kraussiana* (Kunze) A. Br. (Erdtman 1957, p. 93, Fig. 179f) showing two smooth-walled microspores with a 'common cover'. This specimen was later described (Erdtman and Sorsa 1971, p. 169) as 'the "non-perinous" parts of two microspores confined within a supposedly perinous cover common to them both'.

Pettitt (1979, fig. 12) provides a similar example for microspores of *Marsilea drummondii*. Each spore of the dyad that results from failure of cytokinesis following meiosis has an individual exine, but the dyad shares a common perispore. The shared perispore (perine) can be traced to circumstances where the tapetal 'periplasmodium is unable to penetrate between the closely apposed exines at the proximal faces of the spores. In a normal tetrad the perine encloses each spore separately . . .' (Pettitt 1979, p. 237).

Chaloner (1958, p. 198) describes a Carboniferous spore tetrad, attributed to coenopterid ferns, as 'enclosed in a fine meshwork of cuticular substance which projects locally beyond the margin of the flattened fertile spores, and leaves a reticulate pattern where it passes over their surface'. He interprets this loose, resistant layer as the probable remains of tapetum (see also Pettitt 1966, p. 249), or one might suggest remnants of perispore, a tapetal derivative.

These examples, fossil and modern, suggest that if individual spores of an obligate tetrad are closely appressed or tightly conjoined, a supraexinous perispore will synthesize around the entire tetrad. Consequently, it is possible to conclude that obligate spore tetrads with enveloping perispores seldom occur, very largely because obligate spore tetrads are also rare among higher cryptogams.

Perispore also has developmental or ontogenetic implications. Thus, Bower (1959) identified perispore as formed from the tapetal periplasmodium of the sporangium, that is, as derived from the tissues of the sporophyte sporangium rather than from spore protoplast. Lacking such information for the tetrad envelope, it cannot be determined whether perispore of modern fern (and moss) spores is a possible developmental homologue for the tetrad envelope of fossil tetrads. Bower also attributed an evolutionary and potential phylogenetic significance to perispore by suggesting that plasmodium (tapetal cytoplasm) remained as a deposit on the outer exine of certain more advanced ferns, but that it was absorbed into the developing spores of more primitive ferns.

With improved techniques for studying spore-wall ultrastructure and morphogenesis, it has become clear that structures that appear to be morphological counterparts may have had very different ontogenetic paths, while what appear to be different morphological structures may have shared a similar ontogeny. Consequently, it is now believed that a perispore forms the outer spore wall layer of *all* modern fern spores 'with few exceptions' (Lugardon 1972, 1974; Pettitt 1979; Lloyd 1981, written communication July 1983; Tryon and Tryon 1982, p. 9), regardless of whether or not it fits Bower's morphological concept.

In vascular cryptogam spores, perispore ultrastructure (and chemistry) is distinct from that of exine (for example, Southworth and Myles 1984). Perispore is formed from products of the tapetal cytoplasm. At least in some cases, exine is formed from products of both tapetal periplasmodium and spore protoplast (Pettitt 1979). Perispore may be visually obvious with the light microscope, as Bower noted, or it may be difficult to impossible to detect with the light microscope. In more 'primitive' fern taxa, perispore is thin and conforms to the contours of the thick underlying spore exine, giving the appearance that it is absent (Tryon and Tryon 1982, p. 10; Tryon 1986, 1987, p. 292); in spores of more derived taxa, perispore is elaborate and obvious above the thin spore wall, obscuring the spore exine and spore sutures, conclusions opposite to Bower's (1959).

The source of ornament on vascular cryptogam spores may be exine, perispore, or both may be ornamented (see, for example, illustrations in Lloyd 1981; Tryon and Tryon 1982; Tryon 1986). Consequently, in tracheophyte spores, wall ornamentation may be synthesized by spore protoplast, or perhaps more usually by spore protoplast and periplasmodium (tapetal cytoplasm), or solely by periplasmodium. The morphological analogue among modern fern spores and Ordovician/Silurian ornamented spore tetrads lies with those with a visually obvious ornamented perispore that overlies a smooth-walled spore.

The spore-wall terminology of vascular cryptogams has been applied to bryophyte spores (Denizot 1974; Sorsa 1976), although equivalency has not been demonstrated for all features and in all groups of taxa customarily included among the bryophytes. Boros and Jarai-Komlodi (1975, p. 15) suggest, for example, that a perispore is not common or characteristic of bryophytes. They apparently were thinking of a readily identifiable topographic layer in the sense of Bower's (1959) fern perispore, since ultrastructure studies (McClymont and Larson 1964; Mueller 1974; Sorsa 1976; Neidhart 1979, p. 261; Mogensen 1983) indicate that, in mosses, the 'outermost part of the sporoderm . . . is a true "perine" ', derived from sporophyte as in most tracheophyte spores, rather than from spore protoplast. In most mosses, perispore forms the spore wall ornament and 'is the sole wall layer responsible for spore sculpturing' (Mogensen 1983, p. 328), although exine is also involved in the ornamentation of mosses with thick exine, either by forming the base of ornament elements or by forming the entire element (McClymont and Larson 1964; Sorsa 1976). A visual counterpart to the obvious supraexinous perispore of many vascular cryptogam spores is, however, unknown among moss spores, even though there is an ontogenetically equivalent structure. Fossil tetrads with optically undetectable psilate envelopes might find a morphological counterpart among modern fern spores with thin, conformable perispore or among moss spores with inconspicuous perispore.

Exine and tetrads. Hornwort and hepatic spores are believed to lack a perispore corresponding to that of vascular cryptogam and moss spores (Denizot 1971*a*, *b*, 1974, 1976; Sorsa 1976; Neidhart 1979).

This applies to all presently investigated Hepaticae, including the Marchantiales and Sphaerocarpales and possibly the Metzgeriales and Jungermanniales (Neidhart 1979). Mogensen (1983) believes perispore absence in hepatics to be in conformity with absence of tapetum and tapetal periplasmodium. Bower (1908, pp. 104–5) first articulated this idea, suggesting that tapetum is not a morphological constant, not being differentiated at all 'throughout the Bryophyta'. Pacini *et al.* (1985, p. 168), on the other hand, believe that tapetum 'is a constant for sporangia and anthers: it is a specialized tissue for spore nutrition and probably the source of sporopollenin precursors; it must be present in all groups of land plants, even if only for a very short time or if it is (especially in lower plants) difficult to recognize'. Whether perispore is genuinely absent or only briefly present, spore-wall ornament in hepatics is believed to be derived solely from exine synthesized by the cytoplasm of young spores.

In hepatics such as *Sphaerocarpos* where the dispersal unit is an obligate tetrahedral tetrad, there may be a superficial ornamented wall, continuous across the juncture of the spores, that visually resembles perispore (see Haynes 1910). Spores of other species of *Sphaerocarpos* have an ornamented outer wall that 'breaks' at the juncture of the individual spores of the tetrad, with the pattern contiguous, but not continuous, from spore to spore. In one genetic race of *S. donnellii*, where the dispersal unit is a single spore, fusion of the outer wall is incomplete (Doyle 1975).

Ontogeny of shared wall in *Sphaerocarpos* tetrads was uncertain to early investigators (for example, McAllister 1916; Siler 1934). They noted that the spore walls themselves often remained undeveloped and thin until the 'retaining wall' had differentiated or had begun to differentiate. Siler (1934, p. 587) suggested that when this special wall first appears, it 'might be called an exine', although she hesitated to do so. Doyle (1975) suggested that its source was callose synthesized by the spore mother cell. Denizot (1971*a*, 1974, 1976) concluded from ultrastructure studies that the sculpture pattern in *Sphaerocarpos* tetrads, despite its topographic similarity to classically defined perispore of fern spores, is developed in an exine layer with a spore cytoplasmic source. In some species of *Sphaerocarpos*, the outermost exine layer, which appears to be deposited first, does not intrude between the individual spores of the tetrad. In these circumstances, each spore is subsequently encased by its own wall, and the outermost exine forms a continuous tetrad envelope.

Major evolutionary trends

MA Zone I spore assemblages show two major evolutionary trends for which preliminary evidence has been presented elsewhere (Gray *et al.*, 1982, 1986; Gray, 1985*b*, 1989). One appears to be a progressive increase in spore tetrad size from an average of less than 27 μm in the Ordovician to over 40–50 μm in the late Llandovery. Although insufficient data points are available to indicate other than the trend's major outline, it may be non-linear, e.g. closely spaced, adjacent populations may temporarily reverse the overall trend shown by the entire lineage toward overall increased size, merely because of between-sample variation.

Tetrad-size increase appears to provide an example of Cope's rule, which posits that during phylogeny taxa may show an evolutionarily significant size increase. Evolutionary size increase must, of course, be kept conceptually distinct from size increase related to environmental changes. The rule was originally devised with the aid of Mesozoic reptiles and Tertiary mammals. It has since been extended to many organisms, including varied marine invertebrates (Boucot 1976). Evidence for evolutionary size increase commonly involves closely related genera belonging to a single family or subfamily, as with certain Neogene proboscideans or Silurian/Devonian brachiopods. If all Ordovician/early Silurian spore tetrads, regardless of other morphological differences, increase in size, it can be reasonably deduced that they represent closely related members, lineages, within a single family or subfamily. However, until vegetative organs belonging to the tetrads are known, it would be premature to be very positive about this possibility.

Tetrad-dominated to trilete-spore-dominated assemblages

Principal advantages of the land plant spore/pollen record in comparison with the megafossil record, are its comparative stratigraphic continuity through time, and the greater abundance of remains. In the pre-Devonian, we are treated to a particularly elegant example of these differences. The stratigraphic continuity of spore remains beginning in the mid-Ordovician has no megafossil counterpart. This continuity provides a record of major morphological changes among spore assemblages; in this case, a change from spore-tetrad-dominated assemblages into trilete-spore-dominated assemblages. The latter represent spore types that cannot be discriminated from *in situ* or dispersed spores of late Silurian/early Devonian tracheophytes. Change from tetrad-dominated to trilete-spore-dominated assemblages is dramatic. Few evolutionary events in plant history can be traced so clearly in such detail.

The broad outlines of this event, recognized for some time (Gray *et al.* 1982; Gray and Boucot 1983; Gray 1985*a*, *b*), provided the basis for MA Zone I and II (Gray 1985*b*), although the boundary between the Zones could only be positioned near late Llandovery Zones C_2–C_3 in the absence of sufficient stratigraphic control and a continuous set of spore-bearing samples. Spore tetrads could clearly be seen to dominate Ordovician and pre-late Llandovery, MA Zone I, spore assemblages. Trilete spores, that provide evidence of independent dissociation, occurred rarely, if at all. None appeared before about mid-way through the Llandovery, although 'trilete spores' broken out of tetrads may be found earlier and potentially even into the Ordovician. There seems little doubt, for example, that 'single' trilete spores from the Caradoc and Llandovery of Libya (Gray *et al.* 1982; Richardson 1988) as well as essentially all of those from the Llandovery of Pennsylvania (Johnson 1985, p. 339, pl. X), some indicated to be 'very rare', did not normally dissociate from tetrads. Other 'trilete spores' illustrated by Johnson (1985, p. 339, pl. X) are indicated to be merely thin-walled bodies 'with folds or scars that resemble trilete marks'. Post-late Llandovery, MA Zone II, assemblages are dominated by smooth-walled trilete spores. The boundary between MA Zones II and III was tentatively positioned in the late Wenlock (Gray 1985*b*) and was tentatively suggested to correlate with a general increase in spore-coat ornamentation in trilete spores.

In the North Atlantic Region in North America, with continuous, close-spaced samples from the Millerstown Section (also Gray 1988) near Harrisburg, Pennsylvania, in the central Appalachians, the change from tetrad-dominated to trilete-spore-dominated assemblages can be seen to occur in the latest late Llandovery, in the base of the Middle Rose Hill Formation, in the lower part of C_5 equivalent age strata (Gray 1989).

Thus, good reason exists to emphasize the evolutionary nature of the boundary between MA Zone I and MA Zone II, as marked by a change from spore-tetrad- to trilete-spore-dominance. Good reason also exists to emphasize a major extinction event involving plants whose spore tetrads bore reticulate and rugulate tetrad envelopes. Nevertheless, there are special circumstances, discussed below, where local dominance of spore-tetrad-producing plants could continue into the Wenlock. Similar circumstances are found among other groups of fossil organisms. Globally speaking, it is to be expected that these circumstances will form an exception.

Relict taxa or a relict flora. The possibility exists that in some part of the world in post-MA Zone I time, there remained a relict taxon or even a relict flora of spore-tetrad-producing plants that dominated the local vegetation. Relict assemblages are recognized in the modern biota among both plants and animals, and often indicate restriction to specific habitats, locally retained, previously more widespread (Gray, submitted).

Pre-Devonian spore assemblages to date come from nearshore marine, estuarine, and nearshore fluvial deposits, and represent the products presumably of both wind- and waterborne spores, the sweepings of the regional vegetation from all environments. No known spore assemblages represent continental fluvial deposits some distance from shoreline, or swamp, bog, or limnic deposits in any environment. The local vegetation adjacent to a small continental water-body or living on a bog surface could conceivably have been dominated by spore-tetrad-producing plants and could have shed spores in greater abundance into depositing sediments than the regional vegetation dominated by trilete-spore-producing plants. In analogous modern circumstances, differential pollen production and limited diaspore dispersal for some vegetation types as compared with others, or for some important taxa within some vegetation types as compared with others, may provide for a biased representation of the regional vegetation. In some circumstances, diaspores of the regional vegetation may be swamped out in local communities of high spore/pollen production. Even in circumstances where taxa do not dominate in vegetation local to depositional basins, it is conceivable that their spores/pollen might come to dominate assemblages through differential concentration related to transport mechanisms. Waterborne spores/pollen from riverine vegetation may distort the representation of windborne spores/pollen in sediments, even of small lakes with limited catchments, to the extent that quantitive correlations are limited between the pollen/spores of bottom sediments and vegetation that surrounds the lake (see Tauber 1977).

If spore-tetrad-producing plants were members of lake shoreline and gallery vegetation of streams leading into interior lakes, or swamp or bog plants living adjacent to open-water lakes (habitats most readily available to water transportation of diaspores), their spores might be disproportionately incorporated into lake sediments in comparison with spores potentially poorly air-dispersed into limnic sediments from plants of drier habitats.

Redeposition or recycling. The propensity for recycling, even polyrecycling, of organic-wall microfossils, makes it tempting to suggest that most biostratigraphically disharmonious assemblages involving remains in the micron size-range, are due to this cause.

It is axiomatic that recycling is most difficult to recognize, the more closely related in age are the sediments into which fossils are reworked, and the reworked fossils. Penecontemporaneous redeposition may be virtually impossible to distinguish in spore/pollen assemblages. Thus Cretaceous-age pollen/spores in Quaternary age sediments in offshore Oregon are readily determinable. It is inconceivable that reworked Tertiary pollen does not also occur in the same beds, considering that the outcrop area drained into the Pacific and the direct erosion of pollen-bearing Tertiary-age coastal rocks. Occasional pollen of taxa now exotic to the Pacific Northwest, such as *Carya, Pterocarya, Juglans*, and *Ilex*, indicate this to be the case. Yet Tertiary-age *Alnus, Quercus,* and *Pinus* cannot be disentangled from Quaternary/Holocene pollen of the same taxa on the basis of preservation alone, and may notably affect the pollen frequency of these taxa in offshore sediments. Circumstances of essentially penecontemporaneous redeposition apparently account for the continuous, relatively high incidence of *Castanea* pollen in lake deposits in eastern North America (Davis 1967; Davis and Webb 1975), although the trees themselves have been largely decimated, and pollen production from young trees before they are killed by blight must be relatively minor.

Both marine and non-marine beds can occur both above and below an unconformity, and water, as noted above, is an efficient pollen/spore conveyer, sometimes accounting for a more significant proportion of spores/pollen in depositing sediments, even of windborne taxa, than wind. Wind distributes predominantly spores/pollen during anthesis; water continues to contribute diaspores to sedimentary basins independently of 'flowering' season, through reflotation from soils and vegetation surfaces by rain and sheet-flooding (cf. Crowder and Cuddy 1973; Peck 1973) and from erosion of previous deposits. Lake Ontario bottom deposits, for example, 'regularly [have] up to 5% Paleozoic spores' (McAndrews and Power 1973, p. 785, Tables 1, 2), a proportion that exceeds the frequency of 88 per cent of extant woody and herbaceous plants whose pollen occurs in surface sediments.

Fossiliferous outcrops any place within the drainage basin of a lake, or at its shoreline, may be the source of redeposited spores. Under normal circumstances it is to be expected that redeposited spores will represent a minor component of contemporary assemblages, but this will depend, among other things, on their outcrop concentration, extent of erosion, lithification of the eroded deposit, and relative frequency of contemporaneous spores entering depositing sediments. Turner (1982) notes, for example, that in 200 specimen counts of late Ordovician acritarchs as much as 94 per cent of them may be reworked from the early Ordovician, totally swamping out the contemporary taxa.

In some pre-Devonian spore assemblages of limited diversity, where only redeposited tetrads might be added to an assemblage consisting predominantly of tetrads and trilete spores, even a relatively minor frequency of redeposited tetrads (10–20 per cent) could cause a considerable skewing in the frequencies of tetrads and single spores. Should this redeposited tetrad element be combined with contemporaneous tetrads from vegetation in a relict situation, it seems clear that tetrad frequency could easily exceed that of contemporary trilete spores.

Lineages of MA Zone I spore-tetrad-producing plants and trilete spores of MA Zones I and II

Several potential lineages of spore-tetrad-producing plants are suggested by these assemblages. It is impossible to say at what taxonomic level these lineages might be related. It is also impossible to say how, or if, any tetrad-producing taxa are related to trilete spore producers whose spores become predominant in the latest early Silurian.

Lineages of plants producing spore tetrads with tetrad envelopes. Plants whose tetrads bore ornamented tetrad envelopes may have become extinct in the late Llandovery, Wenlock, or early Ludlovian. No ornamented envelope-bearing tetrads are known in latest Llandoverian or post-Llandoverian rocks in the Appalachian region, or in Wenlockian/Ludlovian rocks of Norway, Sweden, UK, Libya or North America, by which time they were surely extinct. Plants whose tetrads bore psilate envelopes apparently survived into the latest Wenlockian/mid-Ludlovian, where their spores are a minor element. But the gap in time between Ordovician/late early Silurian tetrads with psilate envelopes and tetrads with psilate envelopes in latest Wenlockian and mid-late Silurian assemblages, may be more than a sampling problem, i.e. separate lineages may be involved. Tetrads with psilate envelopes are a potential source of smooth-walled tetrads in late Wenlock and Ludlovian assemblages of Norway and Sweden.

No evidence suggests that envelope-bearing tetrads ultimately dissociate at maturity. Spores broken from tetrads always display a triradiate mark, but are always damaged.

Plants that produced spores with ornamented tetrad envelopes may have briefly coexisted, in the latest late Llandovery and into the early Wenlock, with plants that produced trilete spores. However, no available evidence suggests that they did. Only smooth-walled tetrads and those with psilate envelopes are now known to coexist with trilete spores. Plants that produced spores with ornamented envelopes potentially coexisted throughout the late Ordovician and early Silurian and into the earliest late Silurian, with plants that produced smooth-walled tetrads and tetrads with psilate envelopes. There is no evidence for envelope-bearing tetrads beyond the late Silurian.

Lineage(s) of plants producing non-envelope-bearing smooth-walled spore tetrads. The incidence of non-envelope-bearing smooth-walled tetrads throughout the late Ordovician and Llandovery and into the late Silurian suggests a potentially significant and long-ranging 'group' of plants. As detailed above, it is impossible to be certain whether smooth-walled tetrads were all originally smooth, all originally bore envelopes, or came from both sources. What ultimately appears in the fossil record as smooth-walled tetrads may have had multiple plant origins. In the absence of tetrads with ornamented envelopes of Llandovery type from the middle and late Wenlockian of Norway (Smelror 1987; Gray and Boucot, in preparation) and the mid-late Silurian of Gotland, where smooth-walled tetrads are present, it is clear that there need not be a 1:1 relation between envelope-bearing and non-envelope-bearing tetrads. Smooth-walled tetrads are a minor element in late Silurian assemblages, and may belong to a different lineage or lineages than smooth-walled tetrads of the Llandovery and Ordovician. Plants producing smooth-walled tetrads may have coexisted with and out-survived plants whose tetrads bore ornamented detachable envelopes. There is no evidence for smooth-walled tetrads beyond the late Silurian. No smooth-walled tetrads dissociate normally in MA Zone I, as contrasted with being torn apart. Their potential relation to early trilete spores is discussed below.

Lineage(s) of late Llandovery and post-Llandovery plants producing smooth-walled, trilete spores. The earliest dissociated trilete spores of the latest late Llandovery and Wenlock, are smooth-walled and, unless broken, covered by a thin, tightly adherent psilate envelope, similar in appearance to the closely adherent perispore of primitive ferns, both only observable with SEM. These unornamented trilete spores are derived from meiotic tetrads, ultimately separated during sporogenesis by periplasmodium that penetrated between the closely apposed exines at the proximal faces of the spores to lay down the 'perisporal' envelope. Trilete spore envelopes closely resemble the thin, closely adherent, psilate tetrad envelopes. Consequently, of the spore tetrads with which they co-occur in the Llandovery and late Silurian, they seem more likely to have been derived by phyletic replacement from psilate envelope-bearing tetrads than from non-envelope-bearing smooth-walled tetrads, all of which are held together by exine bridges on proximal spore faces. This assumes that smooth-walled, non-envelope-bearing tetrads represent an authentic tetrad type and not one merely derived from psilate envelope-bearing tetrads by envelope loss. If early trilete spores are not the result of evolutionary dissociation of smooth-walled spore tetrads, this group of obligate tetrad-producing plants also became extinct before the Devonian.

The trilete spore is a morphology that is considered phylogenetically ancestral or primitive in vascular cryptogams. It probably is in bryophytes as well. Consequently these 'perispore-bearing' trilete spores, and potentially some among the tetrads with psilate envelopes, could have come from vascular plant progenitors, possibly from vascular plants themselves, and/or from bryophyte lineages. They could, of course, have come from an independent lineage not directly ancestral to either group of embryophytes, although potentially sharing a common ancestry with them. If these early trilete spores are from ancestral tracheophytes, the smooth-walled, 'perispore-bearing' trilete spore is clearly the 'primitive' tracheophyte spore type.

Unresolved questions

There have been some challenges to the major evolutionary timing of change from tetrad- to trilete-spore-dominated assemblages.

The first has to do with the age of Hoffmeister's (1959) Libyan trilete spores. Berry (in Hoffmeister 1959) first suggested a late Llandovery age; later (in Gray and Boucot 1971, p. 919) he suggested a much earlier Llandovery age. Berry only had limited graptolitic fossils available to him, as well as limited time to examine them, which may account for the discrepency. The late Llandovery age determination is compatible with events described above, but until more graptolites from the spore horizon become available the age of these beds will be uncertain.

Another challenge is provided by the recent description and illustration of single trilete spores from the late Ordovician of Greenland (Nøhr-Hansen and Koppelhus 1988). Nøhr-Hansen was kind enough to send me a sample from the Troedsson Cliff Formation, Washington Land, North Greenland, which was extracted in my laboratory (G1615). I have been unable to confirm, either with the light microscope or with SEM, the presence of obligate spore tetrads (potentially their Pl. 1 specimens) or single trilete spores. I did see folded *Lophosphaeridia* that resemble the ornamention claimed for the trilete spores, which might also be construed as trilete. Since trilete spores have not been otherwise observed in this time-interval throughout a vast geographic region, I discount as artefactual the trilete spores of Nøhr-Hansen and Koppelhus (1988) until such a morphology can be verified and documented elsewhere. Artefactual 'trilete scars' are not uncommon. A remarkable example from the Precambrian (cf. Schopf and Blacic 1971) was discounted by Knoll and Golubic (1979) (see pp. 65–66).

Conclusions

The traditional conception that the land surface lay fallow until the late Silurian is incorrect. Simplicity of the earliest vascular plants led to the assumption that they were not far removed, morphologically or chronologically, from their algal prototypes. Consequently, the conclusion became fixed in biological thinking that tracheophytes were the first land plants. Assuming rapid evolution, expectations for finding fossil evidences of a pre-vascular complex from which tracheophytes evolved seemed remote.

Fossil spore tetrads provide the first evidences for one group of pre-vascular land plants. Spore tetrads were doubtless both dispersal and reproductive devices. The spore-tetrad producers of the mid-Ordovician may have arisen from some fully terrestrial, soil-inhabiting green algal complex, from semi-aquatic green algae, or from emergent freshwater green algae. The earliest evidences for these pre-vascular land plants (spore tetrads) provide no basis for connecting them phylogenetically with any group of extant algae. The vegetative nature of these plants, deduced only from their spores, necessarily remains speculative. Spore characters provide reason to interpret them as the earliest fossil evidence for ancestral embryophytes, or lineages that included ancestral embryophytes. These plants sound the death knell for the concept that late Silurian/early Devonian vascular plants were the first post-microbial land plants, as well as to the concept of a rapid evolutionary transformation from alga to vascular plant.

The first recognized adaptive radiation of these land plants, potentially related to acquisition of preservable spores, begins in the earlier mid-Ordovician (Llanvirn–Llandeilo) of the Malvinokaffric Realm, 40–50 Ma before evidence of tracheophytes (Gray *et al.* 1982; Gray, unpublished). By the late Ordovician/earliest Silurian, spore-tetrad-producing plants were widespread in central and eastern North America, Europe, South America, North and South Africa, Arabia (Gray 1988), and Australia (Foster and Williams 1991). Several potential lineages can be recognized on the basis of spore types.

By the late early Silurian, plants that were potentially direct progenitors to vascular and non-vascular embryophytes were well established on land. This radiation is traced from trilete spores, a phylogenetically ancestral morphology in vascular cryptogams and probably in bryophytes as well. The first vascular plants are already known by the late

Silurian (Ludlovian of Australia) and plants with similar morphology, but lacking 'true tracheids' are known in the North Atlantic Region. The adaptive radiation of trilete-spore-producing plants took place explosively in probably less than 500 000 years. From the time of their dominance of the fossil record, about 10 Ma elapsed until the evolution of lignified tracheids in vascular megafossils. During this time-interval some of the earliest embryophyte progenitors were already becoming extinct. Extinction among this group of spore-tetrad producers continued in the late Silurian.

No available data suggest whether the change from obligate tetrads to trilete spores is a developmental change of neutral or selective nature. The trend in some angiosperms from single pollen back to tetrads and other multiple units might be a minor developmental, neotenic change of no selective value.

Acknowledgements

I wish to thank Darlene Southworth, John Rowley, and Don Wimber for reading an early version of this paper, as well as S. Blackmore and S. Barnes for many editorial suggestions. Southworth and Rowley both provided additional reference citations. My colleague A.J. Boucot read a number of versions of the manuscript and has been a valuable sounding-board for various ideas.

References

Allen, C.E. (1923). An apparently sex-linked sporophytic character in *Sphaerocarpos*. *Anatomical Record* **26**, 388–9.

Allen, C.E. (1925). The inheritance of a pair of sporophytic characters in *Sphaerocarpos*. *Genetics* **10**, 72–9.

Allen, C.E. (1945). The genetics of bryophytes. II. *Botanical Review* **11**, 260–87.

Arneson, R.D. (1973). *Pseudotetracystis*, a new chlorosarcinacean alga. *Journal of Phycology* **9**, 10–14.

Atkinson, A.W. Jr., Gunning, G.E.S., and John, P.C.L. (1972). Sporopollenin in the cell wall of *Chlorella* and other algae: ultrastructure, chemistry, and incorporation $g^{14}G$ acetate, studies in synchronous cultures. *Planta (Berl.),* **107**, 1–32.

Ayensu, E.S. and Skvarla, J.J. (1974). Fine structure of Velloziaceae pollen. *Bulletin of the Torrey Botanical Club* **101**, 250–66.

Barnes, S.H. and Blackmore, S. (1986). Some functional features in pollen development. In *Pollen and spores: form and function*, (ed. S. Blackmore and I.K. Ferguson), pp. 71–80. Academic Press, London.

Blackmore, S. and Crane, P. R. (1988). The systematic implications of pollen and spore ontogeny. In *Ontogeny and systematics*, (ed. C.J. Humphries), pp. 83–115. Columbia University Press, New York.

Boros, A. and Jarai-Komlodi, M. (1975). *An atlas of recent European moss spores*. Akademiai Kiado, Budapest.

Boucot, A.J. (1976). Rates of size increase and of phyletic evolution *Nature* **261**, 694–6.

Boucot, A.J. and Gray, J. (1982). Geologic correlates of early land plant evolution. In *Proceedings of the Third North American Paleontological Convention*, Vol. I, (ed. B. Mamet and M.J. Copeland), pp. 61–6.

Bower, F.O. (1908). *The origin of a land flora: a theory based upon the facts of alternation*. Macmillan and Company, London.

Bower, F.O. (1959). *Primitive land plants: also known as the Archegoniatae*. Hafner Publishing Company, New York. (Originally published in 1935.)

Brown, M. Jr and Bold, H.C. (1964). *Phycological studies. V. Comparative studies of the algal genera* Tetracystis *and* Chlorococcum. The University of Texas Publication 6417, pp. 1–214.

Chaloner, W.G. (1958). Isolated megaspore tetrads of *Stauropteris burntislandica*. *Annals of Botany* **22**, 197–204.

Chamberlain, A.H.L. and Evans, L.V. (1973). Aspects of spore production in the red alga *Ceramium*. *Protoplasma* **76**, 139–59.

Chantanachat, S. and Bold, H.C. (1962). *Phycological studies. II. Some algae from arid soils*. The University of Texas Publication 6218, pp. 1–76.

Cranwell, L.M. (1953). New Zealand pollen studies: the monocotyledons. *Bulletin of the Auckland Institute and Museum* **3**, 1–91.

Crowder, A.A. and Cuddy, D.G. (1973). Pollen in a small river basin: Wilton Creek, Ontario. In *Quaternary plant ecology*, (ed. H.J.B. Birks and R.G. West), pp. 61–77. John Wiley & Sons, New York.

Davis, R.B. (1967). Pollen studies of near-surface sediments in Maine lakes. In *Quaternary paleoecology*, (ed. E.J. Cushing and H.E. Wright, Jr), pp. 143–73. Yale University Press, New Haven.

Davis, R.B. and Webb, T. III (1975). The contemporary distribution of pollen in eastern North America: a comparison with the vegetation. *Quaternary Research* **5**, 395–434.

Denizot, J. (1971*a*). Sur la présence d'ensembles exinique élémentaires dans le sporoderme de quelques Marchantiales et Sphaerocarpales. *Comptes Rendus Hebdomadaires des Séances de l'Académie des Sciences, Paris, Série D* **272**, 2166–9.

Denizot, J. (1971*b*). Recherches sur l'origine des ensembles exiniques élémentaires du sporoderme de quelques Marchantiales et Sphaerocarpales. *Comptes Rendus Hebdomadaires des Séances de l'Académie des Sciences, Paris, Série D* **272**, 2305–8.

Denizot, J. (1974). Genèse des parois sporocytaires et sporales chez *Targionia hypophylla* (Marchantiales). Justification de la terminologie utilisée. *Pollen et Spores* **16**, 303–71.

Denizot, J. (1976). Remarques sur l'édification des differentes couches de la paroi sporale à exine lamellaire de quelques Marchantiales et Sphaerocarpales. In *The evolutionary significance of the exine*, (ed. I. K. Ferguson and J. Muller), pp. 185–210. Academic Press, London.

Doyle, W. T. (1975). Spores of *Sphaerocarpos donnellii*. *Bryologist* **78**, 80–4.

Duffield, S. L. (1985). Land-derived microfossils from the Jupiter Formation (Upper Llandoverian), Anticosti Island, Quebec. *Journal of Paleontology* **59**, 1005–10.

Dykstra, R. F. (1971). *Borodinellopsis texensis*: gen. et. sp. nov. A new alga from the Texas Gulf Coast. In *Contributions in Phycology*, (ed. B. C. Parker and R. M. Brown, Jr.), pp. 1–8. Allen Press, Lawrence Kansas.

Erdtman, G. (1943). *An introduction to pollen analysis*. Chronica Botanica Company, Waltham, Massachusetts.

Erdtman, G. (1945). Pollen morphology and plant taxonomy. V. On the occurrence of tetrads and dyads. *Svensk Botaniska Tidskrift* **39**, 286–97.

Erdtman, G. (1952). *Pollen morphology and plant taxonomy. Angiosperms (An Introduction to palynology. I.)* Almqvist and Wiksell, Stockholm.

Erdtman, G. (1957). *Pollen and spore morphology/plant taxonomy. Gymnospermae, Pteridophyta, Bryophyta (An introduction to palynology, II.)* Almqvist and Wiksell, Stockholm.

Erdtman, G. (1969). *Handbook of palynology. An introduction to the study of pollen grains and spores*. Hafner Publishing Company, New York.

Erdtman, G. and Sorsa, P. (1971). *Pollen and spore morphology/plant taxonomy. Pteridophyta: text and additional illustrations (An introduction to palynology. IV.)* Almqvist and Wiksell, Stockholm.

Faegri, K. and Iversen, J. (1950). *Text-book of modern pollen analysis*. Munksgaard, Copenhagen.

Ford, J. H. (1971). Ultrastructural and chemical studies of pollen wall development in Epacridaceae. In *Sporopollenin*, (ed. J. Brooks, P. R. Grant, M. Muir, P. Van Gijzel, and G. Shaw), pp. 130–73. Academic Press, London.

Foster, C. B. and Williams, G. E. (1991). Late Ordovician–Early Silurian age for the Mallowa Salt of the Carribuddy Group, Canning Basin, Western Australia, based on occurrences of *Tetrahedraletes medinensis* Strother & Traverse 1979. *Australian Journal of Earth Sciences* **38**, 223–8.

Fritsch, F. E. (1935). *The structure and reproduction of the algae*, Vol. I. MacMillan, New York.

Fritsch, F. E. (1945) *The structure and reproduction of the algae*, Vol. II. Cambridge University Press, Cambridge.

Furch, B. and Gooday, G. W. (1978). Sporopollenin in *Phycomyces blakesleeanus*. *Transactions of the British Mycological Society* **70**, 307–9.

Furch, B. and Pambor, L. (1978). Cell wall constituents of *Phycomyces blakesleeanus*. 2. Localisation of sporopollenin in the zygospores and sporangiospores. *Microbios Letters* **4**, 211–19.

Gooday, G.W., Fawcett, P., Green, D., and Shaw, G. (1973). The formation of fungal sporopollenin in the zygospore wall of *Mucor mucedo*: a role for the sexual carotenogenesis in the Mucorales. *Journal of General Microbiology* **74**, 233–9.

Gooday, G.W., Green, D., Fawcett, P., and Shaw, G. (1974). Sporopollenin formation in the ascospore wall of *Neurospora crassa*. *Archiv für Mikrobiologie* **101**, 145–51.

Gray, J. (1985*a*). Ordovician–Silurian land plants: the interdependence of ecology and evolution. In *Autecology of Silurian organisms*, Special Papers in Palaeontology 32, (ed. M.G. Bassett and J.D. Lawson), pp. 281–95. The Palaeontological Association, London.

Gray, J. (1985*b*). The microfossil record of early land plants: advances in understanding of early terrestrialization, 1970–1984. In *Evolution and environment in the Late Silurian and Early Devonian*, (ed. W. Chaloner and J.D. Lawson), pp. 281–95. *Philosophical Transactions of the Royal Society of London* **B309**, 167–95.

Gray, J. (1988). Land plant spores and the Ordovician–Silurian boundary. In *A global analysis of the Ordovician–Silurian boundary*, Bulletin of the British Museum (Natural History) Geology 43, (ed. L.R.M. Cocks and B. Rickards), pp. 351–8. British Museum (Natural History), London.

Gray, J. (1989). Adaptive radiation of early land plants. *Abstracts of the 28th International Geological Congress, Washington, DC*, Vol. I, I-582-I-583.

Gray, J. and Boucot, A.J. (1971). Early Silurian spore tetrads from New York: earliest New World evidence for vascular plants? *Science* **173**, 918–21.

Gray, J. and Boucot, A.J. (1972). Palynological evidence bearing on the Ordovician–Silurian paraconformity in Ohio. *Geological Society of America Bulletin* **83**, 1299–314.

Gray, J. and Boucot, A.J. (1975). Color changes in pollen and spores: a review. *Geological Society of America Bulletin* **86**, 1019–33.

Gray, J. and Boucot, A.J. (1977). Early vascular land plants: proof and conjecture. *Lethaia* **10**, 145–74.

Gray, J. and Boucot, A.J. (1983). A spore-based first order biostratigraphy for the pre-Devonian of the Appalachian region. *Geological Society of America, Abstracts with Programs* **15**, 586.

Gray, J., Massa, D., and Boucot, A.J. (1982). Caradocian land plant microfossils from Libya. *Geology* **10**, 197–201.

Gray, J., Colbath, G.K., de Faria, A., Boucot, A.J., and Rohr, D.M. (1985). Silurian-age fossils from the Paleozoic Paraná Basin, southern Brazil. *Geology* **13**, 521–5.

Gray, J., Theron, J. N., and Boucot, A.J. (1986). Age of the Cedarberg Formation, South Africa and early land plant evolution. *Geological Magazine* **123**, 445–54.

Griffiths, D.A. and Griffiths, D.J. (1969). The fine structure of autotrophic and heterotrophic cells of *Chlorella vulgaris* (Emerson strain). *Plant and Cell Physiology* **10**, 11–19.

Gupta, A. and Udar, R. (1986). Palyno-taxonomy of selected Indian liverworts. *Bryophytorium Bibliotheca* **29**, 1–127.

Hassel de Menendez, G.G. (1986). *Leiosporoceros* Hassel n. gen. and *Leiosporocerotaceae* Hassel n. fam. of *Anthocerotopsida*. *Journal of Bryology* **14**, 255–9.

Haynes, C.C. (1910). *Sphaerocarpos hians*, sp. nov., with a revision of the genus and illustrations of the species. *Bulletin of the Torrey Botanical Club* **37**, 215–30.

Hemsley, A.R. (1991). *Parka decipiens* and land plant spore evolution. *Historical Biology*. (In press.)

Hickok, L.G. (1979). Apogamy and somatic restitution in the fern *Ceratopteris*. *American Journal of Botany* **66**, 1074–8.

Hickok, L.G. and Klekowski, E.J. Jr (1973). Abnormal reductional and nonreductional meiosis in *Ceratopteris*: alternatives to homozygosity and hybrid sterility in homosporous ferns. *American Journal of Botany* **60**, 1010–22.

Hoffmeister, W.S. (1959). Lowest Silurian plant spores from Libya. *Micropaleontology* **5**, 331–4.

Honneger, R. and Brunner, U. (1981). Sporopollenin in the cell walls of *Coccomyxa* and *Myrmecia* phycobionts of various lichens: an ultrastructural and chemical investigation. *Canadian Journal of Botany* **59**, 2713–34.

Huynh, K.-L. (1973). L'arrangement des spores dans la tetrade chez les Pteridophytes. *Botanische Jahrbucher für Systematik Pflanzengeschichte und Pflanzengeographie* **93**, 9–24.

Huynh, K.-L. (1976). Arrangement of some monosulcate, disulcate, trisulcate, dicolpate, and tricolpate pollen types in the tetrads, and some aspects of evolution in the angiosperms. In *The evolutionary significance of the exine*, (ed. I.K. Ferguson and J. Muller), pp. 101–24. Academic Press, London.

Iversen, J. and Troels-Smith, J. (1950). *Pollenmorfologiske definitioner og typer*. C.A. Reitzels Forlag, Axel Sandal, København.

Johnson, N.G. (1985). Early Silurian palynomorphs from the Tuscarora Formation in central Pennsylvania and their paleobotanical and geological significance. *Review of Palaeobotany and Palynology* **45**, 307–60.

Klekowski, E.J. Jr (1973*a*). Genetic load in *Osmunda regalis* populations. *American Journal of Botany* **60**, 146–54.

Klekowski, E.J. Jr (1973*b*). Sexual and subsexual systems in homosporous pteridophytes: a new hypothesis. *American Journal of Botany* **60**, 535–44.

Knoll, A.H. and Golubic, S. (1979). Anatomy and taphonomy of a Precambrian algal stromatolite. *Precambrian Research* **10**, 115–51.

Knox, R.B. (1984). The pollen grain. In *Embryology of angiosperms*, (ed. B.M. Johri), pp. 197–271. Springer-Verlag, Berlin.

Knox, R.B. and McConchie, C.A. (1986). Structure and function of compound pollen. In *Pollen and spores: form and function*, (ed. S. Blackmore and I.K. Ferguson), pp. 265–82. Academic Press, London.

Kremp, G.O.W. (1965). *Morphologic encyclopedia of palynology*. The University of Arizona Press, Tucson.

Le Thomas, A., Morawetz, W., and Waha, M. (1986). Pollen of palaeo- and neotropical Annonaceae: definition of the aperture by morphological and functional characters. In *Pollen and spores: form and function*, (ed. S. Blackmore and I.K. Ferguson), pp. 375–88. Academic Press, London.

Levan, A. (1942). A gene for the remaining in tetrads of ripe pollen in *Petunia*. *Hereditas* **28**, 429–35.

Lloyd, R.M. (1981). The perispore in *Polypodium* and related genera (Polypodiaceae). *Canadian Journal of Botany* **59**, 175–89.

Lugardon, B. (1972). La structure fine de l'exospore et de la périspore des Filicinees isosporées. I. Généralités. Eusporangiées et Osmundales. *Pollen et Spores* **14**, 227–61.

Lugardon, B. (1974). La structure fine de l'exospore et de la périspore des Filicinées isosporées. II. Filicales. Commentaires. *Pollen et Spores* **16**, 161–226.

McAllister, F. (1916). The morphology of *Thallocarpus Curtisii*. *Bulletin of the Torrey Botanical Club* **43**, 117–26.

McAndrews, J.H. and Power, D.M. (1973). Palynology of the Great Lakes: the surface sediments of Lake Ontario. *Canadian Journal of Earth Sciences* **10**, 777–92.

McClymont, J.W. and Larson, D.A. (1964). An electron-microscope study of spore wall structure in the Musci. *American Journal of Botany* **51**, 195–200.

McGlone, M.S. (1978*a*). Pollen wall structure of the New Zealand species of *Epacris* (Epacridaceae). *New Zealand Journal of Botany* **16**, 83–9.

McGlone, M.S. (1978*b*). Pollen structure of the New Zealand members of the Stypheliaceae (Epacridaceae). *New Zealand Journal of Botany* **16**, 91–101.

Maheshwari, P. (1950). *An introduction to the embryology of angiosperms*. McGraw-Hill, New York.

Melville, R. (1981). Surface tension, diffusion and the evolution and morphogenesis of pollen aperture patterns. *Pollen et Spores* **23**, 179–203.

Meyen, S.V. (1987). *Fundamentals of palaeobotany*. Chapman and Hall, London.

Miller, M.A. and Eames, L.E. (1982). Palynomorphs from the Silurian Medina Group (Lower Llandovery) of the Niagara Gorge, Lewiston, New York, U.S.A. *Palynology* **6**, 221–54.

Mishler, B.D. and Churchill, S.P. (1984). A cladistic approach to the phylogeny of the 'bryophytes'. *Brittonia* **36**, 406–24.

Mishler, B.D. and Churchill, S.P. (1985). Transition to a land flora: phylogentic relationships of the green algae and bryophytes. *Cladistics* **1**, 305–28.
Mogensen, G.S. (1983). The spore. In *New manual of bryology*, (ed. R.M. Schuster), Vol. I, pp. 325–42. The Hattori Botanical Laboratory.
Mueller, D.M.J. (1974). Spore wall formation and chloroplast development during sporogenesis in the moss *Fissidens limbatus*. *American Journal of Botany* **61**, 525–34.
Nayar, B.K. (1963). Spore morphology of *Loxogramme*. *Grana Palynologica* **4**, 388–92.
Neidhart, H.V. (1979). Comparative studies of sporogenesis in bryophytes. In *Bryophyte systematics*, (ed. G.C.S. Clarke and J.G. Duckett), pp. 251–80. Academic Press, London.
Nøhr-Hansen, H. and Koppelhus, E.B. (1988). Ordovician spores with trilete rays from Washington Land, North Greenland. *Review of Palaeobotany and Palynology* **56**, 305–11.
Oehler, J.H., Oehler, D.Z., and Muir, M.D. (1976). On the significance of tetrahedral tetrads of Precambrian algal cells. *Origins of Life* **7**, 259–67.
Pacini, E., Franchi, G.G., and Hesse, M. (1985). The tapetum: its form, function, and possible phylogeny in Embryophyta. *Plant Systematics and Evolution* **149**, 155–85.
Peck, R.M. (1973). Pollen budget studies in a small Yorkshire catchment. In *Quaternary plant ecology*, (ed. H.J.B. Birks and R.G. West), pp. 43–60. John Wiley & Sons, New York.
Pettitt, J.M. (1966). Exine structures in some fossil and recent spores and pollen as revealed by light and electron microscopy. *Bulletin of the British Museum (Natural History) Geology* **13**, 223–57.
Pettitt, J.M. (1979). Ultrastructure and cytochemistry of spore wall morphogenesis. In *The experimental biology of ferns*, (ed. A.F. Dyer), pp. 213–52. Academic Press, London.
Richardson, J.B. (1988). Late Ordovician and Early Silurian cryptospores and miospores from northeast Libya. In *Subsurface palynostratigraphy of northeast Libya*, (ed. A. El-Arnauti *et al.*), pp. 89–109.
Roland, F. (1965). Précision sur la structure et l'ultrastructure d'une tetrade calymmée. *Pollen et Spores* **7**, 5–8.
Roland, F. (1971). The detailed structure and ultrastructure of an acalymmate tetrad. *Grana* **11**, 41–4.
Sampson, F.B. (1977). Pollen tetrads of *Hedycarya arborea* J.R. et G. Forst. (Monimiaceae). *Grana* **16**, 61–73.
Satterthwait, D.F. and Playford, G. (1986). Spore tetrad structures of possible hepatic affinity from the Australian Lower Carboniferous. *American Journal of Botany* **73**, 1319–31.
Schopf, J.W. and Blacic, J.M. (1971). New microorganisms from the Bitter

Springs Formation (Late Precambrian) of the North-Central Amadeus Basin, Australia. *Journal of Paleontology* **45**, 925–59.

Schuster, R. M. (1966). *The Hepaticae and Anthocerotae of North America*, Vol. I. Columbia University Press, New York.

Selling, O. H. (1944). Studies in the recent and fossil species of *Schizaea*, with particular reference to their spore characters. *Meddelanden Goteborgs Botaniska Tradgard* **XVI**, 1–112.

Selling, O. H. (1946). *Studies in Hawaiian pollen statistics. Part I. The spores of the Hawaiian pteridophytes*. Bernice P. Bishop Museum Special Publication 37, pp. 1–87. Bishop Museum, Honolulu, Hawaii.

Selling, O. H. (1947). *Studies in Hawaiian pollen statistics. Part II. The pollens of the Hawaiian phanerogams*. Bernice P. Bishop Museum Special Publication 37, pp. 1–430. Bishop Museum, Honolulu, Hawaii.

Siler, M. B. (1934). Development of spore walls in *Sphaerocarpos donnellii*. *Botanical Gazette* **95**, 563–91.

Skvarla, J. J. and Larson, D. A. (1963). Nature of cohesion within pollen tetrads of *Typha latifolia*. *Science* **140**, 173–5.

Skvarla, J. J., Raven, P. H., and Praglowski, J. (1975). The evolution of pollen tetrads in Onagraceae. *American Journal of Botany* **62**, 6–35.

Skvarla, J. J., Raven, P. H., and Praglowski, J. (1976). Ultrastructural survey of Onagraceae pollen. In *The evolutionary significance of the exine*, (ed. I. K. Ferguson and J. Muller), pp. 447–79. Academic Press, London.

Smelror, M. (1987). Llandovery and Wenlock meiospores and spore-like microfossils from the Ringerike district, Norway. *Norsk Geologisk Tidsskrift* **67**, 143–50.

Sorsa, P. (1976). Spore wall structure in Mniaceae and some adjacent bryophytes. In *The evolutionary significance of the exine*, (ed. I. K. Ferguson and J. Muller), pp. 211–29. Academic Press, London.

Southworth, D. and Myles, D. G. (1984). Ultraviolet absorbance spectra of megaspore and microspore walls of *Marsilea vestita*. *Pollen et Spores* **26**, 481–8.

Strohl, W. R., Larkin, J. M., Good, B. H., and Chapman, R. L. (1977). Isolation of sporopollenin from four myxobacteria. *Canadian Journal of Microbiology* **23**, 1080–3.

Strother, P. K. (1982). Non-marine palynomorphs from Llandoverian and Wenlockian strata. *Palynology* **6**, 292.

Strother, P. K. and Traverse, A. (1979). Plant microfossils from Llandoverian and Wenlockian rocks of Pennsylvania. *Palynology* **3**, 1–21.

Takahashi, H. and Sohma, K. (1980). Pollen development in *Pyrola japonica* Klenze. *Science Reports of the Tohoku University, 4th Series, Biology* **38**, 57–71.

Takahashi, H. and Sohma, K. (1982). Pollen morphology of the Droseraceae and its related taxa. *Science Reports of the Tohoku University, 4th Series, Biology* **38**, 81–156.

Takahashi, H. and Sohma, K. (1984). Development of pollen tetrads in *Typha latifolia* L. *Pollen et Spores* **26**, 5–18.
Tauber, H. (1977). Investigations of aerial pollen transport in a forested area. *Dansk Botanisk Arkiv* **32**, 120 pp.
Taylor, W.A. and Taylor, T.N. (1987). Spore wall ultrastructure of *Protosalvinia*. *American Journal of Botany* **74**, 437–43.
Tryon, A.F. (1986). Stasis, diversity and function in spores based on an electron microscope survey of the Pteridophyta. In *Pollen and spores: form and function*, (ed. S. Blackmore and I.K. Ferguson), pp. 233–49. Academic Press, London.
Tryon, A.F. (1987). Evolutionary levels and ecological impact on fern spores. *XIV International Botanical Congress, Berlin (West), Germany, Abstracts*, p. 292.
Tryon, R.M. and Tryon, A.F. (1982). *Ferns and allied plants with special reference to Tropical America*. Springer-Verlag, New York.
Turner, R.E. (1982). Reworked acritarchs from the type section of the Ordovician Caradoc Series, Shropshire. *Palaeontology* **25**, 119–43.
Van Campo, M. and Guinet, P. (1961). Les pollens composées. L'exemple des Mimosacées. *Pollen et Spores* **3**, 201–18.
Wagner, W.H. Jr (1974). Structure of spores in relation to fern phylogeny. *Annals of the Missouri Botanical Garden* **61**, 332–53.
Walker, J.W. (1971). Pollen morphology, phytogeography, and phylogeny of the Annonaceae. *Contributions from the Gray Herbarium of Harvard University* **202**, 1–132.
Walker, J.W. and Doyle, J.A. (1975). The bases of angiosperm phylogeny: palynology. *Annals of the Missouri Botanical Garden* **62**, 664–723.
Wodehouse, R.P. (1935). *Pollen grains: their structure, identification and significance in science and medicine*. McGraw-Hill, New York.

5. The control of spore wall formation

GERDA A. van UFFELEN
Rijksherbarium/Hortus Botanicus, PO Box 9514, Leiden, The Netherlands

Abstract

In a study of the formation of the spore wall in several ferns of the family Polypodiaceae (Filicales), different series of surface patterns that are superimposed upon each other during exospore formation have been described (van Uffelen 1985, 1986, 1990, 1991). This study has led to the contemplation of some of the factors that may play a role in establishing surface patterns.

Introduction

SEM observation of mature fern spores (e.g. Tryon and Lugardon (1991); van Uffelen and Hennipman 1985) has shown an astounding variety of surface patterns. This has led to a study of the formation of these patterns in one family of ferns, the Polypodiaceae, in which several other characters have already been studied in a comprehensive way (Hennipman 1984, 1985).

Transmission electron microscopy (TEM) and scanning electron microscopy (SEM) were used, in the latter case using the freeze-fracturing technique adapted at the Natural History Museum for the study of pollen and spores (Blackmore and Barnes 1985). Application of this technique to the three fern species, *Drynaria sparsisora* (Desv.) Moore, *Microgramma ciliata* (Willd.) Copel., and *Belvisia mucronata* (Fée) Copel., showed a different series of patterns superimposed upon each other during exospore formation in each species. The results of this study are or will be published in the near future (van Uffelen 1990, 1991).

Following observation of all these different patterns on spore walls, both during and after spore wall formation, it was noted that some of these patterns closely resemble patterns encountered in spores of plants

Pollen and Spores (ed. S. Blackmore and S. H. Barnes), Systematics Association Special Volume No. 44, pp. 89–102, Clarendon Press, Oxford, 1991.

that are not closely related at all, or in pollen, or even resemble patterns encountered in other living or non-living systems.

An example of a close resemblance in perispore surface between quite distantly related fern spores is shown in Fig. 5.1(a) (spore of *Pyrrosia boothii* (Hooker) Ching, Polypodiaceae) and Fig. 5.1(b) (spore of *Lygodium heterodoxum* Kuntze, Schizaeaceae); the resemblance also pertains to perispore structure: both species have a perispore containing spherical bodies, which apparently consist of the same material as the exospore.

An example of resemblance between a fern perispore and a tectate pollen wall is shown in Figs 5.1(c) and (d): the perispore of *Asplenium ledermanni* Hier. consists of two layers that are separated by columellar structures in many places.

An instance of a rather close resemblance between a pattern found on a spore wall and a pattern developed in a non-living system is shown in Fig. 5.2. Figure 5.2(a) shows a SEM picture of the perispore surface of *Drynaria sparsisora* and Fig. 5.2(b) shows the traced basic pattern (the spines have been omitted). Figure 5.2(c) shows a light micrograph of the pattern found on a glass slide on which glycerine jelly had precipitated after it had evaporated from some preparations that had been kept in a closed box for a long time; this pattern has been traced and is shown in Fig. 5.2(d), to facilitate comparison with the perispore pattern of Fig. 5.2(a) and (b).

These resemblances may be explained in different ways. In the next four sections the following aspects of the formation of surface patterns in sporogenesis will be discussed:

(1) how spore shape and size may influence and constrain surface patterns;

(2) whether pattern formation is controlled by haploid or diploid parts of the system, i.e. by the haploid spore (gametophytic control), by the spore mother cell (sporocytic control), or by other diploid tissues (sporophytic control);

(3) whether spore wall formation may be regarded as a condensation process, which is influenced by relatively simple physico-chemical factors, such as material, condensation surface, and condensation circumstances; and

(4) how prepatterning may be established and how materials may be supplied.

At first sight it seems reasonable to suppose that establishment of pattern only needs to be explained in the case of a non-random pattern (for ways of defining randomness, see Sachs 1984). However, even when

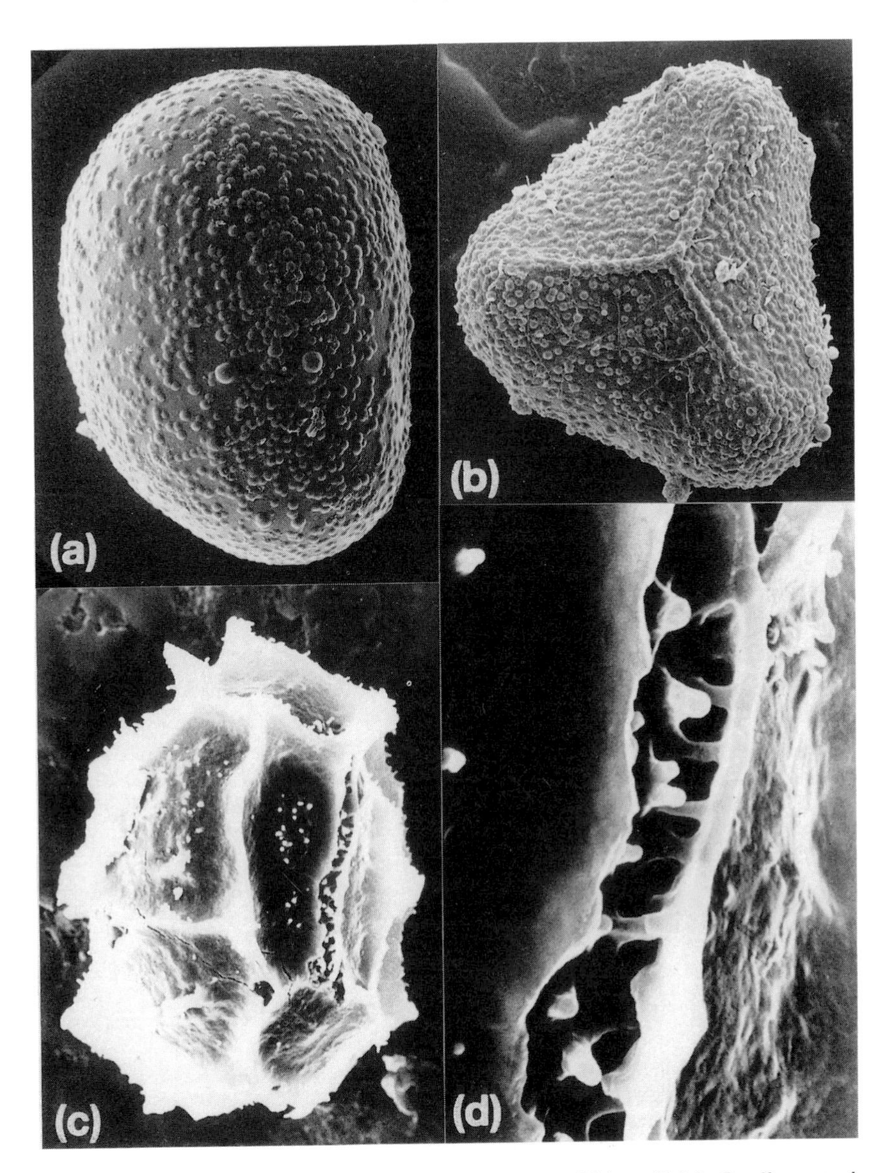

Fig. 5.1 (a) A spore of *Pyrrosia boothii* (Hooker) Ching (BM, Ludlow *et al.* 18646). SEM, 10 μm = 7 mm. (b) A spore of *Lygodium heterodoxum* Kuntze (L, Vregel 12905). SEM, 1 μm = 8.5 mm. (c) A spore of *Asplenium ledermanni* Hier. (L, Brass 24598). SEM, 10 μm = 18 mm. (d) Detail of the same spore of *Asplenium ledermanni* Hier., showing the columellar structure of the perispore. SEM, 1 μm = 11 mm.

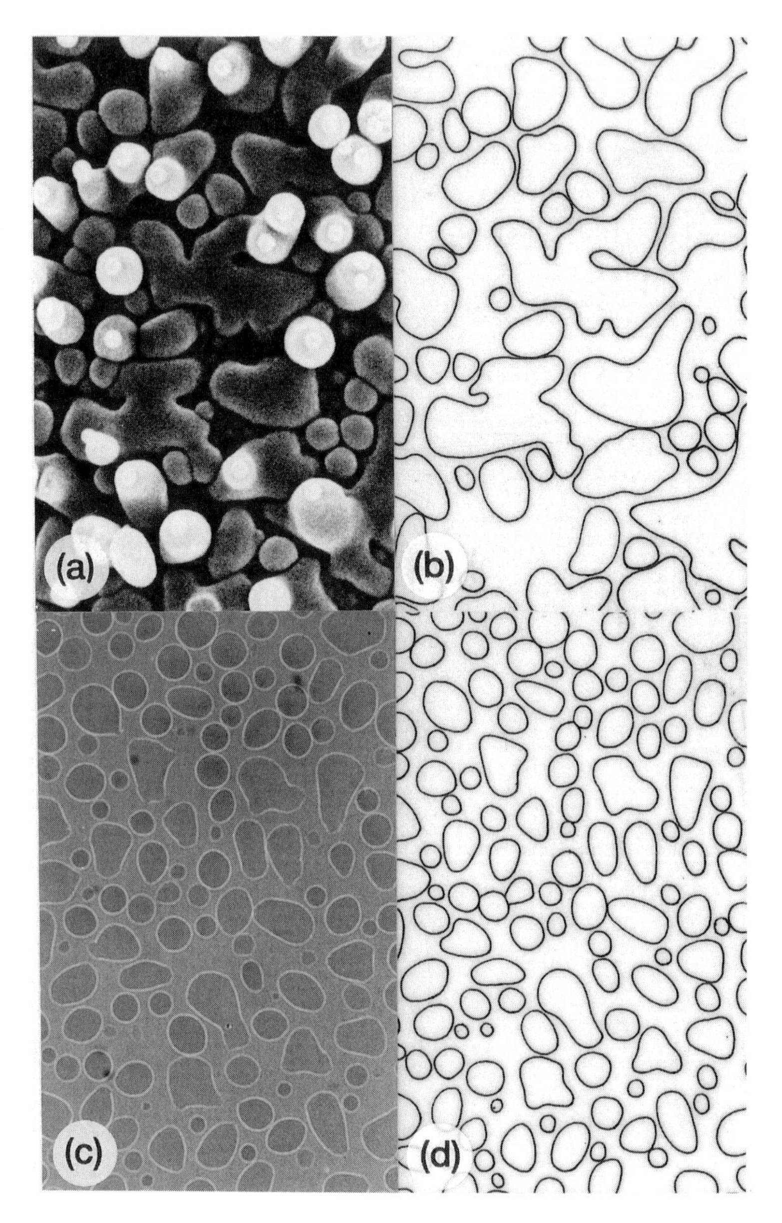

Fig. 5.2 (a) Detail of the perispore surface of *Drynaria sparsisora* (Desv.) Moore (LEI 20339), showing the basic pattern and superimposed spines. SEM, 1 μm = 12 mm. (b) Tracing of the basic pattern on the perispore surface as depicted in (a). (c) Micrograph of the pattern formed by glycerine jelly, which has precipitated on a glass slide. Light microscopy, × 175. (d) Tracing of the pattern depicted in (c).

features are distributed at random over the spore surface, something must govern their form and deposition on the spot where they are found. Therefore, both structure and spacing of surface features need some governing principle, even if the features are randomly spaced.

Functional aspects will not be considered. Functional demands exploit the possibilities yielded by the existence of similar or different patterns; they do not govern the basic processes of pattern formation: function operates on an evolutionary level, and not on pattern formation processes *per se*. Similar functional demands made upon a system will often be the driving force behind the development of similarities, and so may lead to evolutionary convergence and parallelisms.

Pattern, shape, and size

Muller (1979) pointed out that physical constraints are very important in explaining architecture and surface patterns in pollen. These constraints reduce the number of patterns that may be realized under certain circumstances, and may therefore lead to similarities in pattern not caused by a similarity in formation process, but by simple lack of 'choice'.

There are only a few regular patterns consisting of one or two different units capable of completely filling flat surfaces (Stevens 1974, p. 15). The shape of the spores and their haptotypic features (e.g. the laesura) pose spatial constraints upon surface pattern. The differences in curvation angle between different parts of the spore surface may also cause slightly different versions of an otherwise uniform pattern.

The presence of a larger number of channels on the proximal than on the distal side, a situation encountered in many spores, including those of *Drynaria sparsisora*, may also cause some differences in pattern between proximal and distal poles.

Expansion of the growing spore also influences pattern, either by stretching and thus enlarging existing pattern units, or by intercalary growth, which involves the addition of new units.

Control of pattern formation

Basically, genetic control of pattern formation is a question of which genes are switched on at a certain moment during the process. The issue is made more complicated by the existence of different phases in the life cycle of ferns: the diploid sporophyte produces haploid spores; after germination the spores grow into haploid gametophytes, which produce haploid gametes, which after fertilization again produce a diploid sporophyte. Apart from the possibility of turning certain genes on and off, these different phases in the fern life cycle lead to the availability of only one genome in the haploid phase, and the availability of two, possibly different ones, in the diploid phase (Bell 1981).

The following terms are currently used in the designation of a control function during the development of spores and pollen (see Fig. 5.3):

Sporocytic control. Controlled by the diploid spore mother cell (found in bryophytes by Brown *et al.* 1986; Brown and Lemmon 1988), or by sporocytic 'remains' (mRNA, enzymes) in the cytoplasm of the haploid spore (see Schneller 1989).

Sporophytic control. Controlled by other diploid tissues, for example the tapetum or sporangium wall. It is proposed by many authors and supported by experimental evidence (Schneller 1989; perispore in *Athyrium*). Knox (1984) states, 'As the pollen grain is seen as a conserved structure, with group- or species-specific patterning and little infraspecific variation [*sic*], exine form and structure will be under sporophytic control.'

Gametophytic control. Controlled by the haploid spore. Lugardon (1971, p. 191) states that formation of the inner exospore and the laesural fold is directed by the spore cytoplasm; in *Blechnum spicant* (L.) Roth he found that both these spore features, and maturation of the entire exospore (reduction of thickness and the establishment of a uniform outer exospore layer), are controlled by the haploid spore, and do not occur normally in aborted spores.

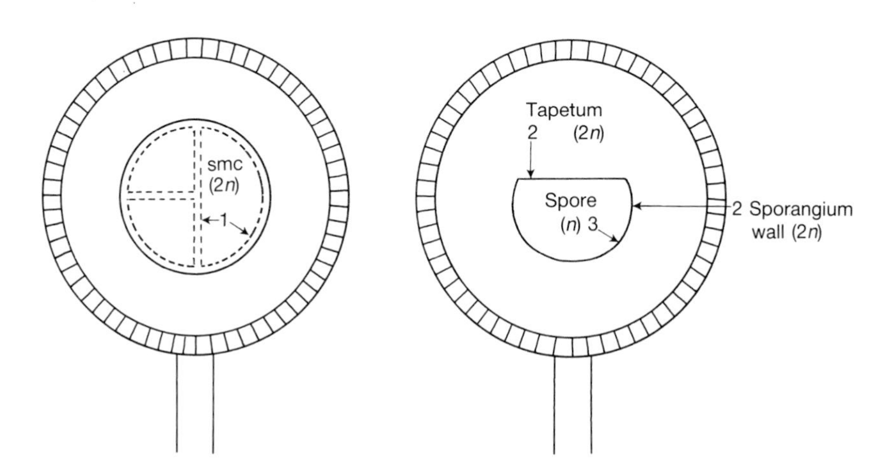

Fig. 5.3 Different types of control possibly playing a role in spore wall formation in ferns. *n*, Haploid; 2*n*, diploid; smc, spore mother cell; 1, sporocytic control; 2, sporophytic control; 3, gametophytic control.

One of the functions ascribed to the callose wall in pollen is that of a barrier between genetically different tetraspores, and between the haploid tetraspores and the diploid parental tissues (Heslop-Harrison 1964, p. 39; Knox 1984, p. 197). If this were true, it would indicate exclusively gametophytic control of pollen wall formation, at least during the tetrad stage. There is one important argument against the barrier theory: in a tetrad, spores are pairwise genetically identical, and the pairs are genetically maximally different; genetic differences between tetrads may also be considerable; nevertheless, there are no great differences in spore wall morphology between spores of the same sporangium. This indicates sporophytic control as the main governing mechanism in spore wall formation (see also Schneller 1989, who finds that perispore formation in *Athyrium* is under sporophytic control).

In *Drynaria sparsisora* only the material of the inner part of the inner exospore layer is deposited from inside the spore, certainly under gametophytic (or residual sporocytic) control. As the rest of the spore wall material is provided by the tapetum, and as I regard the formation of the rest of the exospore as a simple condensation process (see next section), sporophytic control may take over quite early in the sporogenesis of *Drynaria sparsisora*.

As for the timing of the control of prepatterning, Heslop-Harrison (1971) conducted a series of experiments by applying spindle poisons and centrifugation to developing pollen of *Lilium*. He concluded that the meiotic spindle determines the main symmetries of the pollen grain, and that the basic surface pattern is already established before the meiotic division, that is by the genome of the pollen mother cell.

Although most researchers are looking for 'the' way of control of microspore wall formation, I believe that the three ways of control mentioned above may each play a role in the process, and that the extent to which each form of control contributes to wall formation may differ between taxa at quite a low level.

Spore wall formation as a condensation process

It is difficult to comprehend that the beautifully patterned perispore is deposited by the then almost completely degenerated tapetum, which cannot possibly contribute anything but simple precursors. This and the sudden deposition of the perispore suggest a quick and simple condensation process. This is supported by the remarkable resemblance between the basic perispore pattern in *Drynaria sparsisora* (Fig. 5.2a, b) and a spontaneous condensation pattern, as shown in Fig. 5(c) and (d).

The perisporal spines may result from the fusion of highly viscous droplets (van Uffelen 1990), again suggesting a relatively simple physico-chemical explanation for their development (W. Linnemans, personal communication). Despite perispore morphology, which may be typical on the species level, the blobs that play a role at the start of perispore formation seem to be an almost universal feature.

The much slower deposition of exospore material may also be explained as a condensation process. Arguments in favour of such an explanation are:

(1) the absence of an obvious transport system from tapetum to growing exospore; and

(2) the accretion of exosporal material in the form of tapetal spherical bodies that show, at least in mature sporangia and if the scale of the pattern is small enough, the same surface pattern as the fully formed exospore (Lugardon 1981).

There are two types of condensation processes: physical condensation, caused by a phase transition, for example from water vapour to water, and chemical condensation, where small building-blocks are fused into larger molecules while losing very small molecules (for example water); here it is presumed that both types of condensation process are basically subject to the same set of rules. Condensation processes are governed by several factors:

(1) the material that is to condense—physico-chemical properties, concentration, concentration gradient;

(2) the surface or particles on which condensation occurs;

(3) the circumstances—temperature, viscosity, and nature of the environment, the available space.

They can give rise to very simple patterns, but also to very sophisticated ones, and not only in living systems. In the case of spore wall formation, the question is to what extent the living system contributes to the establishment of pattern. This may occur in several ways:

(1) by supplying the necessary material, and by governing the amount, concentration, and/or concentration gradients;

(2) by prepatterning the condensation surface;

(3) by governing part of the circumstances—the presence of water and of any molecules that do not condense, the available space, and, by growing in certain places and not in others, even temperature.

However, once the living system has contributed its share, many aspects of the condensation process are out of its control, and the remaining part of the process of spore wall formation may be a more or less autonomous process.

The supply of materials and the establishment of pattern

In the previous section it was stated that the contribution of a living system to the formation of its spore wall may be threefold: by way of material, surface patterning, and governing deposition circumstances. Circumstances of sporogenesis are difficult to measure and should be studied experimentally. Therefore, only the first two factors can be illustrated here by some examples from the literature and the author's observations.

Material

It has been established beyond doubt that all perisporal material is of tapetal origin, with the possible exception of spherical bodies as found in *Pyrrosia* (van Uffelen and Hennipman 1985), which probably are of the same origin as the exospore.

The exospore material may originate in two sites:

1. In the haploid spore cytoplasm, therefore under gametophytic control (as Mepham and Lane claimed in 1970 for *Tradescantia*, gametophytic control then being more popular than sporophytic control), although a slight residual sporocytic influence may still be present in the form of enzymes or mRNA. In *Drynaria sparsisora*, the inner exospore layer is formed by the accretion of sporopollenin on both sides of the white line (Rowley 1988); the sporopollenin is almost certainly derived from the spore cytoplast on the inner side of this line, and from outside the spore on the outer side. This means that only the innermost part of the inner exospore is made up of material provided by the haploid spore.
2. In the tapetum, therefore under sporophytic control, as in *Equisetum* (Lehmann *et al.* 1984), and as implicitly proposed as a possibility by Pettitt (1976) in his publication about channels in the exospore wall in *Lycopodium gnidioides* L.; also in *Tradescantia virginiana* L. (Tiwari and Gunning 1986*a*, *b*, *c*).

Many others have also found a combination of sporal and tapetal provenance of exine/exospore material (Bell 1985, p. 488; Lugardon 1971). As both spore cytoplasm and tapetum in ferns can secrete sporopollenin precursors, the extent to which cytoplasm and tapetum contribute to exospore formation may vary widely between groups of ferns.

In *Drynaria sparsisora* no involvement of the sporangium wall in sporopollenin secretion has been found (van Uffelen 1990), so the only remaining tissue to provide the necessary material for most of the exospore is the tapetum.

The succession of patterns during exospore formation can be

explained under the condensation model by the existence of sporopollenin precursors of different chain length and slightly different physico-chemical properties, perhaps aided by slightly different condensation circumstances, which result in different grain sizes and different patterns.

Prepatterning

Some suggestions have been made concerning the factors that may be involved in prepatterning condensation surfaces:

(1) sporocytic control during meiotic prophase by cytoplasmic projections of the sporocyte wall capped by an extracellular material, described for several liverworts (Brown *et al.* 1986);
(2) the mucopolysaccharide coat in ferns, which may also act as a template (Pettitt and Jermy 1974);
(3) electrostatic phenomena, as proposed by Locquin (1981), and ionic currents (Jaffe 1981);
(4) microtubules, as described by Brown and Lemmon (1988) for bryophytes;
(5) the plasma membrane, as described by Takahashi (1989) for the formation of the reticulate pattern on the pollen of *Caesalpinia japonica* Sieb. et Zucc.

A different way of prepatterning may be provided by a template around the sporoderm in the course of its formation. The mucopolysaccharide coat as proposed by Pettitt and Jermy (1974) has already been mentioned. Another obvious candidate for this function is the callosic wall present around meiocytes and young tetrads during pollen formation. Some debate has been going on about whether the surface coats around pteridophyte meiocytes and young tetraspores contain callose or not. Callose has been found only in *Selaginella* (Bienfait and Waterkeyn 1976). It has been searched for in several other pteridophytes, e.g. by Sheffield and Bell (1979) who could demonstrate no callose in *Pteridium aquilinum* (L.) Kuhn with a fluorescence method; however, Sheffield and Bell (1987) state that fluorescence may have been suppressed by secondary plant substances, such as tannins and phenols.

In ferns the dual function of the callose wall as a template during wall formation and as a barrier between genetically different cells could be served by the mucopolysaccharide surface coats around sporocytes and spores. However, in *Drynaria sparsisora* there is no sign of the surface coats acting as a template.

Conclusions

In pattern formation during spore wall formation, three important factors may play a role:

1. Universal restrictions. Similarities due to this factor cannot be ascribed to any underlying similarity in formation process, their resemblance is entirely spurious.

2. Universal processes. Usually, the main part of a pattern is directly due to its process of formation. Resemblance between patterns found in living systems with those in non-living systems indicates that the process underlying the formation of many patterns may be rather simple and therefore easy to study experimentally, possibly in non-living systems.

3. A genetic component. Large differences between pattern in closely related groups (e.g. Polypodiaceae) indicate that if the genetic component plays an important role, it governs the formation process in a very simple way. A small genetic change may then result in a very different pattern.

The contribution of gametophytic, sporocytic, and sporophytic control to the process of sporogenesis could lead to differences between taxa at quite a low level, depending on the timing of the change from one system of control to another during sporogenesis.

As Pettitt (1979) mentioned in his publication on spore wall morphogenesis, the elucidation of the relation between cause and effect will not be attained easily by direct observation of normal development. Experimental interference with the process of sporogenesis will be necessary to answer many of the questions posed here.

In addition to the search for ways of surface prepatterning and the template function of callose and other substances, more research should be undertaken on condensation processes in connection with sporogenesis, and to the condensation properties of sporopollenin and its precursors. A very important contribution to our knowledge of biological pattern formation may also be gained by experiments in non-living systems.

Acknowledgements

I want to thank my colleagues of the Rijksherbarium/Hortus Botanicus, Leiden, of the State University of Utrecht, and of the Natural History Museum in London for their help in practical matters and for their participation in many stimulating discussions on sporogenesis.

The investigations were supported by the Foundation for Fundamental

Biological Research (BION), which is subsidized by the Netherlands Organization for Scientific Research (NWO).

References

Bell, P. R. (1981). The phase change in ferns. *Acta Societatis Botanicorum Poloniae* **50**, 307–21.

Bell, P. R. (1985). Maturation of the megaspore in *Marsilea vestita*. *Proceedings of the Royal Society of London* **B223**, 485–94.

Bienfait, A. and Waterkeyn, L. (1976). Sur la nature des parois sporocytaires chez les mousses et chez quelques ptéridophytes. Étude comparative. *Comptes Rendus de l'Académie des Sciences, Paris D*, **282**, 2079–81.

Blackmore, S. and Barnes, S. H. (1985). *Cosmos* pollen ontogeny: a scanning electron miscroscope study. *Protoplasma* **126**, 91–9.

Brown, R.C. and Lemmon, B.E. (1988). Sporogenesis in bryophytes. *Advances in Bryology* **3**, 159–223.

Brown, R.C., Lemmon, B.E., and Renzaglia, K.S. (1986). Sporocytic control of spore wall pattern in liverworts. *American Journal of Botany* **73**, 593–6.

Hennipman, E. (1984). A new approach to the systematics of the Polypodiaceae (Filicales). *Taxon* **33**, 140–1.

Hennipman, E. (1985). Progress report of the Polypodiaceae project. *Taxon* **34**, 354–5.

Heslop-Harrison, J. (1964). Cell walls, cell membranes and protoplasmic connections during meiosis and pollen development. In *Pollen physiology and fertilization*, (ed. H.F. Linskens), pp. 39–47. North Holland Publishing Company, Amsterdam.

Heslop-Harrison, J. (1971). Sporopollenin in the biological context. In *Sporopollenin*, (ed. J. Brooks, P.R. Grant, M. Muir, P. van Gijzel, and G. Shaw), pp. 1–30. Academic Press, London.

Jaffe, L.F. (1981). The role of ionic currents in establishing developmental pattern. *Philosophical Transactions of the Royal Society of London* **B295**, 553–66.

Knox, R.B. (1984). The pollen grain. In *Embryology of Angiosperms*, (ed. B.M. Johri), pp. 197–271, Springer Verlag, New York.

Lehmann, H., Neidhart, H.V., and Schlenkermann, G. (1984). Ultrastructural investigations on sporogenesis in *Equisetum fluviatile*. *Protoplasma* **123**, 38–47.

Locquin, M.V. (1981). Mécanismes physiques de la genèse des formes et des ornements des spores et pollens: chréodes, singularités catastrophiques et attracteurs électrostatiques. *Courier Forschungs-Institut Senckenberg* **50**, 75–88.

Lugardon, B. (1971). Contribution à la connaissance de la morphogenèse et de la structure des parois sporales chez les Filicinées isosporées. Unpublished thesis, Université P. Sabatier, Toulouse.

Lugardon, B. (1981). Les globules des filicinées, homologues des corps d'Ubisch des spermaphytes. *Pollen et Spores* **23**, 93–124.

Mepham, R.H. and Lane, G.R. (1970). Observations on the fine structure and developing microspores of *Tradescantia bracteata*. *Protoplasma* **70**, 1–20.

Muller, J. (1979). Form and function in angiosperm pollen. *Annals of the Missouri Botanical Garden* **66**, 593–632.

Pettitt, J.M. (1976). A route for the passage of substances through the developing pteridophyte exine. *Protoplasma* **88**, 117–31.

Pettitt, J.M. (1979). Ultrastructure and cytochemistry of spore wall morphogenesis. In *The experimental biology of ferns*, (ed. A.F. Dyer), pp. 213–52. Academic Press.

Pettitt, J.M. and Jermy, A.C. (1974). The surface coats on spores. *Biological Journal of the Linnean Society* **6**, 245–57.

Rowley, J.R. (1988). Substructure within the endexine, an interpretation. *Journal of Palynology* **23–24**, 29–42.

Sachs, T. (1984). Controls of cell patterns in plants. In *Pattern formation, a primer in developmental biology*, (ed G.M. Malacinski and S.V. Bryant), pp. 367–91. MacMillan, New York.

Schneller, J.J. (1989). Remarks on hereditary regulation of spore wall pattern in intra- and interspecific crosses of *Athyrium*. *Botanical Journal of the Linnean Society* **99**, 115–23.

Sheffield, E. and Bell, P.R. (1979). Ultrastructural aspects of sporogenesis in a fern, *Pteridium aquilinum* (L.) Kuhn. *Annals of Botany* **44**, 393–405.

Sheffield, E. and Bell, P.R. (1987). Current studies of the pteridophyte life cycle. *Botanical Review* **53**, 442–90.

Stevens, P.S. (1974). *Patterns in nature*. Little, Brown and Company, Boston.

Takahashi, M. (1989). Pattern determination of the exine in *Caesalpinia japonica* (Leguminosae: Caesalpinioidae). *American Journal of Botany* **76**, 1615–26.

Tiwari, S.C. and Gunning, B.E.S. (1986*a*). Cytoskeleton, cell surface and the development of invasive plasmodial tapetum in *Tradescantia virginiana* L. *Protoplasma* **133**, 89–99.

Tiwari, S.C. and Gunning, B.E.S. (1986*b*). An ultrastructural, cytochemical and immunofluorescence study of postmeiotic development of plasmodial tapetum in *Tradescantia virginiana* L. and its relevance to the pathway of sporopollenin secretion. *Protoplasma* **133**, 100–14.

Tiwari, S.C. and Gunning, B.E.S. (1986*c*). Colchicine inhibits plasmodium formation and disrupts pathways of sporopollenin secretion in the anther tapetum of *Tradescantia virginiana* L. *Protoplasma* **133**, 115–28.

Tryon, A.F. and Lugardon, B. (1991). *Spores of the Pteridophyta*. Springer-Verlag, New York.

Uffelen, G.A. van (1985). Spore wall formation in Polypodiaceae (abstract). *Proceedings of the Royal Academy of Edinburgh* **86B**, 443–5.

Uffelen, G. A. van (1986). Sporogenesis in *Drynaria sparsisora* (Desv.) Moore (Polypodiaceae) (abstract). *Acta Botanica Neerlandica* **35**, 116–17.

Uffelen G. A. van (1990). Sporogenesis in Polypodiaceae (Filicales). I. *Drynaria sparsisora* (Desv.) Moore *Blumea* **35**, 177–215.

Uffelen, G. A. van (1990). Sporogenesis in *Polypodiaceae* (Filicales). II. *Belvisia mucronata* (Fée) Copel. and *Microgramma ciliata* (Willd.) Copel., in preparation.

Uffelen, G. A. van and Hennipman, E. (1985). The spores of *Pyrrosia* Mirbel (Polypodiaceae), a SEM study. *Pollen et Spores* **27**, 155–97.

6. Diversification of spores in fossil and extant Schizaeaceae

JOHANNA H.A. VAN KONIJNENBURG-VAN CITTERT
Laboratory of Palaeobotany and Palynology, State University, Utrecht, The Netherlands

Abstract

Data from literature on both extant and fossil schizaeaceous spores have been combined with research on spores of 23 extant species and new scanning electron microscopy (SEM) data on several fossil species. In addition to *in situ* fossil spores, dispersed fossil spores attributed to the Schizaeaceae have been taken into account. In this way some ideas were obtained about the diversification of spores in the Schizaeaceae.

Introduction

The Schizaeaceae are a rather primitive, mainly tropical, fern family with a creeping or rising rhizome and dichotomous or pinnate leaves that sometimes show unlimited growth. The fertile pinnae may or may not have a reduced lamina, but the main characteristic is in the sporangia: the single sporangia (there are no sori) are arranged in two (or four) rows, either marginally or in special sporangiophores. The sporangia are large and have an apical annulus consisting of a single subapical row of cells and an apical plate. The apical plate consists of one cell in *Schizaea*, *Actinostachys*, and *Lygodium*, and several cells in *Anemia* and *Mohria*.

The extant Schizaeaceae comprise five genera: *Schizaea*, *Actinostachys* (sometimes treated as a subgenus of *Schizaea*), *Anemia*, *Mohria*, and *Lygodium*. Some authors combine them into three subfamilies—Schizaeae with *Schizaea* and *Actinostachys*, Lygodiae with *Lygodium*, and

Pollen and Spores (ed. S. Blackmore and S. H. Barnes), Systematics Association Special Volume No. 44, pp. 103–118, Clarendon Press, Oxford, 1991.

Anemiae with *Anemia* and *Mohria*, (for example, de la Sota and Morbelli 1987). Other authors (for example, Reed 1947) distinguish within the order Schizaeales seven families (four fossil ones and three recent: Schizaeaceae, Lygodiaceae, and Anemiaceae). The present author considers the five genera to belong to one family, but that a division into three subfamilies might be useful.

Schizaeaceae spores are usually large and often show a striking, striate exine ornamentation. Since Bolkhovitina's work (1961) on the fossil and extant spores of the Schizaeaceae, many new data have been obtained, especially by SEM work on extant species (Mickel 1962, 1981; Hill 1977, 1979; de la Sota and Morbelli 1987). These data are scattered throughout the literature and the aim of this paper is to combine them with new research (especially SEM) on some fossil and extant Schizaeaceous spores, in order to understand the diversification of the spores within the Schizaeaceae.

Materials and methods

Spores of 23 species (Utrecht herbarium material) of the Schizaeaceae (representing the five genera) were studied both by light microscopy and by SEM (Camscan). In addition, fossil material of the Middle Jurassic *Klukia exilis* and *Stachypteris spicans*, which had been studied previously by the author (1981), was re-studied, and spores of the Early Cretaceous *Pelletixia valdensis* and *Ruffordia goeppertii* (from the collections of the Natural History Museum, London) were studied by SEM.

The spores of the herbarium species are labelled with a number; the slides of the fossil spores kept at Utrecht have a Yor. prefix, the hand specimens an S. prefix, and the material from London is labelled with a V. prefix.

Spore morphology of the extant genera

Schizaea and Actinostachys

These mainly pantropical genera are closely related; the main differences are in the gametophyte, the number of rows of sporangia in the sporangiophores (two in *Schizaea* arranged in a terminal spike, and four in *Actinostachys* arranged in a penicillate tuft), and the spores. The spores of *Schizaea* and *Actinostachys* are monolete—trilete spores have only been recorded from *S. dichotoma* (Selling 1946)—and reniform in outline (only *Schizaea* subg. *Microschizaea* has subglobose to ovoid spores), but there is a striking difference in exine ornamentation.

Actinostachys. Actinostachys (15 species) spores (Fig. 6.1a) usually show patterns of parallel ridges and furrows, giving them a striate appearance. Although the number and orientation of the ridges and the width of the ridges and furrows may vary, the basic pattern is the same in most species. This pattern of ridges and furrows is a property of the exospore. The perispore is usually thin and finely granulate, and can be removed easily (Fig. 6.1b). The size of *Actinostachys* spores varies between 35 and 120 μm (there are three small species; the majority have spores between 60 and 120 μm).

Schizaea. Spores of *Schizaea* (25 species) have a thin, smooth or finely granulate perispore which can be easily removed (Fig. 6.1c). The exospore may be almost smooth but is usually more or less granulate. Sometimes the granules are stalked and fused, giving the spores an alveolate ornamentation (e.g. *S. pusilla*).

Various authors distinguish different subgenera within *Schizaea*: the best known are *Schizaea*, *Microschizaea*, and *Lophidium* (Selling 1944, 1946, 1947; Reed 1947). *Microschizaea* has, as already mentioned, large (60–100 μm), subglobose to ovoid spores with a smooth to granular exospore. *Lophidium* spores are usually small (30–45 μm) and smooth or punctate (Fig. 6.1e), while larger spores (45–100 μm) with larger granules occur in *Schizaea* (Fig. 6.1d, f).

Mohria and Anemia

Mohria, the smallest genus in the family, with three species in Madagascar and SE Africa, and *Anemia*, the largest with *c.* 80 species, mainly in America, some in Africa, and one in India, are closely related. Both have pinnate fronds; all the pinnae in a fertile frond of *Mohria* bear sporangia, while only the lower two pinnae bear sporangia in *Anemia*. The spores of *Anemia* and *Mohria* are always trilete and show a pattern of ridges and furrows.

Mohria. Mohria caffrorum (the most common *Mohria* species) has large spores (*c.* 100 μm) with a fairly simple pattern of broad ridges and narrow furrows (Fig. 6.2b). This spore type is considered to be the most primitive in the *Mohria*/*Anemia* group. The exospore shows some granules, the thin perispore follows the exospore closely and is almost smooth. When examined closely under light microscopy *Mohria* spores show a light, apparently hollow area along the centre of each ridge (Fig. 6.2c; see also Mickel 1962). A broken spore under SEM clearly shows a hollow area in the centre of the ridges (Fig. 6.2a). Furthermore, this figure illustrates that the exospore consists of two layers (comparable with sexine and nexine in pollen).

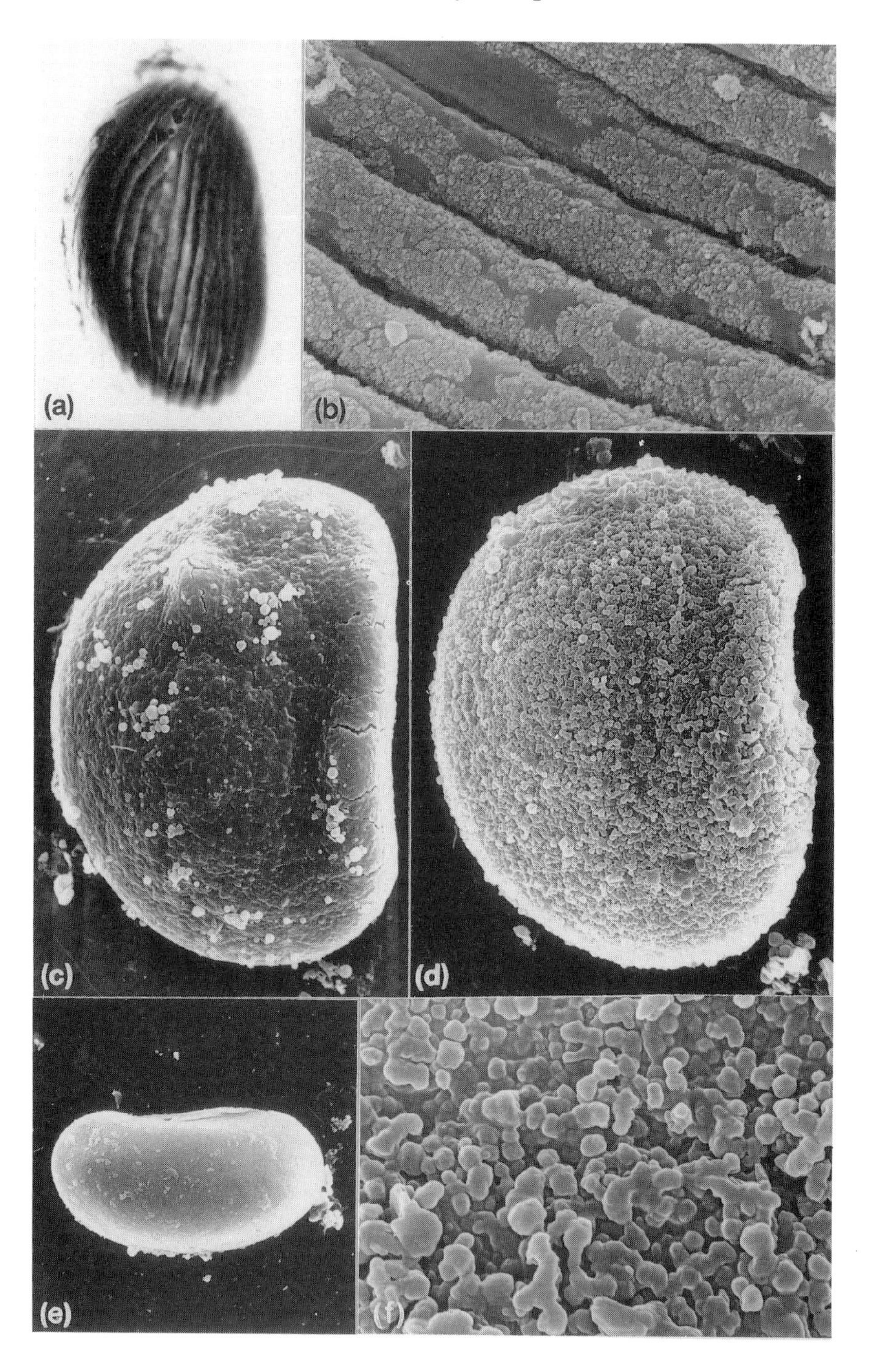
(a)
(b)
(c)
(d)
(e)
(f)

Anemia. The genus *Anemia* consists of three subgenera: *Coptophyllum*, *Anemiorrhiza*, and *Anemia*. Work on the spores of the subgenus *Coptophyllum* (40 species) has been done by Mickel (1962) and Hill (1977). The spores are usually large (65–120 μm) and show a pattern of broad ridges and narrow furrows like *Mohria*. The ridges usually anastomose in the centre of the distal side and at the angles of the tetrahedron, where short, wart-like protuberances (appendices) are formed (Fig. 6.2d). The exospore is almost smooth, the perispore usually shows microspines (Fig. 6.2f), sometimes even forked spines or baculae (Hill 1977). The ridges, unlike-those of *Mohria*, are solid in section.

In *Anemiorrhiza* (12 species) the spores are smaller (50–75 μm); the spore ridges show a tendency to become lower, fewer, and further apart; they also tend to become somewhat undulating with near or total anastomoses (Fig. 6.3a). The ridges are hollow in section (Mickel 1962, 1981). The exospore is smooth, the perispore smooth to scabrate.

The subgenus *Anemia* (25 species) has the most derived spores. The ridges are, as in *Anemiorrhiza*, further apart than in *Coptophyllum*, but in addition are covered with processes: bacula or spines (Atkinson 1962; Hill 1979). A good example is *A. phyllitidies*: the exospore is covered with spines (Fig. 6.3c); the perispore follows the exospore. The most extreme species in this respect is *A. hirsuta*: the spines are so numerous and large that they conceal the original pattern of ridges almost completely (Fig. 6.3b). Only when the perispore is removed, is the pattern visible. The ridges in this subgenus are usually solid, but sometimes show a different layer in the centre: the medulla (Mickel 1962). The spores are rather large (60–100 μm) but not as large as in *Coptophyllum*.

Lygodium

Lygodium stands slightly apart from the other genera in sometimes having fronds of unlimited growth which may even be up to 10 m long. Also palmate and pinnate fronds are common. The sporangia are borne separately on marginal lobes or wholly fertile segments, each covered by a laminar outgrowth. There are about 45, mainly tropical species divided over three subgenera: *Gisopteris* (with palmate fronds), *Lygodium* (with those long fronds), and *Odontopteris*. This division into three subgenera is not clearly supported by spore morphology. *Lygodium* spores

Fig. 6.1 *Actinostachys laevigata*, 10728; light micrograph, × 500. (b) *Actinostachys laevigata*, 10728, exospore and perispore; SEM, × 5000. (c) *Schizaea pectinata*, 10730, perispore; SEM, × 1000. (d) *Schizaea pectinata*, 10730, exospore; SEM, × 1000. (e) *Schizaea elegans*, 10729; SEM, × 1000. (f) *Schizaea pectinata*, 10730, detail of exospore; SEM, × 5000.

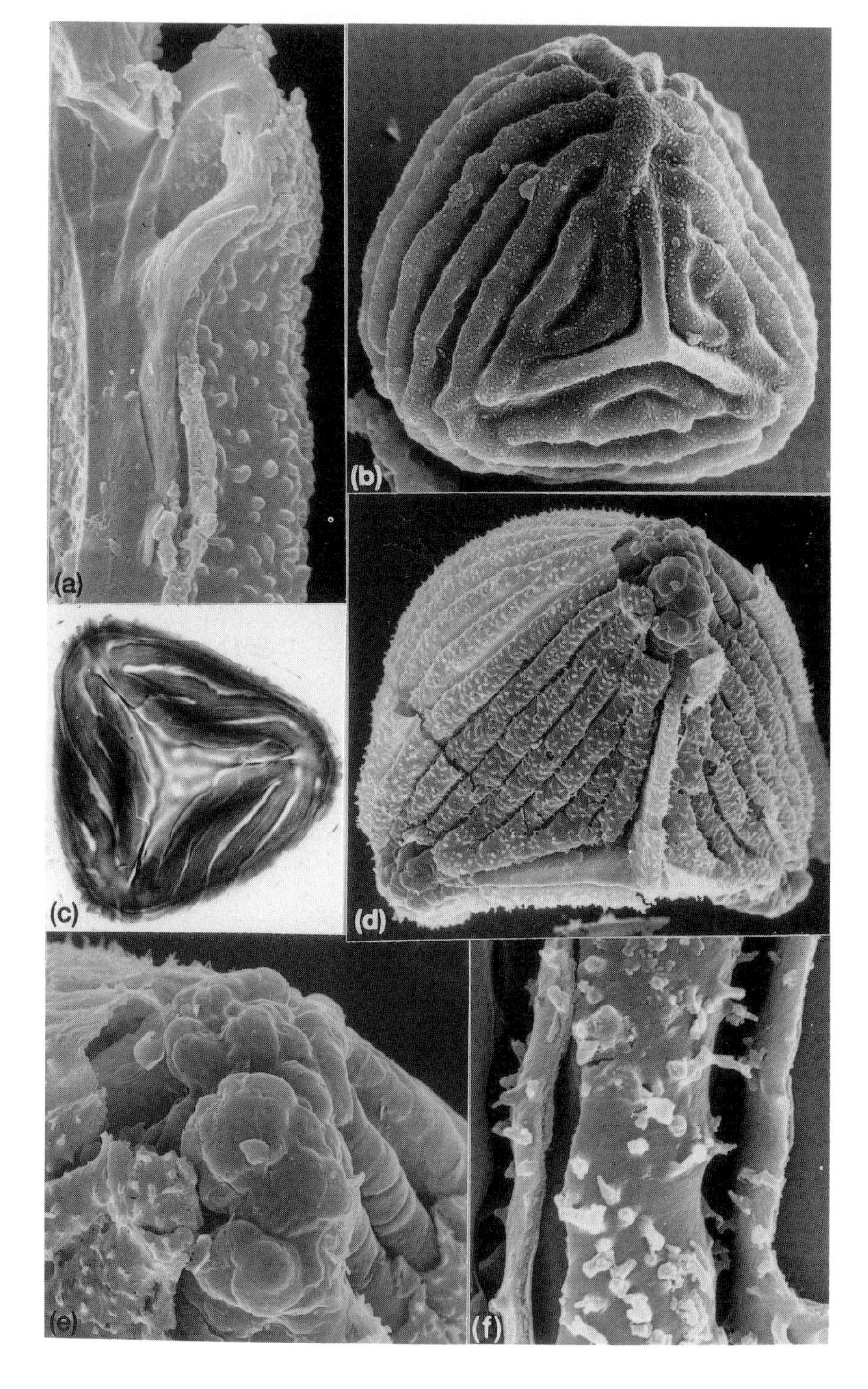
(a)
(b)
(c)
(d)
(e)
(f)

are always trilete, usually show a margo along the laesurae, and unlike the other genera, often have a thick perispore that is usually scabrate or verrucate (Fig. 6.3e; Lugardon 1974). The ornamentation on the proximal surface of the exospore is often less pronounced than on the distal surface. There is no pattern of ridges and furrows. In some species the two layers of the exospore split easily; a completely smooth 'central' or 'inner' body (comparable to nexine) and an outer ornamented layer (sexine) can be seen. These inner bodies are often found dispersed in the fossil record (Kräusel and Weyland 1950; Pocock 1964).

The subgenera *Gisopteris* (20 species) and *Lygodium* (15 species) have indistinguishable spores; the exospore may be smooth but is usually more or less granulate or verrucate, and the ornamentation near the laesura-ends is sometimes more pronounced than on the remainder of the proximal surface. The spore size varies between 50 and 130 μm (Fig. 6.3d).

In the subgenus *Odontopteris* (10 species; spore size 75–130 μm) the ornamentation is either coarsely verrucate (Fig. 6.3f) or the verrucae have fused to form a reticulum (Fig. 6.4b). Reticulate spores are only found in this subgenus.

Spore morphology of fossil *in situ* genera

The oldest fossil that has been attributed to the Schizaeaceae is the Carboniferous *Senftenbergia* with two species: *S. plumosa* and *S. pennaeformis* (Radforth 1938, 1939). The fossils show a great many characters of the Schizaeaceae but the difference is that the sporangia are stalked and that the apical annulus consists of 2–4 rows of cells rather than the one row characteristic of the Schizaeaceae. Therefore, *Senftenbergia* cannot be placed in the Schizaeaceae but is certainly related to the family. The spores of *S. plumosa* are 40–70 μm and show a pattern of anastomosing ridges sometimes with short spines on them. *S. pennaeformis* spores are 50–70 μm and show a pattern of fused ridges forming a reticulum (like certain *Lygodium* species and the Jurassic *Klukia* and *Stachypteris*).

The oldest fossils that definitely can be attributed to the Schizaeaceae

---

Fig. 6.2 (a) *Mohria caffrorum*, 10732, section of hollow ridges; SEM, × 5000. (b) *Mohria caffrorum*, 10732; SEM, × 1000. (c) *Mohria cafforum*, 10732; hollow ridges; light micrograph, × 500. (d) *Anemia tomentosa*, 10740, appendices; SEM, × 500. (e) *Anemia tomentosa*, 10740, detail of (d), appendices (exospore) and perispore; SEM, × 2000. (f) *Anemia guatemalensis*, 10735, exo- and perispore; SEM, × 5000.

are the Jurassic Eurasian genera *Klukia* and *Stachypteris*. The spores of the various *Klukia* and *Stachypteris* species (Fig. 6.4a) have been well described (for a summary see van Konijnenburg-van Cittert 1981). They are all trilete, 45–70 μm, distally reticulate, proximally reticulate-granulate (especially *Stachypteris*). They show a great deal of resemblance to spores of some species of *Lygodium* subg. *Odontopteris*.

In the Early Cretaceous, the first *Anemia/Mohria*-like spores occur *in situ*—the well-known *Ruffordia* and *Pelletixia*. *Pelletixia valdensis* spores are about 75 μm and show the typical pattern of broad ridges and narrow furrows (Fig. 6.4e). They even show some thickenings at the end of the laesurae like *Coptophyllum*. The exospore sometimes splits into two layers: a smooth inner and an outer with the ridges (Hughes and Moody-Stuart 1966). SEM study revealed that the ridges are solid (Fig. 6.4f). *Ruffordia goeppertii* spores are smaller (50 μm) and show more narrow and undulating ridges and wider furrows (like *Anemiorrhiza*) but they also show a tendency to ridges with small spines like subg. *Anemia* (Hughes and Moody-Stuart 1966). SEM study revealed again that the ridges are solid in section (Fig. 6.4d). The sculpture on the proximal surface is less pronounced than on the distal surface (Fig. 6.4c).

Another, and very interesting, Early Cretaceous fossil is *Schizaeopsis americana* (Berry 1911)—the first occurrence of a schizaeaceous fossil from America. The macrofossil is like *Schizaea* or *Actinostachys*, but the spores (60–80 μm) are trilete instead of monolete and show the typical striate pattern (Hughes and Moody-Stuart 1966). Thus, this genus combines the macromorphology of *Schizaea/Actinostachys* with the spore morphology of *Mohria/Anemia*.

The first *in situ* fossil that has been attributed to an extant genus occurs in the Late Cretaceous: the american *Anemia fremontii*. The spores are about 40 μm and again show the typical ridge pattern. This species may be compared (also on the basis of the macrofossil) to *A. adiantifolia* (Andrews and Pearsall 1941).

By the Tertiary more fossils can be attributed to extant genera, e.g. the English *Anemia poolensis* and *A. colwellensis* (Chandler 1955). The latter shows the typical ridge pattern (spore size about 45 μm) but *A. poolensis* has 50–60 μm large, smooth spores. The macrofossil can be easily compared with some extant *Anemia* species, but smooth spores

---

Fig. 6.3 (a) *Anemia adiantifolia*, 10736; SEM, × 1000. (b) *Anemia hirsuta*, 10734; SEM, × 1000. (c) *Anemia phyllitides*, 5443, ridges with spines; light micrograph, × 500. (d) *Lygodium japonicum*, 10742; SEM, × 1000. (e) *Lygodium volubile*, 10741, exospore and thick perispore; SEM, × 5000. (f) *Lygodium volubile*, 10741; SEM, × 500.

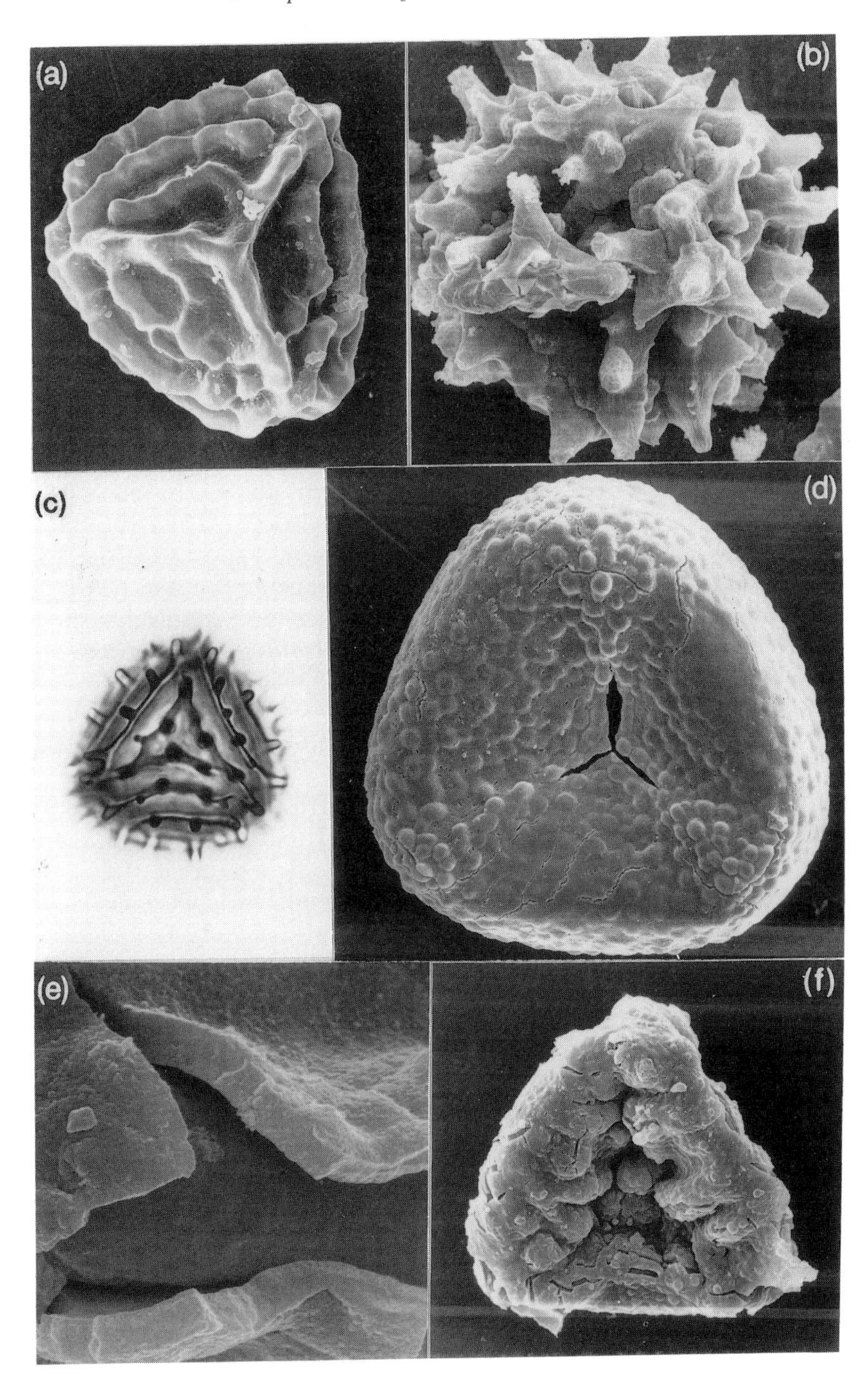
(a)
(b)
(c)
(d)
(e)
(f)

have not been found in extant species except for some immature spores (*A. perrieriana* Chandler 1955 and *A. trichoriza* Pocock 1964). *A. poolensis* spores might be immature; it might also be that only the inner layer of the exospore has been preserved.

In the Tertiary we also find several palmate *Lygodium* fossils, e.g. *L. skottsbergii* from Chili (Halle 1940; Kräusel and Weyland 1950). The spores are about 85 μm, trilete with a granular exospore. Other Tertiary *Lygodium* species of which spores are known are *L. kaulfussii*, with up to 100 μm large, scabrate spores, and *L. gaudinii*, also with scabrate (60–90 μm) spores (Kräusel and Weyland 1950). All these spores show a great resemblance to extant spores of the subg. *Gisopteris* and *Lygodium*.

Lygodium poolensis (Chandler 1955) yielded trilete spores of about 65 μm, with large tubercles that are partly fused, forming a reticulum like *Klukia* spores and those of some extant species in the subg. *Odontopteris*.

Schizaea/Actinostachys-like fossils with monolete spores have not yet been found before the Quaternary. The earliest being *Schizaea skottsbergii* from Hawaii (Selling 1946), with large (over 100 μm) spores that show an alveolate exospore (like the extant *S. pusilla*) and a granulate perispore.

Spore morphology of supposedly Schizaeaceous dispersed spores

The oldest dispersed spores that can be attributed with reasonable certainty to the Schizaeaceae are found in the Jurassic. Before this there were spores that might have belonged to the Schizaeaceae, but they might also have belonged to the Parkeriaceae or Pteridaceae or (common) ancestors (see Filatoff and Price 1988).

From the earliest Jurassic onwards we find dispersed spores like those from *Klukia* and *Stachypteris*, which were described by Balme (1957) as *Ischyosporites* and by Couper (1958) as *Klukisporites* (for discussion on the nomenclature see van Konijnenburg-van Cittert 1981). These spores have been recorded from the beginning of the Jurassic to the end of the Cretaceous, with a maximum during the Middle Jurassic–Early

Fig. 6.4. A, *Klukia exilis*, Yor-8; light micrograph, × 500. B, *Lygodium reticulatum*, 7715; SEM, × 1000. C, *Ruffordia goeppertii*, V.2192 $1, type, proximal surface; SEM, × 1000. D, *Ruffordia goeppertii*, V.2192 $1, type, section of massive ridges; SEM, × 5000. E, *Pelletixia valdensis*, V.2329 $1, type, proximal surface; SEM, × 1000. F, *Pelletixia valdensis*, V.2329 $1, type, section of massive ridges; SEM, × 5000.

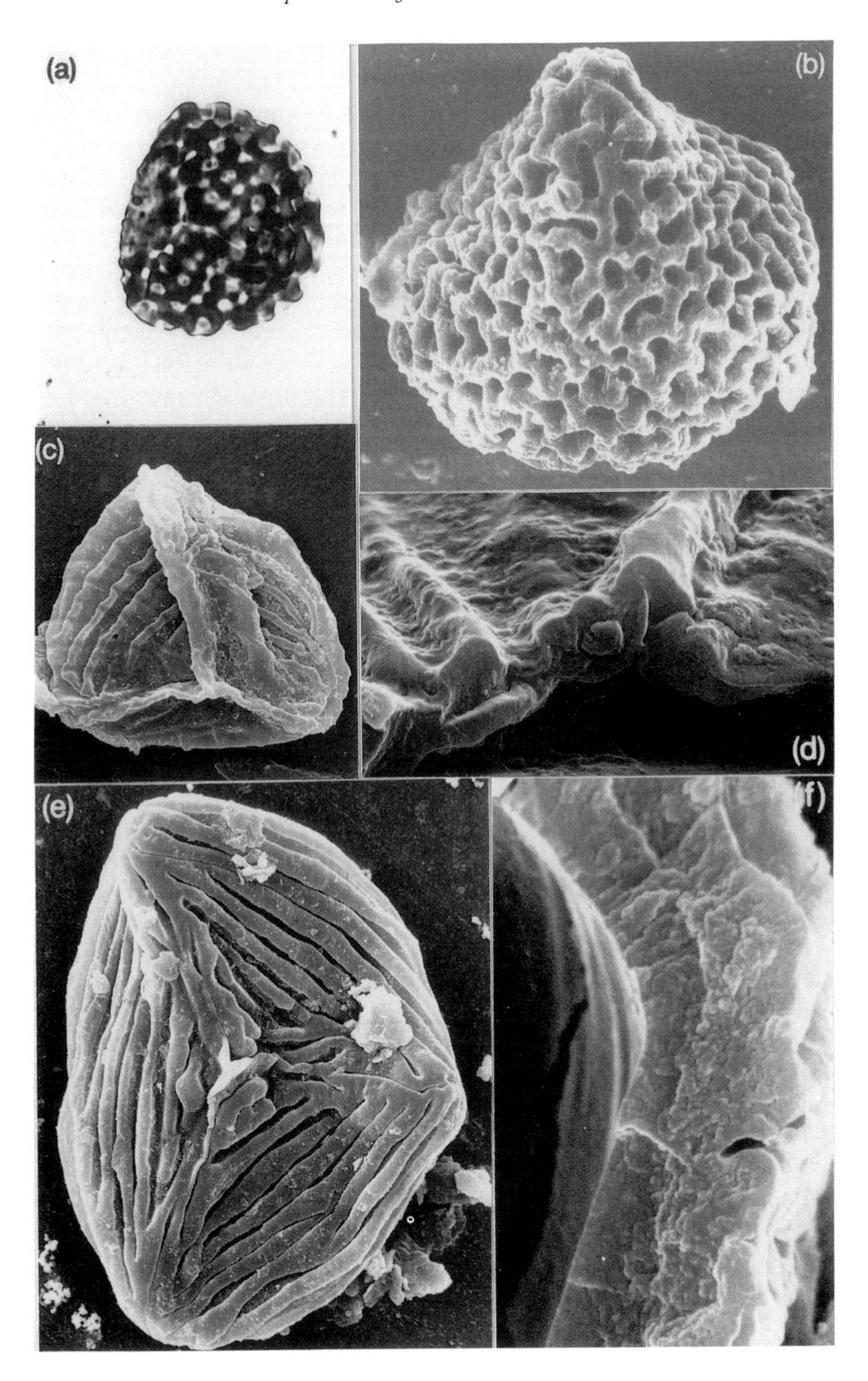
(a)
(b)
(c)
(d)
(e)
(f)

Cretaceous, from all over the world, but the majority of the records come from Eurasia (Bolkhovitina 1961; Filatoff 1975).

Later in the Jurassic, and especially in the Cretaceous, there are a great many records of trilete, striate dispersed spores attributed to the Schizaeaceae (Couper 1958; Davies 1985; Fensome 1987). They usually fall into two genera: *Cicatricosisporites* and *Appendicisporites*. *Cicatricosisporites* has spores in which the ridges are usually broader than the furrows, and they show a great resemblance to the spores of the extant *Mohria* and some *Anemia* species, and the fossil *Pelletixia* and *Ruffordia*. The two layers of the exospore (nexine and sexine) are recorded to have approximately the same thickness (Pocock 1964). These spores are usually 30–65 μm large but some (especially later) are larger (up to 120 μm). *Appendicisporites* has similar spores, but in addition shows appendices of the exospore at the angles of the tetrahedron, like in the extant *Anemia* subg. *Coptophyllum*. Furthermore, the outer layer of the exospore has been reported to be thicker than the inner layer (Pocock 1964). The appendices may have been much longer than in the extant species (Bolkhovitina 1961). *Appendicisporites* species are about 35–100 μm.

Also from the Middle Jurassic onwards, with a maximum in the Tertiary, *Lygodium*-like smooth–verrucate spores have been recorded from all over the world. They have been described under various generic names according to the exospore ornamentation (Bolkhovitina 1961; Filatoff 1975; Fensome 1987), but it is by no means certain that all these spores did really belong to the Schizaeaceae, their spore type, especially the smooth or granulate forms, is too general for definite assignment.

The first *Actinostachys*-like dispersed monolete, striate spores occur in the Aptian (Cretaceous). They are usually placed in the genus *Schizaeosporites* (Pocock 1964; Archangelsky *et al.* 1983). Pocock mentioned that in the Aptian *S. phaseolus* the majority of spores are monolete but trilete spores and intermediate forms occur. During the Tertiary we find several *Actinostachys*-like spores and in the Miocene the first alveolate monolete spores (like the extant *S. pusilla*) occur (Selling 1946). Granulate monolete spores (like *Schizaea*) have also been described from the Tertiary onwards (Selling 1946) but they are of too general a type to attribute them definitely to the Schizaeaceae.

The data concerning the spore morphology of the extant, the fossil *in situ* and dispersed genera are summarized in Table 6.1.

Geographical distribution throughout time

The extant Schizaeaceae occur mainly in tropical areas in the southern hemisphere, although some species are found in the northern

Table 6.1 Comparison of spore morphology of extant and fossil schizaeaceous genera

Genus	Exospore		Perispore	Fossil genus	Dispersed genus
Actinostachys	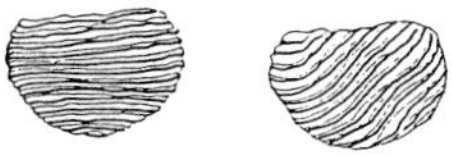		Finely granulate		*Schizaeosporites* (Aptian–Recent)
Schizaea		Smooth Scabrate Granulate Alveolate	Smooth–finely granulate	*Schizaeopsis* (L.Cretaceous)	*Schizaeosporites* (Cret/Tertiary–Recent)
Anemia (*Coptophyllum*)		Cross section ridges	Microspines	*Pelletieria* (L.Cretaceous)	*Appendicisporites* (L.Cretacous–Recent)
(*Anemia*)			Smooth or microspines	*Anemia* (U.Cretaceous –Recent)	*Cicatricosisporites* (Latest Jurassic –Recent)
(*Anemiorrhiza*)			Smooth–scabrate	*Ruffordia* (L.Cretaceous)	*Cicatricosisporites*
Mohria			Smooth	*Pelletieria*	*Cicatricosisporites*
Lygodium (*Gisopteris*) (*Lygodium*)	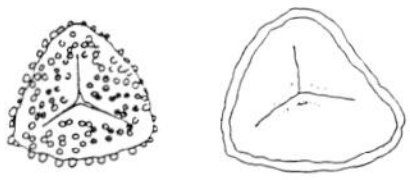		Scabrate	*Lygodium* (Tertiary–Recent)	*Lygodium*-like spores (various genera) (Middle Jurassic–Recent)
(*Odontopteris*)	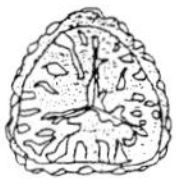		Smooth Scabrate Verrucate	*Klukia* *Stachypteris* (Jurassic) *Senftenbergia* (Carboniferous)	*Klukisporites* (L.Jurassic–end of Cretaceous)

hemisphere (for example *Schizaea pusilla* occurs as far north as Nova Scotia and Newfoundland).

Mesozoic schizaeaceous macrofossils have so far all been found in the northern hemisphere; first in Eurasia (the possibly ancestral *Senftenbergia* in Germany, *Klukia*, *Stachypteris*, *Ruffordia*, and *Pelletixia* in Europe and/or Asia), later also in North America (*Schizaeopsis*, *Anemia fremontii*). The oldest macrofossil recorded from the southern hemisphere was the Tertiary *Lygodium skottsbergii*. The same trend is apparent in the dispersed spores; the various genera appear first in the northern hemisphere and are only later (and first in smaller numbers) found in the southern hemisphere. During the Cretaceous *Cicatricosisporites* and *Appendicisporites* show a world-wide distribution. At the end of the Cretaceous there was a contraction of the distribution area; in the Eocene a second, mainly pantropical, expansion and at the end of the Tertiary a second decline to the relic areas of present distribution in America, Africa, Asia, and Australia. Schizaeaceae no longer occur in Europe (Germeraad *et al.* 1968).

Conclusions and evolutionary trends

Schizaeaceous spores exhibit the well-known trend from trilete to monolete spores (Wagner 1974). All the older spores are trilete. The oldest *Schizaea*/*Actinostachys*-like macrofossil (the Early Cretaceous *Schizaeopsis*) still has trilete spores; and quite a few trilete or intermediate forms occur in the earliest dispersed monolete spore genus (Pocock 1964).

The oldest schizaeaceous spores are heavily ornamented and show a thick exospore, a feature that is supposed to be derived (Wagner 1974), but I believe it to be primitive, at least in this family. Spores with ridges and reticula occur in *Senftenbergia* and *Klukia*/*Stachypteris*, and later spores with parallel ridges in *Pelletixia*, *Ruffordia*, *Schizaeopsis*, and *Anemia*. The more smooth and granulate spores are only found much later. I think that in this family the heavily ornamented exospores are the primitive type and the smooth or granulate ones derived. Sculpture elements like the spines on the ridges in *Anemia* are, of course, derived characters.

Within the striate spores there might be a trend from massive ridges to hollow ridges in *Mohria* and *Anemiorrhiza* via ridges with a kind of medulla in the centre.

I think that, judging from the fossil evidence, the original spore size within the Schizaeaceae is somewhere between 50–70 μm, and that from there we can observe two trends: one leading to very large spores in, e.g. *Anemia* subg. *Coptophyllum*, *Lygodium*, and *Schizaea* (all up to 130 μm); the other to smaller (and usually smooth or granulate) spores, especially in *Schizaea* subg. *Lophidium* and some *Actinostachys* species.

As to the geographical distribution, we see a gradual shifting from an occurrence in the northern hemisphere in Mesozoic (or even Palaeozoic) times to the southern hemisphere in the Tertiary and Quaternary.

Acknowledgements

I wish to thank Dr W. Punt for his help and advice. I am most grateful to the Natural History Museum, Department of Palaeontology (London) for the opportunity to borrow material. I am thankful to Mrs M. Verf for the preparation of the herbarium material, to Mr P. Hoen for his help with the SEM photos and to Mr H. Elsendoorn for his help in the darkroom.

References

Andrews, H.N. and Pearsall, C.S. (1941). On the flora of the Frontier formation of Southwestern Wyoming. *Annals of the Missouri Botanical Garden* **28**, 167–93.

Archangelsky, S., Baldoni, A., Gamerro, J.C., and Seiler, J. (1983). Palinologia estratigraphica del Cretacico de Argentina Australis. II. Descripciones sistematicas. *Ameghiniana* **20**, 199–226.

Atkinson, L.R. (1962). The Schizaeaceae: the gametophyte of Anemia. *Phytomorphology* **12**, 264–88.

Balme, B.E. (1957). Spores and pollen grains from the Mesozoic of Western Australia. *Coal Research CSIRO* **25**, 1–48.

Berry, E.W. (1911). A Lower Cretaceous species of Schizaeaceae from Eastern North America. *Annals of Botany* **25**, 193–9.

Bolkhovitina, N.A. (1961). Fossil and recent spores of the family Schizaeaceae. *Trudy Geologica Instituta, Akademia Nauk SSSR* **40**, 1–176.

Chandler, M.E.J. (1955). The Schizaeaceae of the South of England in Early Tertiary times. *Bulletin of the British Museum (Natural History)* **2**, 291–314.

Couper, R.A. (1958). British Mesozoic microspores and pollen grains. *Palaeontographica* **B103**, 77–175.

Davies, E.H. (1985). The Anemian, Schizaeacean and related spores: An index to genera and species. *Canadian Technical Report of Hydrography and Ocean Sciences* **67**, 1–21.

de la Sota, E.R. and Morbelli, M.A. (1987). Schizaeales. *Phytomorphology* **37**, 365–93.

Fensome, R.A. (1987). Taxonomy and biostratigraphy of Schizaealean spores from the Jurassic–Cretaceous boundary beds of the Aklavik Range, District of Mackenzie, *Palaeontographica Canadiana* **4**, 1–49.

Filatoff, J. (1975). Jurassic palynology of the Perth Basin, Western Australia. *Palaeontographica* **B154**, 1–113.

Filatoff, J. and Price, P.L. (1988). A Pteridean spore lineage in the Australian Mesozoic. *Memoirs of the Association for Australasian Palaeontology* **5**, 89–124.

Germeraad, J. H., Hopping, C. A., and Muller, J. (1968). Palynology of Tertiary sediments from tropical areas. *Review of Palaeobotany and Palynology* **6**, 189–348.

Halle, T. G. (1940). A fossil fertile *Lygodium* from the Tertiary of South Chile. *Svensk Botanisk Tidskrift* **34**, 257–64.

Hill, S. R. (1977). Spore morphology of *Anemia* subgenus *Coptophyllum*. *American Fern Journal* **67**, 11–17.

Hill, S. R. (1979). Spore morphology of *Anemia* subgenus *Anemia*. *American Fern Journal* **69**, 71–9.

Hughes, N. F. and Moody-Stuart, J. (1966). Descriptions of Schizaeaceous spores taken from early Cretaceous macrofossils. *Palaeontology* **9**, 274–89.

Kräusel, R. and Weyland, H. (1950). Kritische Untersuchungen zur Kutikular-Analyse tertiäre Blätter, I. *Palaeontographica* **91B**, 7–92.

Lugardon, B. (1974). La structure fine de l'exospore et de la périspore des Filicinées isosporées. *Pollen et Spores* **16**, 161–226.

Mickel, J. T. (1962). A monographic study of the fern genus *Anemia*, subgenus *Coptophyllum*. *Iowa State Journal of Science* **36**, 349–82.

Mickel, J. T. (1981). Revision of *Anemia* subgenus *Anemiorrhiza* (Schizaeaceae). *Brittonia* **33**, 413–29.

Pocock, S. A. J. (1964). Pollen and spores of the Chlamydospermidae and the Schizaeaceae from Upper Mannville Strata of the Saskatoon Area of Saskatchewan. *Grana Palynologica* **5**, 129–209.

Radforth, N. W. (1938). An analysis and comparison of the structural features of *Dactylotheca plumosa* Artis sp. and *Senftenbergia ophiodermata* Goeppert sp. *Transactions of the Royal Society of Edinburgh* **59**, 385–96.

Radforth, N. W. (1939). Further contributions to our knowledge of the fossil Schizaeaceae; genus *Senftenbergia*. *Transactions of the Royal Society of Edinburgh* **59**, 745–61.

Reed, C. F. (1947). The phylogeny and ontogeny of the *Pteropsida*. I. Schizaeales. *Boletim de Sociedade Broteriana* **21**, 71–173.

Selling, O. H. (1944). A new species of *Schizaea* from Melanesia and some connected problems. *Svensk Botanisk Tidskrift* **38**, 207–25.

Selling, O. H. (1946). Studies in the Recent and fossil species of *Schizaea*, with particular reference to their spore characters. *Meddelelser Göteborgs Botanisk-Trödgard* **16**, 1–113.

Selling, O. H. (1947). Further studies in *Schizaea*. *Svensk Botanisk Tidskrift* **41**, 431–50.

van Konijnenburg-van Cittert, J. H. A. (1981). Schizaeaceous spores *in situ* from the Jurassic of Yorkshire. *Review of Palaeobotany and Palynology* **33**, 169–81.

Wagner, W. H. (1974). Structure of spores in relation to fern phylogeny. *Annals of the Missouri Botanical Garden* **61**, 332–53.

7. Diversification of modern heterosporous pteridophytes

MARGARET E. COLLINSON*
Biosphere Sciences Division, King's College London, Campden Hill Road, London W8 7AH, UK

Abstract

The fossil record of the heterosporous lycopsids (Isoetales and Selaginellales) and water ferns (Marsileales and Salviniales) is reviewed. Fertile whole plants and dispersed megaspores with attached microspores or microspore massulae are the crucial fossils for understanding the evolution, diversification, and phylogeny of these plants. Salviniales are particularly well represented with fertile fossil plants substantiating an extensive dispersed spore record. Azollaceae (*Azolla*) and Salviniaceae (*Salvinia*) are distinct, and the modern genera clearly identifiable, from the earliest Tertiary onwards. *Azolla* and extinct relatives are also known in the late Cretaceous. A mosaic of characters is presented by these late Cretaceous fossils most of which is lost during the early Tertiary. Marsileales are represented by one sporocarp species in the early Tertiary but otherwise only by a patchy record of dispersed spores from the early Cretaceous onwards. Isoetales and Selaginellales have a longer history. *Selaginella* is known from the Palaeozoic from whole fertile plants indistinguishable from modern representatives. However, the potential for *in situ* spores to clarify the more extensive dispersed spore record has not been fully exploited. Associated microspores, spore wall ultrastructure, and other features may help to indicate the affinity of dispersed magaspores. These probable isoetalean and selaginellalean spores are apparently more diverse in the Mesozoic than in the Tertiary. In *Selaginella* the Mesozoic diversity may not exceed that of the modern

*Present Address:
Department of Geology, Royal Holloway and Bedford New College, Egham Hill, Egham, Surrey TW20 0EX UK.

Pollen and Spores (ed. S. Blackmore and S. H. Barnes), Systematics Association Special Volume No. 44, pp. 119–150, Clarendon Press, Oxford, 1991.

genus, for which many spores have not yet been studied in detail. In Isoetales the Mesozoic diversity undoubtedly includes representatives of extinct families, including Pleuromeiaceae. Fertile *Isoetites* plants (probable Isoetaceae, from the Cretaceous and Tertiary) contain *Minerisporites* megaspores (with attached monolete microspores in one case). Dispersed *Minerisporites* spores are more abundant and diverse in the Cretaceous than in the Tertiary. The apparent Tertiary decline of heterosporous lycopsids is particularly inconsistent with the high modern diversity of *Selaginella*. This may possibly be explained if the increasing proportion of angiosperms in Tertiary vegetation influenced the chances of transport to a site of potential fossilization for megaspores produced by low-growing, pteridophyte herbs in non-wetland habitats.

Introduction

Heterosporous plants are most frequently represented in the fossil record by megaspores. These large spores (having one dimension >200 μm) are known from the early Devonian onwards (Traverse 1988*a*). Many Palaeozoic examples have been found *in situ*, and the ecology, functional biology, and phylogeny of their parent plants are comparatively well understood. The more familiar examples are arborescent lycopsids which dominated late Carboniferous coal-forming swamps and other communities (Bartram 1987; Collinson and Scott 1987; Gastaldo 1987; Dimichele and Phillips 1988). The importance of herbaceous heterosporous lycopsids in these communities has been demonstrated recently (Bartram 1987).

In comparison with their Palaeozoic relatives, Mesozoic heterosporous plants, although apparently abundant and diverse, are very poorly understood. Parent plants are unknown for most of the 500 or so Mesozoic megaspore species reviewed by Kovach and Batten (1989). Sweet (1979) gives a general, well-illustrated account of megaspore diversity. In contrast, Cainozoic heterosporous plants are better known but of much lower diversity. Generally, megaspores almost identical with those of modern genera of heterosporous plants are first encountered in Mesozoic or younger strata.

Methods

Spores were removed from fertile plants using fine needles and then treated as detailed below. Dispersed spores were obtained by disaggregating sediment using hot water, hydrogen peroxide, or hydrofluoric acid, and grading the residues through a series of sieves, retaining all material larger than 125 μm. Collinson *et al.* (1985) and Traverse (1988*a*) give further details. Specimens were picked from samples in water using a fine paintbrush and then cleaned in hydrofluoric acid for at least 7 days. Specimens for scanning electron microscopy (SEM) were mounted on stubs using Durofix or Bostick on a cover-glass, sputter-coated with gold, and examined using a Philips 501B SEM. Specimens for transmission electron microscopy (TEM) were removed from stubs using acetone and embedded in Spurr resin. Sections 60 nm thick were cut using a Reichert–Jung ultracut microtome, stained using uranium acetate/lead citrate, and examined using a Philips EM 301S TEM. Semi-thin sections 0.5–1 μm were studied using light microscopy. All samples were sectioned in a near median longitudinal plane, from proximal to distal pole (or along their longest axis when orientation was obscured by hairs). Illustrations are taken from the equatorial rather than the polar zones. Generally, at least two specimens have been examined by TEM and ultrastructural uniformity confirmed by SEM using fractured specimens. Numerous specimens have been studied using SEM. For *Glomerisporites* and *Molaspora* only two specimens were available but results accord well with previous work. For *Ariadnaesporites* and *Arcellites* only one specimen has been studied. This is simply to provide comparative illustrations using identical techniques. Clearly, the full range of species requires future study. This is also true for *Azollopsis* as material of only one (of seven) species was available.

Lycopsids: Isoetales

Introduction

The three families Isoetaceae (Mesozoic–Recent), Pleuromeiaceae (Mesozoic), and Chaloneriaceae (Palaeozoic) were included in Isoetales by Pigg and Rothwell (1983). Thomas and Brack-Hanes (1984) added the Takhtajanodoxaceae (Mesozoic) and excluded Pleuromeiaceae. Knowledge of the extinct families is very limited. Reconstructions show a broadly similar growth form (Fig. 7.1a,b). Megasporophylls and microsporophylls may be grouped in terminal or lateral strobili. Sporophyll morphology may be identical with, or distinct from, the vegetative leaves. In many cases, genera are based only on the strobilus. For further details see Chaloner (1967), Taylor (1981), Meyen (1987), and Thomas and Spicer (1987).

Paurodendron fraipontii (Leclercq) Fry is a small Carboniferous lycopsid with a basal rhizomorph, branching stems, and bisporangiate cones (with *in situ* megaspores and microspores). This reconstructed plant was often included in *Selaginella*, the aerial portions being almost identical to some modern species (Taylor 1981). Rothwell and Erwin (1985) interpret the rooting organ as a modified shoot system, thus resembling *Isoetes* (and other Isoetales and Lepidodendrales) not *Selaginella*. This emphasizes the need for caution when intepreting affinities solely from dispersed spores.

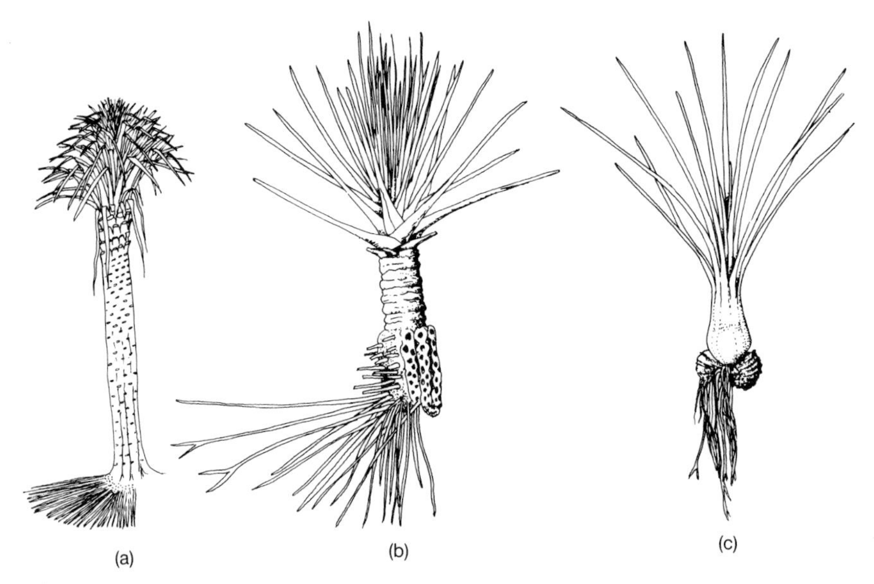

Fig. 7.1 Isoetalean plants: (a) *Pleuromeia*, stem up to 2 m tall; (b) *Nathorstiana*, stem about 10 cm tall; (c) *Isoetes*, stem about 1–3 cm tall. (Redrawn after Thomas and Spicer 1987 and Stewart 1983.)

Mesozoic

The best-known Mesozoic isoetaleans with megaspores and microspores *in situ* are assigned to Pleuromeiaceae. These Triassic fossils contain the dispersed megaspore genera *Banksisporites, Tenellisporites,* and *Horstisporites* in cones named *Cylostrobus, Annalepis,* and *Skilliostrobus,* respectively. Monolete microspores named *Aratrisporites* were produced by all of these (Meyen 1987, p. 79; Traverse 1988*a*, p. 232–5).

Retallack (1981) used a multidisciplinary approach to reconstruct the Triassic vegetation of eastern Australia. According to him a 'Pleuromeia plant' (Fig. 7.1a) probably bore *Cylostrobus* cones and grew in wetland meadows fringing interdistributary bays of lobate deltas entering coastal lagoons. This is the most detailed interpretation of an extinct isoetalean plant. Similar species grew in Europe and Asia.

Several other probable Mesozoic isoetaleans are based on strobili or on isolated spores. Amongst these may be cited the Triassic strobili named *Lycostrobus*, for which *L. scottii* (see Seward 1910; (Potonié 1956 p. 72–3; Chaloner 1967; Taylor 1981) yields spiny megaspores (cf. Fig. 7.2a) and granulate ellipsoid monolete microspores. Dispersed megaspores of *Nathorstisporites*, with monolete *Aratrisporites* microspores entangled in the spines (Scott and Playford 1985), and of *Paxillitriletes* (Fig. 7.2a), also with entangled monolete microspores (Kovach and

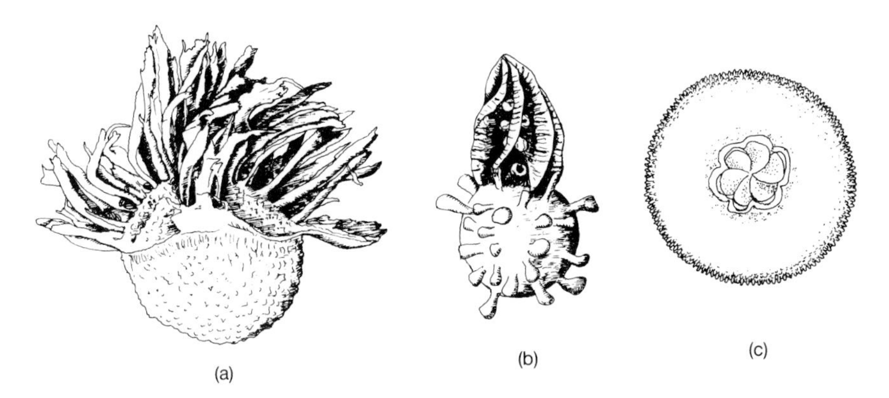

Fig. 7.2 Megaspores: (a) *Paxillitriletes vittatus* Kovach and Dilcher 1985, (extinct Isoetales); equatorial view, small monolete microspores occur amongst the proximal spines; equatorial diameter of spore body is 590 μm. (b) *Arcellites disciformis* Miner emend Ellis and Tschudy 1964, (extinct Marsileales); equatorial view showing spirally twisted acrolamella with adherent microspores; equatorial diameter of spore body is 250 μm. (c) *Regnellidium diphyllum* Lind. (Marsileaceae); apical view showing terminal acrolamella and papillate surface; equatorial diameter is 500 μm. (Redrawn after: A, Kovach and Dilcher 1985; B, Hueber 1982 and Ellis and Tschudy 1964; C, Higinbotham 1941.)

Dilcher 1985), are superficially similar. These spore types probably floated proximal face downwards, just under the water, allowing floating microspores to become entangled in the spines (Kovach and Dilcher 1985). Kovach and Dilcher (1985) support an isoetalean affinity on the basis of monolete microspores, megaspore wall ultrastructure (see also p. 128), and facies occurrence, implying a wetland growth site for the parent plant. Batten (1988) argued that the strongest evidence was the entangled or associated monolete microspores. Koppelhus and Batten (1989) base isoetalean affinity for *Tenellisporites* (Swedish Cretaceous) on wall ultrastructure and associated (non-adherent) monolete microspores. Along with *Minerisporites* (Fig. 7.3n, see below), all these apparently isoetalean spores have equatorial and triradiate flanges, features therefore more typical of the extinct than extant members of the group.

One frequently figured Cretaceous isoetalean is *Nathorstiana* (Fig. 7.1b; Karfalt 1984). Reproductive structures are unknown. Sporophylls may have been like those of modern *Isoetes* (Fig. 7.1c) or borne in strobili. This limited whole-plant information demonstrates that most Mesozoic dispersed spores cannot be included with certainty in any particular family of the order.

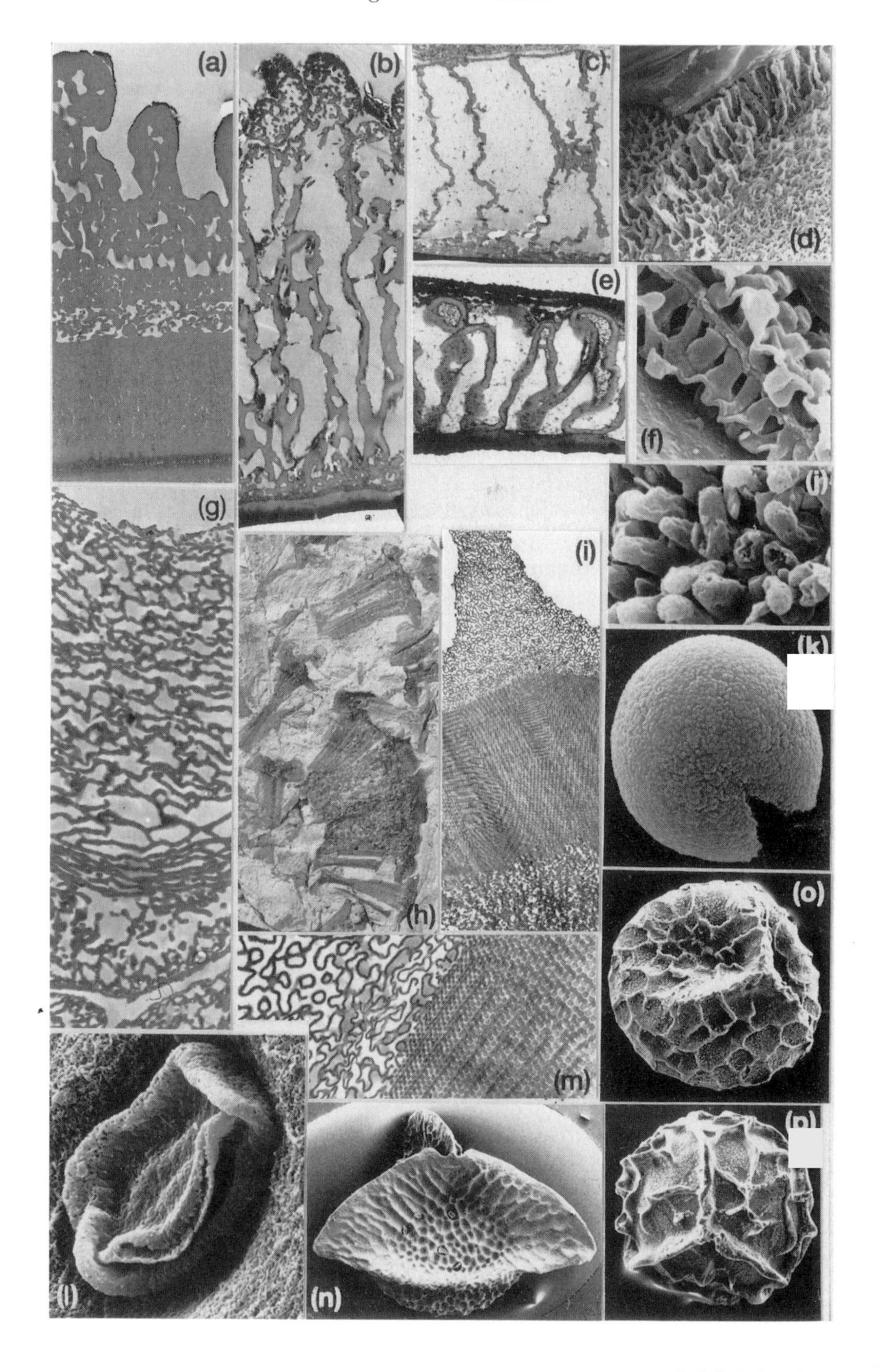
(a)
(b)
(c)
(d)
(e)
(f)
(g)
(h)
(i)
(j)
(k)
(l)
(m)
(n)
(o)
(p)

Isoetaceae

Plant fossils resembling modern *Isoetes* have been assigned to the genus *Isoetites*. This is the only extinct genus of Isoetaceae according to some authors (e.g. Taylor 1981), although others include *Nathorstiana* (Fig. 7.1b) and similar forms (Chaloner 1967, Thomas and Brack-Hanes 1984). *Isoetites* fossils are compressions of sporophylls which are linear or spatulate with a basal sporangium (Chaloner 1967; Taylor 1981). The European Cretaceous *I. choffatii* (Saporta) Seward includes an associated compressed axis reaching a breadth of 1 cm and having radiating appendages (or their scars) resembling the roots of *Isoetes* (Seward 1910). Sporangia were also noted on this species by Dijkstra (1951, p. 17) and Teixeira (1948) but spores were not described. *I. elegans* Walkom 1944 (Jurassic, Australia) includes associated mega- and microsporangia with spores *in situ*, but details were omitted. The Indian Cretaceous *I. indicus* contains megaspores resembling those of the dispersed spore genus *Minerisporites*. *Minerisperites cutchensis* is abundant in the same strata (Bose and Roy 1964; Sukh-Dev 1980). The species *M. mirabilis* is found *in situ* in *Isoetites horridus* (Dawson) Brown (Hickey 1977; Melchior 1977; Melchior and Hall 1983; personal observations) from the Palaeocene of North America. Monolete microspores

Fig. 7.3 (a) *Arcellites medusus* (Knobloch 1984), TEM of megaspore wall, × 3000. (b) *Molaspora lobata* (Batten 1988), TEM of megaspore wall, × 2000. (c) *Marsilea vestita* (cultivated), from sorus fixed with glutaraldehyde after sporocarp germination, TEM of megaspore wall, × 2000. (d) *Marsilea vestita* (cultivated), dry within sporocarp, SEM of fractured megaspore wall, × 500. (e) *Marsilea vestita* (cultivated), from sorus fixed with glutaraldehyde after sporocarp germination, TEM of microspore wall, × 8000. (f) *Marsilea vestita* (cultivated), dry within sporocarp, SEM of fracture of two adjacent microspore walls, × 4000. (g) *Minerisporites mirabilis* (Collinson *et al.* 1985), TEM of megaspore wall, × 4000. (h) *Isoetites horridus*, portion of fertile plant × 1.35 (Field Museum of Natural History, Chicago, Geology Department pp 33936). (i) *Erlansonisporites* sp. (Collinson *et al.* 1985), TEM of megaspore wall, × 1000. (j) *Molaspora lobata* (Batten 1988), SEM of megaspore surface, × 1100. (k) *Marsilea vestita* (cultivated), air-dried from water after germination, SEM of microspore, × 500. (l) *Minerisporites mirabilis* (Collinson *et al.* 1985), SEM of microspore on megaspore surface, × 1300. (m) *Erlansonisporites* sp. (Collinson *et al.* 1985), TEM of megaspore wall, × 4000. (n) *Minerisporites mirabilis* (Collinson *et al.* 1985), SEM of megaspore with surface microspores, × 50. (o) *Erlansonisporites* sp. (Collinson *et al.* 1985), SEM of megaspore, × 50. (p) *Selaginella galeottii* Spring [British Museum (Natural History), Botany Department, J.A. Speymark 49718], SEM of megaspore, × 50. See text and cited references for details of species and localities.

have been found (personal observations) on megaspores extracted from USNM 42785 figured by Hickey (1977). Identical megaspores (Fig. 7.3n), also with adherent monolete microspores (Fig. 7.3l) are common in the Palaeocene of southern England (Collinson *et al.* 1985).

Kovach and Batten (1989) list three Triassic and Jurassic species of *Minerisporites* and the genus is abundant throughout the Cretaceous and Palaeocene (Tschudy 1976; Collinson *et al.* 1985; Kovach and Dilcher 1988) often in facies which imply a wetland habitat for the parent plants. *I. horridus* (see Brown 1962; Hickey 1977) and *I. serratus* Brown (1939) both show a double row of quadrate depressions in the leaves, interpreted as collapsed air chambers; both occur in depositional settings consistent with a wetland habitat (Brown 1939, 1962; Hickey 1977). Brown (1939) interpreted the rosette of sporophylls as floating or close to the soil or mud in a well-watered situation. According to Melchior and Hall (1983) corms of *I. horridus* were part of an emergent aquatic community with water depth a maximum of 30 cm. In both species the sporophylls or leaves radiate, in spiral phyllotaxy, from a central structure said to be corm-like but which is very poorly preserved. In material figured by Hickey (1977) and Brown (1962) and new material (collected by Crane and Collinson in 1984 from Camels Butte), the sporophyll bases (some with, some without sporangia) are distinct dumb-bell shaped units (Fig. 7.3h). They not only seem to have been 'shed', or at least displaced, from the corm-like base but also to have lost the (possibly deciduous) distal vegetative portion. This may result from a mechanism for spore dispersal subtly distinct from the swelling and rotting of modern *Isoetes* sporophylls. In this respect, and in the spatulate tips to the leaves, these *Isoetites* species differ from modern *Isoetes*. Furthermore, the corm-like base is poorly understood and might have been more like *Nathorstiana* (Fig. 7.1b) than *Isoetes* (Fig. 7.1c). Brown (1939) recognized root-like structures on one specimen, which, along with *I. choffatii*, seem to exclude an alternative interpretation of *Isoetites* as detached strobili.

Drinnan and Chambers (1986) assigned an early Cretaceous plant to *Isoetes*. This has a corm-like base and radiating elongate linear leaves with flared tips, but no reproductive structures. Another *Isoetes* is from the early Miocene of Tasmania (Hill 1987). The compressed elongate leaves narrow to a rounded apex, and megaspores very similar to those of modern *Isoetes* occur at their bases. Sporangia are not preserved, microspores are unknown. The leaves lack stomata, suggesting that the plant was submerged. SEM illustrations show a megaspore surface very similar to that of *Minerisporites* (Fig. 7.3l). Whereas these fossils may well represent *Isoetes*, it is impossible to exclude other genera without details of the basal portion of the plant.

Lycopsids: Selaginellales

Fossil plants similar to modern iso- and anisophyllous *Selaginella* species are recorded from the Carboniferous onwards as *Selaginellites* or *Selaginella* (Chaloner 1967; Thomas and Brack-Hanes 1984; Meyen 1987). Dispersed spores of some are important indicators of herbaceous plants in Carboniferous coal-forming communities (Bartram 1987). *Paurodendron* (see above) is a problematic case, but the Permian *S. harrisiana* Townrow 1968 from Australia, does show all the characteristics of the modern genus (Stewart 1983; Meyen 1987). Traverse (1988*b*) cites *Selaginella* as the most conservative land vascular plant, with a history covering some 300 million years for forms closely resembling the modern genus.

Unfortunately, what little is known of *in situ* spores from these fossils is largely unhelpful for recognizing affinites of dispersed spores. *Banksisporites* megaspores with trilete *Lundbladispora* microspores were recorded in species from the Jurassic of Sweden and Triassic of Greenland (Helby and Martin 1965; Chaloner 1967). However, *Banksisporites* occurs in *Cylostrobus* of the Pleuromeiaceae. Townrow (1968) noted several dispersed megaspore genera which could not be distinguished from spores *in situ* in *S. harrisiana*. Fertile *Selaginella* with leafy axes identical to those of modern species are known from the Triassic of Arizona (Ash 1972), Tertiary of the USA (Knowlton 1916; Brown 1962), and early Cretaceous of England (Watson 1969). Only in the last case were details of spores given, but they were not encountered dispersed in the same area and time-interval (Batten 1974).

Dispersed megaspores—recognition of Selaginellales

Without more extensive knowledge of *in situ* spores, this task is problematic. Possible criteria include: presence of trilete versus monolete microspores (preferably adherent or abundant in associated palynological preparations); occurrence of clusters of large numbers of megaspores [Selaginellales typically have 4,8, or 12 (max. 42) spores per sporangium]; and ultrastructural characteristics (Collinson *et al.* 1985; Kovach 1989).

The author knows of no examples of trilete microspores adherent to dispersed Mesozoic or Cainozoic lycopsid megaspores (in a Palaeozoic example from Lepidodendrales the adherence has survived even reworking, see Collinson *et al.* 1985). Trilete microspores are adherent to megaspores whose perispore form and ultrastructure and general morphology indicate affinities with water ferns (see later). Such adhesion is probably greatly facilitated by water and by ornament or mucilage production on megaspores. Thus absence (or rarity) of adherence is to be expected if ancient Selaginellales, like their modern relatives

(Tryon and Tryon 1982), tended not to grow in wetlands and lacked suitably ornamented spores (e.g. Fig. 7.3p). In *Selaginella* species where megaspores are initially retained in a single basal sporangium, they may receive a microspore rain from above. Tryon and Tryon (1982, p. 823) figure a microspore on an *S. galeottii* megaspore. Even if assisted by surface microtopography, such an association might not survive transport to a depositional site. In similarly ornamented *Minerisporites* (Fig. 7.3l,n), by analogy with modern relatives (see earlier), mucilaginous material may have been produced around the sporangia, further assisting the adhesion.

Tetrads are recorded from a variety of dispersed spore types (Hueber 1982; Collinson *et al.* 1985; Kovach and Dilcher 1988; Koppelhus and Batten 1989; and in the collections of material published by Knobloch 1984, 1986). *Minerisporites* and also *Paxillitriletes*, in contrast, are known from clusters of large numbers of spores (Collinson *et al.* 1985; Marcinkiewicz 1989).

Kovach and Dilcher (1985, 1988), Batten (1988), Koppelhus and Batten (1989), and Kovach (1989), quoting personal observations and previous studies, suggested that a loose arrangement of sporopollenin threads, orientated mainly parallel to the spore surface, is typical for isoetalean megaspores, whereas a loose, random or regular, compact arrangement is typical for selaginellalean forms (e.g. cf. Fig. 7.3(g) with Fig. 7.3i,m)

Of the 40 or so species of modern *Selaginella* examined ultrastructurally (TEM and/or SEM) (e.g. Tryon and Lugardon 1978; Minaki 1984; Bajpai and Maheshwari 1986; Taylor and Taylor 1987; Kovach 1989; personal observations), five show a very regular, compact, grid-like arrangement of elements in one layer of the sporoderm, which is otherwise unknown in Recent plant spores. The same structure has been observed (Fig. 7.3i,m) in *Erlansonisporites* megaspores (Fig. 7.3o) from the Palaeocene (Collinson *et al.* 1985) and Cretaceous (Bergad 1978; Hueber 1982; Kovach and Dilcher 1988; Koppelhus and Batten 1989; Kovach 1989). Opalescent iridescence of the spore wall that results from this structure (personal observations; Hueber 1982; Koppelhus and Batten 1989) has also been seen in Jurassic material (Collinson *et al.* 1985, p. 395), in Cretaceous material assigned to *Kerhartisporites* Knobloch 1984 (probably a synonym of *Erlansonisporites*), and in Neogene material assigned to *Selaginella* (Knobloch 1986) (last two, personal observation, Geological Survey collections, Prague). At least two of these fossil assemblages include tetrads (Collinson *et al.* 1985; personal observations). All these megaspores show gross morphological similarity to the modern species *S. galeottii* (Fig. 7.3p) which has the grid-like wall structure.

The same ultrastructure has, however, also been observed in fossils where the gross morphology is rather distinct from modern *Selaginella*, notably in the Cretaceous *Thylakosporites* (Hueber 1982; Koppelhus and Batten 1989) which has a mesh-like perispore but does occur in tetrads (Hueber 1982). Other megaspores with this ultrastructure include species of *Ricinispora, Horstisporites, Rugotriletes,* and *Trileites* (Batten 1988; Taylor and Taylor 1988; Koppelhus and Batten 1989) All of these are generally assumed to have selaginellalean affinities.

Water Ferns: Marsileales

This order is poorly represented in the fossil record (Fig. 7.7), especially compared to Salviniales (see below). Dorofeev (1981) described megaspores of *Marsilea* (four species, Oligocene and Miocene); *Regnellidium* (three species, Eocene and Oligocene), and *Pilularia* (two species, Miocene and Pliocene) from the USSR. The *Marsilea* and *Regnellidium* megaspores had adherent microspores. Adhesion of microspores to megaspores occurs in laboratory dishes of *Marsilea* if these dry out after spore release; and the adhesion persists on rewetting (personal observations). The papillate surfaces of megaspores and microspores (Fig. 7.3f,k) may interlock and mucilage may aid initial contact. *Regnellidium* megaspores were also described by Mai and Walther (1978) from the German Oligocene.

According to Batten (1988, p. 10) megaspores of *Molaspora lobata* (Dijkstra) Hall (Fig. 7.3j) from the late Cretaceous and early Tertiary (Kovach and Batten 1989) are almost identical to *Regnellidium* (Fig. 7.2c). I am not aware of any *Molaspora* with adherent microspores. The ultrastructure of the terminal leaf-like lobes (acrolamella) and of the wall (Fig. 7.3b) is also almost identical to modern *Regnellidium* illustrated by Chrysler and Johnson (1939) and Higinbotham (1941). The megaspores and microspores of the three modern genera have similar wall morphology and ultrastructure (e.g. Fig. 7.3c–f; Chrysler and Johnson 1939; Higinbotham 1941). This includes a central prismatic or vesiculate layer (with thin inner and outer layers) which results in a more or less papillate surface. (Fig. 7.3f,j,k). This combination of features is diagnostic for Marsileaceae.

No vegetative fossils can be reliably assigned to Marsileales (see Seward 1910 for dismissal of some early claims). Sporocarps named *Rodeites* Sahni from the Deccan Intertrappean cherts of India (probably close to the Cretaceous/Tertiary boundary (Courtillot *et al.* 1988) were reinvestigated by Chitaley and Paradkar (1972). The sporocarps resemble those of *Marsilea* in occurrence in groups, in bilateral form, and in soral attachment. They contain microsporangia and megasporangia, with microspores and megaspores of characteristic structure. The megaspores are said to 'agree perfectly' with those of the dispersed spore *Molaspora lobata*. Whereas this spore, as noted above, is almost identical with that of modern *Regnellidium*, it seems that the *Rodeites* sporocarps differ from this modern genus in soral attachment and because microspores lack a terminal papilla (Chitaley and Paradkar 1972).

Megaspores (Fig. 7.2b) of the dispersed spore genus *Arcellites* Miner emend Ellis and Tschudy 1964 (including *Pyrobolospora*), which occurs throughout the Cretaceous (Kovach and Batten 1989), are usually assigned to the Marsileales. Although diverse in their distal spore ornament (Li and Batten 1986), all share the presence of a prominent proximal neck made up of six leaf-like units (termed acrolamella or flame-like organ). These units commonly exhibit a spiral twisting (Fig. 7.2b). Trilete microspores of the *Crybelosporites* type are found between the elements of the acrolamella (Fig. 7.2b) in several species from a variety of locations, as well as being common in palynological preparations from sediments yielding the megaspore (Cookson and Dettmann 1958; Hall 1963; Ellis and Tschudy 1964; Hall and Peake 1968; Heuber 1982; Li and Batten 1986). Wall ultrastructure of *Arcellites* (Fig. 7.3a) and *Crybelosporites* (e.g. Li and Batten 1986) differs from that of Marsileaceae. The microspores are morphologically very similar to those of the modern genera, and the spirally twisted acrolamella on the megaspore (Fig. 7.2b) can be interpreted as an elongate version of that on modern *Regnellidium* (Fig. 7.2c). These spores may belong in Marsileales but should perhaps be considered to represent an extinct family. Like some isoetalean megaspores (see above) *Arcellites* probably floated with the acrolamella downwards in the water, thus trapping microspores. Capture was possibly assisted by mucilaginous material, and retention may have resulted through some degree of re-twisting or closing of the acrolamellar units on drying (Hueber 1982). In the modern genera, microspore adhesion occurs all over the megaspore surface (Dorofeev 1981; personal observations).

A remarkable feature of the fossil record of Marsileales is the absence of sporocarps other than *Rodeites*. These strongly sclerotic structures are borne on plants growing in a subtle variety of ephemerally or permanently wet situations. They are the main organs of dissemination and perennation and have a high longevity (up to 100 years) being very resistant to environmental stress (Kornas 1988). This should enhance their fossilization potential. Perhaps their buoyancy and comparatively rapid germination, 6–10 hours after wetting, means that few are preserved. Careful searches are needed in wetland sites with rapid sedimentation, preferably in areas where parent plant growth has been inferred (Kovach 1988) or where megaspores and microspores are abundant in the same sediments, implying limited transport.

Water Ferns: Salviniales

Salvinia

Hall (1974, p. 354) referred to possible Cretaceous vegetative remains of *Salvinia* but all other records are Tertiary (Fig. 7.7) (Andrews and Boureau 1970). These fossils usually consist of impressions or partial compressions of the surface floating leaves with their diagnostic shape, arrangement, and surface patterning created by a combination of venation, surface hairs, and internal air cavities. Very few of these examples include fertile material.

S. zeillerii Fritel 1908 (French Early Eocene) includes floating leaves, groups of sporocarps, and absorption segments in close association on the bedding plane, consistent with derivation from the same plant being slightly disturbed during fossilization. No details of the sori are known.

S. preauriculata Berry 1925 from the Middle Eocene of the USA includes floating leaves, absorption segments, and possible sporocarps. Details of sori are unknown. Material from the Early Eocene Camels Butte Member, Golden Valley Formation (Hickey 1977) is not fertile, but Jain and Hall (1969) noted abundant dispersed megaspores and massulae (named *S. aureovallis*) in the same strata.

S. intertrappea Mahabale 1950, from the Indian Deccan Intertrappean cherts (near the Cretaceous/Tertiary boundary; Courtillot *et al.* 1988), is represented by megaspores and massulae. Vegetative remains (undescribed) are said to occur in the same strata.

Megaspores from the Late Cretaceous of North America (Hall and Bergad 1971; Jain 1971) have been assigned to three species of *Salvinia*. Hall (1974) considered these 'equivocal' as they lacked the proximal flap-like extensions of perispore characteristic of the genus. Furthermore, no Cretaceous salviniaceous massulae had been reported. The three species were transferred to *Azollopsis* by Sweet and Hills (1974).

Salvinia megaspores have diagnostic ultrastructure and morphology (Figs 7.4b,c; 7.5i; Kempf 1971). Massulae are eglochidiate and are not found attached to megaspores (cf. *Azolla*, see below). In soral clusters (Fig. 7.4b) *Salvinia* can be distinguished as only one massula is formed per sporangium (in contrast to *Azolla* where there are three or more, Fig. 7.6a). Examples of amphisporangiate sori (Fig. 7.4b; Bůžek *et al.* 1971; Mai and Walther 1978) provide conclusive evidence for conspecifity, which must otherwise be based on association. Palaeocene *Salvinia* occurs in southern England at several sites (Martin 1976*a*; Collinson 1986, and in progress) represented by massulae in soral clusters and associated megaspores. *Triletes ? exiguus* Dijkstra 1961 from the Palaeocene of Belgium and Holland is a *Salvinia* massula, also known from soral clusters (Collinson and Batten, in progress).

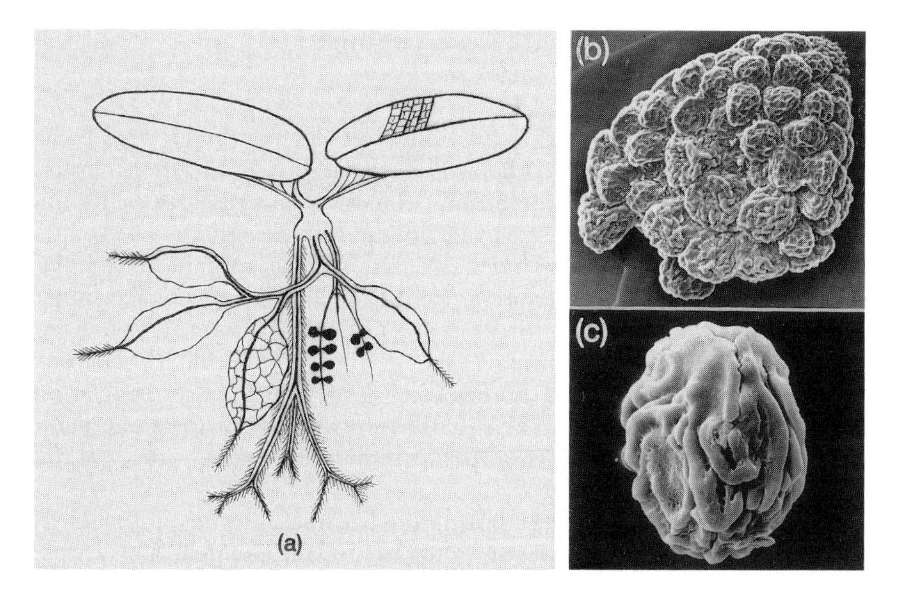

Fig. 7.4 *Salvinia reussii* Ettingshausen, Miocene, Bohemia, (Bůžek *et al.* 1971). (a) A schematic reconstruction of a single node showing two surface floating leaves, four submerged 'flotation structures', submerged hairy absorption segments, and submerged sporocarp-bearing appendage (details of venation only on part of one leaf and one 'flotation structure'); leaves about 2.5 cm long. (b) SEM of amphisporangiate sorus, megaspores in lower and microspore massulae in upper part, × 20. (c) SEM of megaspore, × 80.

Fig. 7.5 TEM sections of wall ultrastructure in salvinialean spores. All except (a), (f), and (g) have exine at the lower edge. (a) *Glomerisporites* (see b,f,h, and Fig. 7.6m) microspore massula, outer portion with perispore prolonged into hair, × 2000. (b) *Glomerisporites* microspore massula, inner portion with exine and innermost perispore (the latter is also on a), × 8000. (c) *Azolla* (see g and Fig. 7.6e) megaspore with exine, two-layered perispore, and surface hairs (suprafilosum), × 4000. (d) *Azollopsis* (see Fig. 7.6q) megaspore with exine, perispore, and hairs, amongst which floats (large vacuoles) are enmeshed, × 2000. (e) *Parazolla* (see Fig. 7.6i) megaspore with exine, perispore, hairs, and vacuolate floats, × 2000. (f) *Glomerisporites* megaspore, single float from outer portion, enmeshed by hairs, × 1000. (g) *Azolla*, detail of small portion of one vacuolate float, × 2000. (h) *Glomerisporites* megaspore, inner portion with exine, perispore, and surface hairs, × 2000. (i) *Salvinia* (see Fig. 7.4c) megaspore with exine and perispore, × 2500. (j) *Ariadnaesporites* (see k and Fig. 7.6h) microspore with exine and perispore, × 8000. (k) *Ariadnaesporites* megaspore with exine and perispore, × 4000.

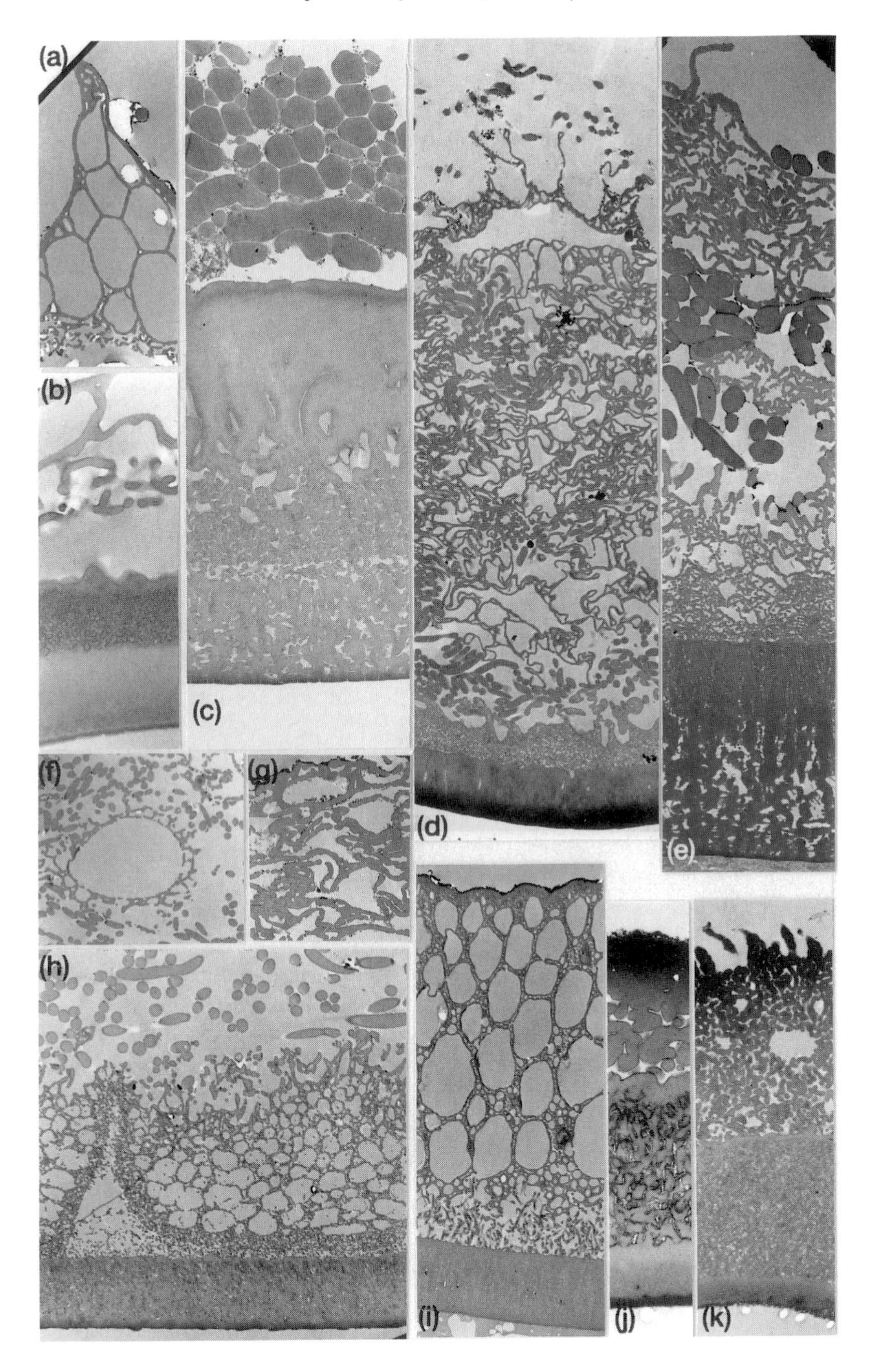
(a)
(b)
(c)
(d)
(e)
(f)
(g)
(h)
(i)
(j)
(k)

Eocene *Salvinia* is more common (seven species noted by Kovach and Batten 1989) and some are represented by associated megaspores and massulae (e.g. Jain and Hall 1969). In the later Tertiary, associated megaspores and massulae are recorded by Friis (1977) and amphisporangiate sori by Mai and Walther (1978).

The best-understood example of fossil *Salvinia* is *S. reussii* from the Miocene of Bohemia (Bůžek *et al.* 1971). This plant is known from floating leaves with details of venation, cuticle, hairs, and air chambers; submerged absorption segments; bisporangiate sori (Fig. 7.4b) found in connection with vegetative material with full details of megaspores (Figs 7.4c, 7.5i) and massulae; and unusual submerged structures interpreted as possible flotation aids. Figure 7.4a is a schematic reconstruction of a single node. This plant shows several subtle differences from modern *Salvinia*, including extensive branching of the absorption segments, presence of bisexual amphisporangiate sori, and the unusual 'flotation aids'.

Azolla

Of the heterosporous plants discussed here, *Azolla* has by far the most extensive fossil record from the Late Cretaceous to Recent (Fig. 7.7) (Collinson 1980). Additional references include Dorofeev (1979, 1980) who reported megaspores of 10 named species of section Azolla (three floats) and 12 named species of Rhizosperma (nine floats), from Oligocene to Pliocene of the USSR; Mai and Walther (1978, 1985) reported *A. prisca* from the German Late Eocene and Oligocene; Zhou (1983) reported *A. pinnata* from the Quaternary of China; and Bůžek *et al.* (1988) reported several species from the Miocene of Bohemia, including whole plants. In addition, Collinson (1980) omitted the revision of Canadian Palaeocene material and description of *A. areolata* (Sweet and Hills 1976).

Azolla megaspores (Fig. 7.6b–e,) can be distinguished by the presence of a number of pseudovacuolate structures, commonly termed floats (Fig. 7.6b–d, 7.5g) (although Dunham and Fowler 1987 have shown that they do not impart buoyancy) grouped to a greater (Fig. 7.6b,c) or lesser (Fig. 7.6d) extent around the proximal pole. The megaspores have a three-layered sporoderm (Fig. 7.5c) comprising exine, endoperine, and exoperine, the latter often baculate or tuberculate and giving rise to hairs. These hairs may arise from the exoperine all over the megaspore (infrafilosum) or from the modified exoperine (including collar and/or columella) within the float aggregation at the proximal pole (suprafilosum) (Fig. 7.6b,d,e). Three to eight massulae are produced per microsporangium, and soral clusters (sporocarps) may reveal the contents of one sporangium, showing preliminary septation

(Fig. 7.6a). The massulae often possess hairs (glochidia) which assist their attachment to megaspores. This is especially the case for those with anchor-shaped tips (Fig. 7.6f) which anchor onto to hairy megaspores (Fig. 7.6e). Smooth megaspores (where hairs scarcely extend outside the floats, Fig. 7.6b) tend to be associated with eglochidiate massulae or simple hair-like glochidia. This is known in two fossil whole-plants (see below and Table 7.1, notes 1 and 6) and here the sori are amphisporangiate, suggesting that massulae and megaspores would have been shed in association (as in *Azinia*, see below).

Well-substantiated latest Cretaceous *Azolla* are recorded from the Hell Creek Formation of Montana and Dakota, USA, and the Edmonton Formation of Canada (Hall and Swanson 1968; Srivastava 1968; Hall and Bergad 1971; Hall 1974). *A. barbata, A. distincta,* and *A. montana* all possess multifloated (cf. Fig. 7.6c,d) megaspore apparatuses (at least 15 up to 24 floats). Attached microspore massulae have anchor-tipped glochidia in the last two species and coiled hair-like glochidia in the first.

Other supposed Cretaceous records (Jain 1971; Collinson 1980; Kovach and Batten 1989) need reassessment. Several are known only from massulae or only from inadequately understood megaspores. *A schopfii* (synonym *A. extincta*) is considered to be Palaeocene (Sweet and Chandrasekharam 1973). Two other Cretaceous species require particular mention as they are claimed to have only one or three floats. The single float is unknown elsewhere in the genus now that *A. primaeva* is known to have nine (Sweet in Sweet and Hills 1976, p. 350). Three floats are only known questionably in the Eocene and securely from the late Oligocene onwards. Of the two Cretaceous species, *A. geneseana* is too poorly known for reliable assessment, and repeated attempts to recollect this form from the type locality have failed (Jain 1971). *Azolla simplex* Hall 1969 was based on corroded material from which the floats had been lost (Martin 1976*b*, footnote p. 166).

Palaeocene species are also typified by large float numbers (15–27) (Sweet and Hills 1976; Collinson 1980); most have anchor-tipped glochidia but *A. schopfii* has coiled or simple hair-like glochidia (Sweet and Chandrasekharam 1973). The youngest multifloated species is *A. colwellensis* from the English late Eocene (Collinson 1980) which also has anchor-tipped glochidia. Other Eocene and younger species have nine (Fig. 7.6b) or three floats and anchor-tipped glochidia or eglochidiate massulae.

Of the modern species, those in section Azolla have three floats and simple, hooked or anchor-tipped glochidia; those in section Rhizosperma have nine floats and glochidia are either absent or simple. The character mosaic revealed by the fossils has resulted in establishment of additional sections (Fowler 1975): Filifera Hall (more than nine floats; coiled glochidia); Kremastospora Jain and Hall (more than nine floats; anchor-tipped glochidia), alternative restricted concept also in Sweet and Hills (1976); Trisepta Fowler (nine floats; anchor-tipped glochidia); Simplicispora Hall (one float) is invalid.

The most valuable aspect of this *Azolla* record is the context and support provided by fertile fossil plants (Fig. 7.6g), i.e. leafy branches with roots and with attached sporocarps from which full details of megaspores and microspore massulae have been obtained. There are

Fig. 7.6 (a) *Azolla* (undescribed Cretaceous/Tertiary boundary material from Teapot Dome, Wyoming), microspore massula cluster (= probable sorus) showing contents of sporangia segregating into massulae in the lower part, SEM, × 15. (b) *A.* aff. *ventricosa* (Bůžek *et al.* 1988), megaspore with nine floats and lacking hairs on the lower part, SEM, × 90. (c) *Azolla* (undescribed Cretaceous/Tertiary boundary material, USGS locality 8504), megaspore revealing many floats, SEM, × 90. (d) *A. velus* (Dijkstra 1961), longitudinal section of megaspore, showing spore wall 'loose' within suprafilosum derived from the proximal perispore, and compressed floats around much of the spore except the distal pole, TEM, × 100. (e) *Azolla* (undescribed Cretaceous/Tertiary boundary material, USGS locality 8504), megaspore with floats and spore hidden by suprafilosum, the microspore massula is attached at lower right, SEM, × 90. (f) *Azolla* (undescribed Cretaceous/Tertiary boundary material, USGS locality 8504), anchor-tip from glochidia of microspore massula, SEM, × 4000. (g) Portion of *Azolla* fertile plant, × 1.5 (see Table 7.1, note 3). (h) *Ariadnaesporites* sp. (D.J. Batten collection 79/103, late Cretaceous, Portugal), megaspore (at right) and two microspores (to left of megaspore) entwined together by perispore hairs, SEM, × 90. (i) *Parazolla* sp. (Hall 1974; new material, same source as c), megaspore with plate-like floats and equatorial flange, SEM, × 45. (j) *Parazolla* sp. (Hall 1974; new material, same source as c), megaspore surface from just below flange on (i) with adpressed hairs and free circinate tips, SEM, × 700. (k) *Parazolla* sp. (Hall 1974; new material, same source as c), detail of a portion of the surface of a microspore massula, SEM, × 400. (l) *Glomerisporites pupus* (Batten 1988), detail of a microspore massula from the apex of (m), showing circinate tips to hairs, SEM, × 250. (m) *Glomerisporites pupus* (Batten 1988), megaspore with hairy surface and four attached massulae at the apex, SEM, × 45. (n) *Parazolla* sp. (Hall 1974; new material, same source as c), microspore massula cluster, SEM, × 45. (o) *Glomerisporites pupus* (Batten 1988) detail of floats seen amongst hairs on the surface of (m), SEM, × 500. (p) *Azollopsis* cf. *pilata* (Sweet and Hills 1974; new material, same source as c); detail of floats seen amongst hairs on the surface of (q), SEM, × 500. (q) *Azollopsis* cf. *pilata* (Sweet and Hills 1974; new material, same source as c), megaspore with attached microspore massula, SEM, × 500. See text and cited references for details of species and localities.

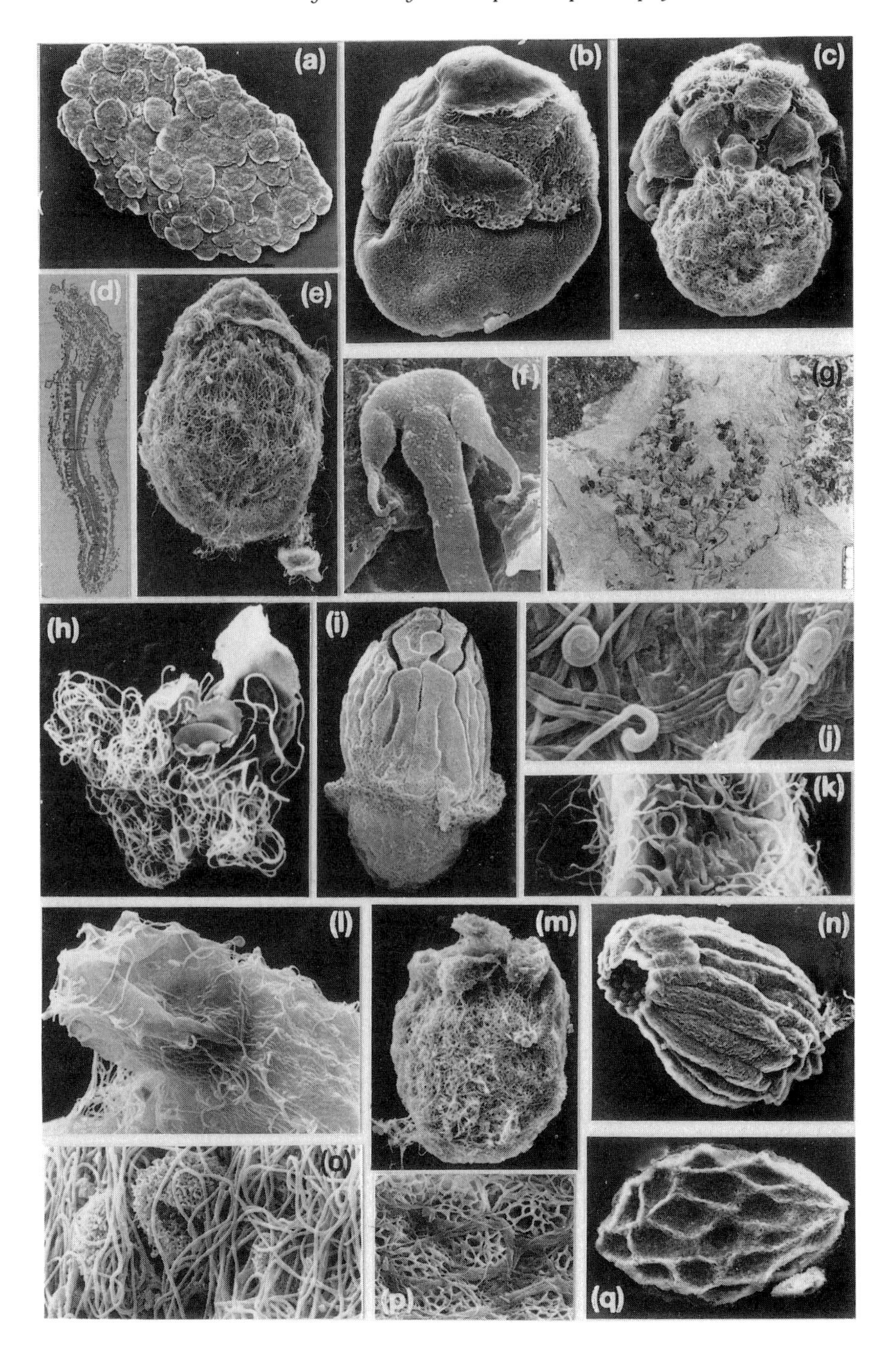
(a)
(b)
(c)
(d)
(e)
(f)
(g)
(h)
(i)
(j)
(k)
(l)
(m)
(n)
(o)
(p)
(q)

Table 7.1. *Azolla* whole plants with leafy branches, rhizomes, roots, ampho-, mega-, and microsporangiate sori and details of megaspores and microspore massulae; only the two features that have been used to define extinct sections of the genus are tabulated here

Species and reference		Age	Floats on megaspore	Glochidia on microspore massula
A. schopfii	(1)	Palaeocene	24 (21–26)	Simple
Undescribed	(2)	Palaeocene	17–24	Anchor
A. stanleyi	(2a)	Palaeocene	20–24	Anchor
Undescribed	(3)	Early Eocene	Not yet known	Anchor
A. primaeva	(4)	Middle Eocene	9	Anchor
A. prisca	(5)	Late Eocene and early Oligocene	9	Anchor
A. aff. *ventricosa*	(6)	Early Miocene	9	Simple
Section Azolla	(5)	Recent	3	Anchor, simple, or hooked
Section Rhizosperma	(5)	Recent	9	Absent or simple

References: (1) Sweet and Chandrasekharam (1973); (2) new material under study from collections of United States Geological Survey, Denver, from the Cretaceous/Tertiary boundary section at Ricks Place, Montana; (2a) Melchior and Hall (1983); (3) new material under study (see Fig. 7.6G), collected by Leo Hickey from the northwestern extension of the Wind River Formation, Freemont County, Wyoming, USA, in the collections of the United States National Museum, Smithsonian Institution, Paleobiology Department number 372411; (4) Hills and Gopal (1967); Sweet in Sweet and Hills (1976); (5) Fowler (1975); (6) Bůžek *et al.* (1988).

seven records from the earliest Palaeocene to the Miocene (Table 7.1) which include representatives of all the sections mentioned above. In addition, associated vegetative and fertile parts are known for *A. indica* and *A. intertrappea* (Trivedi and Verma 1971); leaves of these species are poorly known but the permineralized preservation provides anatomical details of roots and sporocarps. Three other *Azolla* species are based only on sterile leafy shoots (Collinson 1980).

These records demonstrate conclusively that the mosaic character distribution existed within the genus *Azolla* not within otherwise unknown but related plants. A detailed re-examination of fossil *Azolla*, especially of neglected ultrastructural characters and whole-plant material, is currently being undertaken, with the aim of providing a phylogenetic analysis of this and related genera (Collinson, in progress).

Extinct genera

Five form genera, based on megaspores with attached microspore massulae, are commonly assigned to Salviniales (Fig. 7.7). These have formed the basis for discussion of salvinialean phylogeny and origin (Hall 1969, 1974, 1975; Martin 1976*b*).

Azinia Balujeva is based on a single assemblage from the Palaeocene of the USSR (Andrews and Boureau 1970). It comprises a heterosporangiate sporocarp containing a single megasporangium with one smooth-walled multifloated megaspore and 6–8 microsporangia, each containing three eglochidiate massulae. Apart from the lignified sporo-

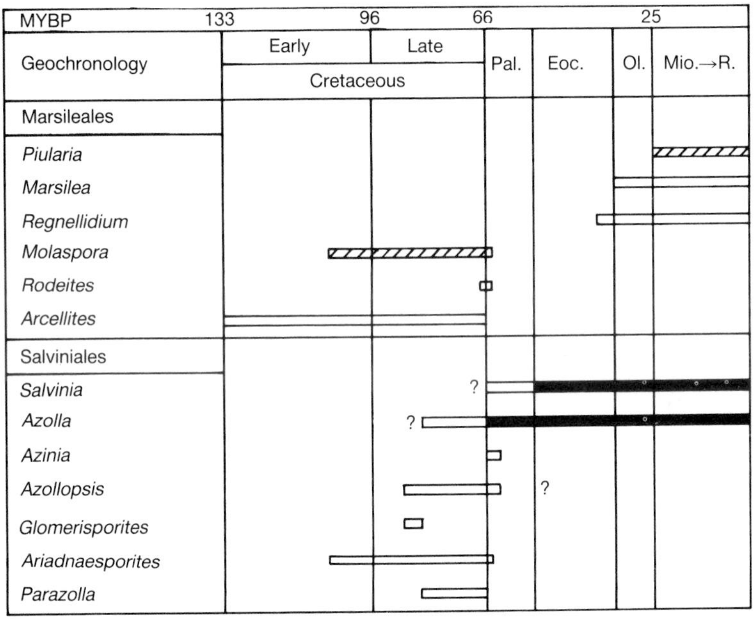

MYBP 133 96 66 25
Geochronology
Early Late
Cretaceous
Pal. Eoc. Ol. Mio.→R.
Marsileales
Piularia
Marsilea
Regnellidium
Molaspora
Rodeites
Arcellites
Salviniales
Salvinia ?
Azolla ?
Azinia
Azollopsis ?
Glomerisporites
Ariadnaesporites
Parazolla

Fig. 7.7 Stratigraphic distribution of water ferns. In Marsileales, hatched bar = megaspores only; open bar = megaspores and adherent microspores, associated in sporocarps in *Rodeites*. In Salviniales, solid bar = whole fertile plants plus characters of open bar; open bar = megaspores with attached or associated (*Salvinia*) microspore massulae. ? Indicates doubtful records based on inadequately understood megaspores (*Azolla*), undescribed vegetative material (*Salvinia*), and a single, probably reworked, massula (*Azollopsis*). MYBP, time in millions of years before present; Pal, Palaeocene; Eoc, Eocene; Ol, Oligocene; Mio, Miocene; R, Recent.

carp wall, this material is very like some fossil species of *Azolla* (see earlier; Martin 1976*b*).

Azollopsis Hall emend Sweet and Hills is based on seven species from the Late Cretaceous and Palaeocene of Canada and the USA (Sweet and Hills 1974). [An Eocene record based on a single massula (Jain and Hall 1969) may well be reworked.] Megaspores have more than 32 floats (often 100+) of equal size distributed evenly around the spore and enmeshed within hairs derived from a fibrous exoperine (Figs 7.6p,q, 7.5d). The endoperine is granular (Fig. 7.5d) and extended into a proximal acrolamella. Attached massulae possess glochidia with either circinate or multibarbed tips. In *A. tomentosa* the massulae occur in groups of 22–30 within sporangia; in *A. intermedia* sporocarps occur (with a thin indusial wall) containing either a single-spored megasporangium or a single microsporangium with 7–10 massulae. Microspores range from one to seven per massula, each averaging more than 40 μm in diameter.

Glomerisporites Potonié is based on a single assemblage from the Dutch Late Cretaceous (Hall 1975; Batten 1988). The megaspore (Fig. 7.6m) is similar to those of *Azollopsis* but the perispore is much more complex, having three or four layers and with extension of the innermost layer into thick hairs that ornament the surface (Fig. 7.5h). The outermost perispore layer gives rise to narrower hairs which enmesh the vacuolate floats from which hairs also arise (Figs 7.5h,f, 7.6o). The central perispore zone is also strongly vacuolate (Fig. 7.5h). Attached massulae/microspores (Fig. 7.6l,m) contain a single microspore with a three-layered perine (Fig. 7.5a,b) having a vacuolate central zone. This zone continues into surface hairs which have circinate tips (Figs 7.5a, 7.6l).

Ariadnaesporites Potonié emend Tschudy 1966 included 12 species from the Late Cretaceous of Canada, USA, Greenland, Peru, and Chile when last revised by Hall (1975). Kovach and Batten (1989) list 24 species, of which one extends just into the Early Cretaceous (Albian) and another into the Early Palaeocene. The genus shows only weak heterospory and the megaspores and microspores (Fig. 7.6h). show little morphological difference. Each has a prominent tripartite acrolamella and long hair-like processes arising from the perispore. The innermost perispore layer is partly extended into these hairs but they are mainly hollow. The sporoderm is two-layered (Fig. 7.5j,k) with exine and perine of almost equal thickness. The perispore has two zones distinguished primarily by the density of packing of fibrils (inner layer less dense) and sometimes by a vacuolate structure in the outer portion (Fig. 7.5j,k). In the species examined by the author, the exine also appears to be two-layered (see also Lugardon *et al.* 1984). Surface ornamentation ranges from a reticulum to spinules to almost smooth. In *A.*

varius several float-like structures occur enmeshed in the hairs at the proximal end of the megaspore. The floats have a two-layered wall, partly resembling the perispore, and an irregularly foveolate outer surface (Hall 1975). As far as the author is aware, floats have not been recorded in other species. Microspores occur in clusters (Hall 1975) but these are not massulae as the spores do not share a common perispore.

Parazolla Hall (Hall 1969, 1974) is known from two species from the Late Cretaceous of Montana. Megaspores (Fig. 7.6i) have plate-like 'floats' attached to the proximal megaspore surface. Hairs arise from the perispore surface where they are adpressed (Fig. 7.6j) and extend across the equator onto the floats. At maturity (presumably) the hairs form a collar-like equatorial flange (Fig. 7.6i) from which they radiate. These hairs have circinate tips (Fig. 7.6j). The sporoderm is two-layered (Fig. 7.5e) with exine and perine of almost equal thicknesses. Both have a complex structure. The perine becomes vacuolate outwards and the outer region gives rise to the surface hairs. The plate-like 'floats' spread apart at maturity around a trilete suture. The microspore massulae are described as triangular by Hall (1969). Hall (1974) and Collinson (herein) record elongate oval 'banana-shaped' massulae grouped into clusters of 20 or more (Fig. 7.6n). Having found no evidence of sporangium wall between massulae in TEM observations, I consider these clusters to be the contents of one sporangium rather than one sporocarp. I have found several megaspores with these massulae attached. The massula surface has simple glochidia (Fig. 7.6k), the attachment being encouraged by the circinate tips (Fig. 7.6j) on the megaspore surface hairs. Microspores have a proximal acrolamella and they share a common perispore. There are more than 10 per massula.

Conclusions

Isoetites is the only fertile plant fossil resembling modern *Isoetes*. Cretaceous and Early Tertiary examples contain *Minerisporites* megaspores, some of which bear monolete microspores on their surface. Other Mesozoic dispersed megaspores may be recognized as isoetalean rather than selaginellalean on the basis of associated monolete microspores, spore wall ultrastructure, etc. Even when recorded *in situ* from strobili, their affinity cannot always be ascertained and they are likely to represent several families.

Selaginella has a long fossil history but fertile plant fossils have not been fully exploited. One dispersed megaspore genus *Banksisporites* has been recorded in a *Selaginella*-like plant as well as in *Cylostrobus*, a strobilus assigned to Pleuromeiaceae (Isoetales). Thus the dispersed spore affinity remains unknown if associated microspores are absent. Wall ultrastructure and occurrence in dispersed tetrads can be taken to imply selaginellalean affinities. In particular, a grid-like arrangement in one layer of the sporoderm is unique to a few species of *Selaginella* amongst modern megaspores. This occurs in several extinct spore taxa which are therefore assumed to be selaginellalean.

Marsileales have a poor fossil record (Fig. 7.7) with only one species based on sporocarps and others based on megaspores, some of which have adherent microspores. If similarity of megaspores (*Molaspora*) alone is sufficiently indicative, then the Marsileaceae extend to the Early Cretaceous. Spore wall ultrastructure is distinctive for the order, but *Arcellites* differs significantly in this character. It may represent an extinct family.

The distinction between Azollaceae and Salviniaceae extends back at least to the beginning of the Tertiary (Fig. 7.7) and almost certainly into the Late cretaceous. The record of *Azolla*, based upon fertile plants, is particularly well substantiated. No extinct genera show significant similarities with *Salvinia*. The late Cretaceous genera *Azollopsis* and *Glomerisporites*, based on megaspores with attached massulae, should probably be included in Azollaceae as they share many features with *Azolla*. *Parazolla* is a distinctive genus which may represent an extinct family of Salviniales. *Ariadnaesporites* is distinct from all of these, perhaps representing an extinct order of heterosporous plants, though it is here retained in Salviniales. Future study is needed to establish the nature and occurrence of floats in the different species.

The hairy perispore is unique to Azollaceae amongst modern plant families. Similar, but distinct, hairy surfaces occur only in one homosporous ant-fern, *Lecanopteris*, Polypodiaceae (Walker 1985) and in some angiosperm pollen (Hesse 1986). *Ariadnaesporites* has a hairy perispore whereas *Arcellites* has an ornament of appendages, otherwise the two megaspores are rather similar, especially if the number of units in the acrolamella (three in the former, six in the latter) are hard to distinguish. Megaspores of *Balmeisporites* and *Styx* (not previously mentioned herein) resemble small *Arcellites* but with only three acrolamella lobes. Microspores are unknown. The megaspores have been considered members of the Salviniaceae, Marsileaceae, or merely as 'water ferns' (Hall 1963, 1974; Sweet 1979). I do not consider that the similarities of these megaspores imply a common ancestry for the two orders of water ferns. They probably reflect convergence in functional adaptation for aquatic dispersal and this emphasizes the crucial importance of features such as associated microspores, and preferably whole-plant fossils, for the understanding of relationships in the early members of the group.

In the three better-known orders discussed in this paper, the Late Cretaceous represents a period of higher diversity than the Tertiary (Fig. 7.7). The extensive dispersed spore record shows a considerably higher diversity in the Mesozoic than in the Tertiary (Kovach and Batten 1989). In the water ferns a reduced Tertiary diversity is consistent with a low modern diversity. This is less true for the lycopsids, especially *Selaginella*. It is possible that the increasing proportions of angiosperms in post-Cretaceous vegetation influenced the ease with which spores produced by low-growing herbs in non-wetland habitats (like *Selaginella*) were transported to sites of deposition.

The Mesozoic diversity, coupled with the combination of novel and mosaic characters on fossils, suggests that the fossil record contains representatives of extinct families or orders of heterosporous plants. Attempts to include these in one of the four extant orders may confuse rather than clarify relationships.

It is necessary to distinguish the synapomorphies from the mosaic of characters represented in the modern and well-known fossil material. Further understanding of the diversification patterns in heterosporous plants must await the results of this analysis. In addition, more detailed knowledge is needed of *in situ* spores from fertile whole plants whose familial affinities can be determined with accuracy. These are crucial, even though they are rare, because they form 'key points' into which the dispersed spore record can be linked. Recognition of sporocarps and megaspores with adherent or associated microspores/microspore massulae also provide critical evidence. Palaeoecological techniques, such as those applied by Kovach (1988), can be used to help predict suitable areas to search for more complete fossils.

Acknowledgements

My studies of heterosporous pteridophytes have benefited from the help of many people who have generously provided fossil and modern material, guidance for field collecting, and helpful discussion. Very little of their generosity is directly evident in this short contribution but all are gratefully thanked: D.J. Batten, P. Bell, C. Bůžek, P.R. Crane, K. Fowler, L.J. Hickey, C. Jermy, E. Knobloch, H. Schorn, R.A. Stockey, M. Stoneberg, A.R. Sweet, G. van Uffelen, W. Wehr, and J. Wolfe. I am grateful for access to material from the following collections: British Museum (Natural History), Palaeontology and Botany Departments; The Burke Museum, Washington State; The Field Museum of Natural History, Chicago; The Princeton Museum, Canada; The University of California Museum of Palaeontology at Berkeley; United States Geological Survey, Denver; United States National Museum of Natural History, Smithsonian Institution, Paleobiology Department. I thank Tony Brain, EM-unit, King's College London, for his patient and skilful help with TEM studies, especially the sectioning of this difficult material. Jerry Hooker drew the line figures and I also thank him for his continued support and encouragement in many ways. This work was undertaken during the tenure of a Royal Society 1983 University Research Fellowship which is gratefully acknowledged.

References

Andrews, H.N. and Boureau, E. (1970). VI—Classe des Leptosporangiopsida. In *Traité de Paléobotanique*, (ed. É. Boureau), Vol. IV, Fasc. 1, *Filicophyta*, (ed. H.N. Andrews, C.A. Arnold, É. Boureau, J. Doubinger, and S. Leclercq), pp. 17–406, Masson et Cie, Paris.

Ash, S.R. (1972). Late Triassic plants from the Chinle Formation in North-Eastern Arizona. *Palaeontology* **15**, 598–618.

Bajpai, U. and Maheshwari, H.K. (1986). SEM study of megaspore sporoderm of some Indian Selaginellas. *Phytomorphology* **36**, 43–51.

Bartram, K.M. (1987). Lycopod succession in coals: an example from the Low Barnsley Seam (Westphalian B), Yorkshire, England. In *Coal and coal-bearing strata: recent advances, Geological Society Special Publication,* No **32**, (ed. A.C. Scott), pp. 187–99. Blackwell Scientific Publications, Oxford.

Batten, D.J. (1974). Wealden palaeoecology from the distribution of plant fossils. *Proceedings of the Geologists' Association* **85**, 433–58.

Batten, D.J. (1988). Revision of S.J. Dijkstra's Late Cretaceous megaspores and other plant microfossils from Limburg, the Netherlands. *Mededelingen Rijks Geologische Dienst.* **41–3**, 1–55.

Bergad, R.D. (1978). Ultrastructural studies of selected North American Cretaceous megaspores of *Minerisporites, Erlansonisporites, Horstisporites,* and *Ricinispora*, n.gen. *Palynology* **2**, 39–51.

Berry, E. (1925). A new *Salvinia* from the Eocene of Wyoming and Tennessee. *Torreya* **25**, 116–18.

Bose, M.N. and Roy, S.K. (1964). Studies on the Upper Gondwana of Kutch - 2. Isoetaceae. *Palaeobotanist* **12**, 226–8.

Brown, R.W. (1939). Some American fossil plants belonging to the Isoetales. *Journal of the Washington Academy of Sciences* **29**, 261–9.

Brown, R.W. (1962). Paleocene flora of the Rocky Mountains and the Great Plains. *United States Geological Survey Professional Paper* **375**, 1–119.

Bůžek, Č., Konzalová, M., and Kvaček, Z. (1971). The genus Salvinia from the Tertiary of the North-Bohemian Basin. *Sborník Geologických Věd. Praha, Řada P. Paleontologie* **13**, 179–222.

Bůžek, Č., Konzalová, M., and Kvaček, Z. (1988). *Azolla* remains from the lower Miocene of the North-Bohemian Basin, Czechoslovakia. *Tertiary Research* **9**, 117–32.

Chaloner, W.G. (1967). Lycophyta. In *Traité de Paléobotanique*, (ed. É.

Boureau), Vol. II, *Bryophyta, Psilophyta, Lycophyta*, (ed. É. Boureau, S. Jovet-Ast, O.A. Höeg, and W.G. Chaloner), pp. 435–802. Masson et Cie, Paris.

Chitaley, S.D. and Paradkar, S.A. (1972). *Rodeites* Sahni reinvestigated - I. *Botanical Journal of the Linnean Society* **65**, 109–17.

Chrysler, M.A. and Johnson, D.S. (1939). Spore production in *Regnellidium*. *Bulletin of the Torrey Botanical Club* **66**, 263–79.

Collinson, M.E. (1980). A New multiple-floated *Azolla* from the Eocene of Britain with a brief review of the genus. *Palaeontology* **23**, 213–29.

Collinson, M.E. (1986). The Felpham flora: a preliminary report. *Tertiary Research* **8**, 29–32.

Collinson, M.E. and Scott, A.C. (1987). Implications of vegetational change through the geological record on models for coal-forming environments. In *Coal and coal-bearing strata: recent advances*, Geological Society Special Publication, No **32**, (ed. A.C. Scott), pp. 67–85. Blackwell Scientific Publications, Oxford.

Collinson, M.E., Batten, D.J., Scott, A.C., and Ayonghe, S.N. (1985). Palaeozoic, Mesozoic and contemporaneous megaspores from the Tertiary of southern England: indicators of sedimentary provenance and ancient vegetation. *Journal of the Geological Society of London* **142**, 375–95.

Cookson, I.C. and Dettmann, M.E. (1958). Cretaceous 'megaspores' and a closely associated microspore from the Australian region. *Micropaleontology* **41**, 39–49.

Courtillot, V., Féraud, G., Maluski, H., Vandamme, D., Moreau, M.G., and Besse, J. (1988). Deccan flood basalts and the Cretaceous/Tertiary boundary. *Nature* **333**, 843–6.

Dijkstra, S.J. (1951). Wealden megaspores and their stratigraphical value. *Mededelingen Geologische Stichting, New Series* **5**, 7–22.

Dijkstra, S.J. (1961). Some Paleocene megaspores and other small fossils. *Mededelingen Geologische Stichting. New Series* **13**, 5–11.

Dimichele, W.A. and Phillips, T.L. (1988). Paleoecology of the Middle Pennsylvanian-age Herrin Coal Swamp (Illinois) near a contemporaneous river system, the Washville paleochannel. *Review of Palaeobotany and Palynology* **56**, 151–76.

Dorofeev, P.I. (1979). On the taxonomy of Tertiary *Azolla* Lam. E Sect *Azolla* Lam. *Botanicheskii Zhurnal SSSR* **64**, 1259–72. (In Russian.)

Dorofeev, P.I. (1980). On the taxonomy of Tertiary *Azolla* Lam. of the section *Rhizosperma* Meyen. *Botanicheskii Zhurnal SSSR* **65**, 297–310. (In Russian.)

Dorofeev, P.I. (1981). On the taxaonomy of the Tertiary Marsileaceae. *Botanicheskii Zhurnal SSSR* **66**, 792–801. (In Russian.)

Drinnan, A.R. and Chambers, T.C. (1986). Flora of the Lower Cretaceous Koonwarra Fossil Bed (Korumbarra Group) South Gippsland, Victoria. In *Plants and invertebrates of the Lower Cretaceous Koonwarra Fossil Bed (Korumbarra*

Group) South Gippsland, Victoria, (ed. P.A. Jell and J. Roberts), pp. 1–77, Memoir 3 of the Association of Australasian Palaeontologists, AAP, Sydney.

Dunham, D.G. and Fowler, K. (1987). Megaspore germination, embryo development and maintenance of the symbiotic association in *Azolla filiculoides* Lam. *Botanical Journal of the Linnean Society* **95**, 43–53.

Ellis, C.H. and Tschudy, R.H. (1964). The Cretaceous megaspore genus Arcellites Miner. *Micropaleontology* **10**, 73–9.

Fowler, K. (1975). Megaspores and massulae of *Azolla prisca* from the Oligocene of the Isle of Wight. *Palaeontology* **18**, 483–507.

Friis, E.M. (1977). EM-studies on salviniaceae megaspores from the Middle Miocene Fasterholt flora, Denmark. *Grana,* **16**, 113–28.

Fritel, P.-H. (1908). Note sur une espèce fossile nouvelle du genre *Salvinia. Journal de Botanique, Paris, Series 2,* **1**, (8), 190–8.

Gastaldo, R.A. (1987). Confirmation of Carboniferous clastic swamp communities. *Nature* **326**, 869–71.

Hall, J.W. (1963). Megaspores and other fossils in the Dakota Formation (Cenomanian) of Iowa, (U.S.A.) *Pollen et Spores* **5**, 425–43.

Hall, J.W. (1969). Studies on fossil Azolla: primitive types of megaspores and massulae from the Cretaceous. *American Journal of Botany* **56**, 1173–80.

Hall, J.W. (1974). Cretaceous Salviniaceae. *Annals of the Missouri Botanical Garden* **61**, 354–67.

Hall, J.W. (1975). Ariadnaesporites and Glomerisporites in the Late Cretaceous: ancestral Salviniaceae. *American Journal of Botany* **62**, 359–69.

Hall, J.W. and Bergad, R.D. (1971). A critical study of three Cretaceous salviniaceous megaspores. *Micropaleontology* **17**, 345–56.

Hall, J.W. and Peake, N.M. (1968). Megaspore assemblages in the Cretaceous of Minnesota. *Micropaleontology* **14**, 456–64.

Hall, J.W. and Swanson, N.P. (1968). Studies on fossil *Azolla*: *Azolla montana*, a Cretaceous megaspore with many small floats. *American Journal of Botany* **55**, 1055–61.

Helby, R. and Martin, R.A. (1965). *Cylostrobus* gen. nov., cones of lycopsidean plants from the Narrabean group (Triassic) of New South Wales. *Australian Journal of Botany* **13**, 389–404.

Hesse, M. (1986). Nature, form and function of pollen-connecting threads in angiosperms. In *Pollen and spores form and function*, Linnean Society Symposium Series, No. 12, (ed. S. Blackmore and I.K. Ferguson), pp. 109–18. Academic Press, London.

Hickey, L.J. (1977). Stratigraphy and paleobotany of the Golden Valley Formation (Early Tertiary) of Western North Dakota. *Memoirs of the Geological Society of America* **150**, 1–183.

Higinbotham, N. (1941). Development of the gametophytes and embryo of *Regnellidium diphyllum. American Journal of Botany* **28**, 282–300.

Hill, R.S. (1987). Tertiary *Isoetes* from Tasmania. *Alcheringa* **12**, 157–62.

Hills, L.V. and Gopal, B. (1967). Azolla primaeva and its phylogenetic significance. *Canadian Journal of Botany* **45**, 1179–91.

Hueber, F.M. (1982). Megaspores and a palynomorph from the Lower Potomac Group in Virginia. *Smithsonian Contributions to Paleobiology* **49**, 1–69.

Jain, R.K. (1971). Pre-Tertiary records of Salviniaceae. *American Journal of Botany* **58**, 487–96.

Jain, R.J. and Hall, J.W. (1969). A contribution to the Early Tertiary fossil record of Salviniaceae. *American Journal of Botany* **56**, 527–39.

Karfalt, E.E. (1984). Further observations on *Nathorstiana* (Isoetaceae). *American Journal of Botany* **71**, 1023–30.

Kempf, E.K. (1971). Elektronenmikroskopie der sporodermis von mega- und mikrosporen der pteridophyten-gattung *Salvinia* aus dem Tertiär und Quatär Deutschlands. *Palaeontographica* **B136**, 47–70.

Knobloch, E. (1984). Megapsoren aus der Kreide von Mitteleuropa. *Sborník Geologických Věd. Praha Řada P. Paleontologie* **26**, 157–95.

Knobloch, E. (1986). Megasporen der gattung Selaginella Beauv. aus dem Neogen von Mähren und der Slowakei. *Časopis pro mineralogii a geolgii* **31**, 113–24.

Knowlton, F.H. (1916). A new fossil *Selaginella* from the Lower Tertiary of Montana. *Torreya* **16**, 201–4.

Koppelhus, E.B. and Batten, D.J. (1989). Late Cretaceous megaspores from southern Sweden: morphology and paleoenvironmental significance. *Palynology* **13**, 91–120.

Kornas, J. (1988). Adaptive strategies of *Marsilea* (Marsileaceae: Pteridophyta) in the Lake Chad Basin of N.E. Nigeria. *Fern Gazette* **13**, 231–43.

Kovach, W.L. (1988). Quantitative palaeoecology of megaspores and other dispersed plant remains from the Cenomanian of Kansas, USA. *Cretaceous Research* **9**, 265–83.

Kovach, W.L. (1989). Quantitative methods for the study of lycopod megaspore ultrastructure. *Review of Palaeobotany and Palynology* **57**, 233–46.

Kovach, W.L. and Batten, D.J. (1989). Worldwide stratigraphic occurrences of Mesozoic and Tertiary megaspores. *Palynology,* **13**, 247–77.

Kovach, W.L. and Dilcher, D.l. (1985). Morphology, ultrastructure, and paleoecology of *Paxillitriletes vittatus* sp. nov. from the Mid-Cretaceous (Cenomanian) of Kansas. *Palynology* **9**, 85–94.

Kovach, W.L. and Dilcher, D.L. (1988). Megaspores and other dispersed plant remains from the Dakota Formation (Cenomanian) of Kansas, U.S.A. *Palynology* **12**, 89–119.

Li, W.-B. and Batten, D.J. (1986). The Early Cretaceous megaspore *Arcellites* and closely associated *Crybelosporites* microspores from northeast inner Mongolia, P.R. China. *Review of Palaeobotany and Palynology* **46**, 189–208.

Lugardon, B., Diníz, F., and Moron, J.-M. (1984). Structure fine d'Ariadnaesporites du Portugal. *Revue de Paleobiologie* **Volume Special**, 139–48.

Mahabale, T.S. (1950). A species of fossil *Salvinia* from the Deccan Intertrappean Series, India. *Nature* **165**, 410–11.

Mai, D.H. and Walther, H. (1978). Die floren der Haselbacher Serie im Weisselster-Becken (Bezirk Leipzig, DDR). *Abhandlungen des Staatlichen Museums für Mineralogie und Geologie zu Dresden* **28**, 1–200.

Mai, D.H. and Walther, H. (1985). Die obereozänen Floren des Weisselster-Beckens und seiner Randegebiete. *Abhandlungen des Staatlichen Museums für Mineralogie und Geologie zu Dresden* **33**, 1–260.

Marcinkiewicz, T. (1989). Remarks on agglomerations of megaspores *Minerisporites institus* Marc. *Acta Palaeobotanica* **29**, 221–4.

Martin, A.R.H. (1976*a*). Upper Palaeocene Salviniaceae from the Woolwich/Reading Beds near Cobham, Kent. *Palaeontology* **19**, 173–84.

Martin, A.R.H. (1976*b*). Some structures in *Azolla* megaspores, and an anomolous form. *Review of Palaeobotany and Palynology* **21**, 141–69.

Melchior, R.C. (1977). On the occurrence of *Minerisporites mirabilis in situ. Scientific publications of the Science Museum, Saint Paul, Minnesota, New Series* **3**, 1–11.

Melchior, R.C. and Hall, J.W. (1983). Some megaspores and other small fossils from the Wannagan Creek Site (Paleocene), North Dakota. *Palynology* **7**, 133–45.

Meyen, S.V. (1987). *Fundamentals of Palaeobotany.* Chapman and Hall, London.

Minaki, M. (1984). Macrospore morphology and taxonomy of *Selaginella* (Selaginellaceae). *Pollen et Spores* **26**, 421–80.

Pigg, K.B. and Rothwell, G.W. (1983). *Chaloneria* gen. nov., heterosporous lycophytes from the Pennsylvanian of North America. *Botanical Gazette* **144**, 132–47.

Potonié, R. (1956). Synopsis der Gattungen der Sporae dispersae. 1. Sporites. *Beihefte zum Geologischen Jahrbuch* **23**, 1–103.

Retallack, G.J. (1981). Two new approaches for reconstructing fossil vegetation with examples from the Triassic of Australia. In *Communities of the past*, (ed. J. Gray, A.J. Boucot, and W.B.N. Berry), pp. 271–95. Hutchinson Ross, Stroudsburg, Pennsylvania.

Rothwell, G.W. and Erwin, D.M. (1985). The rhizomorph apex of *Paurodendron*; implications for homologies among the rooting organs of Lycopsida. *American Journal of Botany* **72**, 86–98.

Scott, A.C. and Playford, G. (1985). Early Triassic megaspores from the Rewan Group, Bowen Basin, Queensland. *Alcheringa* **9**, 297–323.

Seward, A.C. (1910). *Fossil plants*, Vol. II. Cambridge University Press, Cambridge.

Srivastava, S.K. (1968). *Azolla* from the Upper Cretaceous Edmonton Formation, Alberta, Canada. *Canadian Journal of Earth Sciences* **5**, 915–19.

Stewart, W.N. (1983). *Palaeobotany and the evolution of plants*. Cambridge University Press, Cambridge.

Sukh-Dev (1980). Evaluation of *in situ* spores and pollen grains from the Jurassic-Cretaceous fructifications. *Proceedings of the IV International Palynological Conference, Lucknow (1976–77)* **2**, 753–68.

Sweet, A.R. (1979). Jurassic and Cretaceous megaspores. *American Association of Stratigraphic Palynologists Contributions Series*, No. **5B**, pp. 1–30.

Sweet, A.R. and Chandrasekharam, A. (1973). Vegetative remains of *Azolla schopfii* Dijkstra from Genesee, Alberta. *Canadian Journal of Botany* **51**, 1491–6.

Sweet, A.R. and Hills, L.V. (1974). A detailed study of the genus *Azollopsis*. *Canadian Journal of Botany* **52**, 1625–42.

Sweet, A.R. and Hills, L.V. (1976). Early Tertiary species of *Azolla* subg. *Azolla* sect. *Kremastospora* from western and artic Canada. *Canadian Journal of Botany* **54**, 334–51.

Taylor, T.N. (1981). *Paleobotany, an introduction to fossil plant biology*. McGraw-Hill, New York.

Taylor, W.A. and Taylor, T.N. (1987). Subunit construction of the spore wall in fossil and living lycopods. *Pollen et Spores* **29**, 241–8.

Taylor, W.A. and Taylor, T.N. (1988). Ultrastructural analysis of selected Cretaceous megaspores from Argentina. *Journal of Micropalaeontology* **7**, 73–87.

Teixeira, C. (1948). *Flora Mesozóica Portuguesa*. Serviços Geológicos de Portugal, Lisbon.

Thomas, B.A. and Brack-Hanes, S.D. (1984). A new approach to family groupings in the lycophytes. *Taxon* **33**, 247–55.

Thomas, B.A. and Spicer, R.A. (1987). *The evolution and palaeobiology of land plants*. Croom Helm, London.

Townrow, J.A. (1968). A fossil *Selaginella* from the Permian of New South Wales. *Botanical Journal of the Linnean Society* **61**, 13–23.

Traverse, A. (1988*a*). *Paleopalynology*. Unwin Hyman, Boston.

Traverse, A. (1988*b*). Plant evolution dances to a different beat. *Historical Biology* **1**, 277–301.

Trivedi, B.S. and Verma, C.L. (1971). Contributions to the knowledge of *Azolla indica* sp. nov. from the Deccan Intertrappean series M.P., India. *Palaeontographica* **136**, 71–82.

Tryon, A.F. and Lugardon, B. (1978). Wall structure and mineral content in *Selaginella* spores. *Pollen et Spores* **20**, 315–40.

Tryon, R.M. and Tryon, A.F. (1982). *Ferns and allied plants, with special reference to tropical America*. Springer Verlag, New York.

Tschudy, R.H. (1966). Associated megaspores and microspores of the Cretaceous genus *Ariadnaesporites* Potonié, 1956, emend. *United States Geological Survey Professional Paper*, 550-D, D76–D82.

Tschudy, R.H. (1976). Stratigraphic distribution of species of the megaspore genus *Minerisporites* in North America. *United States Geological Survey Professional Paper*, 743-E, E1–E11.

Walker, T.G. (1985). Spore filaments in the ant-fern *Lecanopteris mirabilis* – an alternative viewpoint. *Proceedings of the Royal Society of Edinburgh* **86B**, 111–14.

Walkom, A.B. (1944). Fossil plants from Gingin, W.A. *Journal of the Royal Society of Western Australia* **28**, 201–7.

Watson, J. (1969). A revision of the English Wealden flora, I. Charales-Ginkgoales. *Bulletin of the British Museum (Natural History) Geology* **17**, 207–54.

Zhou, Z. (1983). Quaternary record of *Azolla pinnata* from China and its sporoderm ultrastructure. *Review of Palaeobotany and Palynology* **39**, 109–29.

8. Heterospory: cul-de-sac or pathway to the seed?

WILLIAM G. CHALONER
ALAN R. HEMSLEY*
Biology Department, Royal Holloway and Bedford New College, University of London, Egham, Surrey, UK

Abstract

The three types of life cycle seen in vascular plants—homospory, heterospory, and the 'seed habit'—appear in that order within the Devonian period. It is widely believed that this sequence represents an evolutionary series, with each succeeding step derived from its predecessor. It is clear from the fossil evidence that this series was repeated separately and in parallel in several vascular plant lineages. However, some authors have suggested that the seed might have arisen directly from homospory, without an intermediate heterosporous phase. The ultrastructure of the megaspore exines of four Palaeozoic plants relevant to this controversy (the Devonian genera *Archaeopteris* and *Archaeosperma*, and the Lower Carboniferous *Bensonites* and cf. *Stamnostoma*) is described, and their bearing on this possibility is discussed. We see no support for the derivation of early seeds directly from homosporous stock in the fossil evidence available to us.

Introduction

Three distinct types of life cycle are represented among vascular plants—homospory, heterospory, and the seed habit. It is widely, but not universally, believed that these three means of reproduction represent surviving members of what has been an evolutionary series;

*Present address:
Laboratoire de Paléobotanique, USTL, Place E. Bataillon, 34060 Montpellier Cedex, France

Pollen and Spores (ed. S. Blackmore and S. H. Barnes), Systematics Association Special Volume No. 44, pp. 151–167, Clarendon Press, Oxford, 1991.

that homospory preceded heterospory, and gave rise to it, and that this type of life cycle was the antecedent to the seed habit. The belief that this is indeed the course of events is particularly defended by palaeobotanists (see, for example, Andrews 1963; Pettitt 1970; Stewart 1983; Chaloner and Pettitt 1987), perhaps because the fossil evidence is so unequivocal. In any event, the wide use of such terms as 'microspore mother cells' in connection with pollen formation in living seed plants, and the use of the 'megaspore mother cell' for the pre-meiotic source of the embryo sac (see, for example, Sporne 1974; Gifford and Foster 1989) are symptomatic of the general acceptance of this relationship. None the less, the possibility that homospory might have given rise directly to the seed, omitting a heterosporous intermediate, was first seriously raised by Thomson (1927), and subsequently revived by Joseph Doyle (1953). Indeed, Doyle refers to the idea that the seed is a derivative of a heterosporous stage as a 'bland assumption'. Both these authors based their case essentially on the claim that if measured at the appropriate stage, the embryo-sac-to-be was, in many conifers, no larger than the pollen grain at the corresponding stage. They attached no significance to the subsequent growth that occurred in gymnosperm embryo sacs, even though these were still enclosed inside a sporopollenin membrane and were in this sense still endosporic. Neither of those authors claimed that they had any stronger evidence than this for the actual evolutionary pathway of seed origins, but suggested that the possibility of a seed being derived directly from homospory should not be ignored. Doyle merely argued that a 'seed origin from homosporous sources cannot, on present evidence, be ruled out'. He went on to support Thomson's suggestion that the term megaspore in seed plants should be replaced by 'gynospore', so as to avoid begging the question of origin, but few later authors seem to have adopted the term.

Over the intervening years, much evidence has accrued from fossil plant studies which seems to reinforce the suggestion that seeds evolved from heterosporous ancestors, and that this happened not once, but probably several times—that 'the seed' is indeed polyphyletic. The study of dispersed spores in the Palaeozoic shows a steady rise in spore sizes, followed by a sharp segregation into a bimodel size distribution once heterospory was established (Chaloner 1967). The fossil record is quite unambiguous on the homosporous nature of the earliest vascular plants, both from a limited number of *in situ* occurrences and dispersed spores, from mid-Silurian to early Devonian time (420 Ma to 380 Ma BP). By the mid-Devonian, evidence of heterospory begins to emerge, and this was to become a major feature of land plant life from the late Devonian until the close of the Carboniferous at around 290 Ma BP. By that time, seed-bearing plants, which first appear right at the end of the Devonian (around 360 Ma BP), were playing an increasingly important role in land plant life, and gymnosperms were an abundant feature of land floras from then onwards. The timing of the homospory–heterospory–seed sequence is thus well documented, but of course this does not prove the derivation of one stage from another. However, it is appropriate to note that by the Carboniferous at least four separate classes of vascular plants had become heterosporous (Chaloner and Pettitt 1987), and that each one of these (progymnosperms, ferns *sensu lato*, lycopods, and sphenopsids) had homosporous forms which predated the heterosporous members. Thus far the fossil record certainly points to a polyphyletic origin of heterospory, in these four lines at least.

The early appearance of both platyspermic and radiospermic seeds more or less synchronously at the end of the Devonian is a strong argument for a polyphyletic seed origin, (Chaloner *et al.* 1977; Beck 1981). More recently, the discovery of an anomalous seed-like body from the Lower Carboniferous of France has brought new evidence favouring the same conclusion (Chaloner 1989; Galtier and Rowe 1989). However, other authors, e.g. Rothwell (1986), regard the gymnosperms as monophyletic; that author reviews the evidence for these two conflicting interpretations.

Despite the considerable body of evidence for the evolution of seeds via heterospory, the idea of a direct derivation of seeds from a homosporous ancestral group has been raised again by DiMichele *et al.* (1989). They bring a new perspective to Thomson's and Doyle's thesis, in arguing that we may have misunderstood the relationship between unisexual gametophytes, endospory, and a heterosporous form of reproduction. In spite of the fact that the former two are fixed features of all heterosporous plants (both living and fossil, where the relevant evidence is available in the fossils), DiMichele *et al.* suggest that we should remain open to the possibility that endospory was the first step. Indeed, they argue that unisexuality of the (two) gametophytes may have been 'a consequence of endospory, not its antecedent'. This would require an intermediate stage in which bisexual gametophytes became endosporic, and so would have needed to carry the minimum food reserve required by the ensuing sporophyte, as do extant heterosporous plants. As a separate but related issue, these authors further argue that 'it is as likely that seed plants were derived directly from homosporous ancestors . . . as from heterosporous "intermediates" '.

Before pursuing this theme it is perhaps worth noting some of the features of the role of these three life cycle strategies in living plants. As pointed out by Chaloner and Pettitt (1987) their relative 'success' as judged by the number of species in each category shows the seed plants (angiosperms plus gymnosperms) as far outnumbering all spore-bearing plants put together. However, the heterosporous plants seem to represent a kind of valley in the topography of success, between the highlands of homospory on one side and seed plants on the other, since they are in turn outnumbered by homosporous plants by roughly five to one. If there has been an evolutionary succession in the line of homospory–heterospory–seed, this relationship may seem odd to the point of being contradictory. We can only suppose that when heterospory first successfully challenged homospory in the late Devonian and early Carboniferous, the former type of life cycle showed its 'competitive edge' most effectively.

The rising seed plants through the Mesozoic may have challenged their heterosporous contemporaries more directly than the homosporous ones, which occupied different ecological niches. Tiffney (1981) offers a useful discussion of the kinds of factors that may have been instrumental in the changing status of these life cycles through geological time. Whatever the true course of events, it is clear that homospory was in no sense 'ousted' by the evolution of heterospory, although in certain situations, such as those of the great Carboniferous coal-forming swamp forests, heterospory was the successful strategy adopted by the dominant lepidodendrid trees of those communities. Similarly, although the seed seems to have successfully challenged heterospory in the majority of habitats, plants with heterosporous reproduction have held their own in a scattered range of habitats, from the submerged *Isoetes* species to epiphytic selaginellas. We must accept that the status of heterospory as we see it now in an angiosperm-dominated plant world is likely to be a poor guide to its role before the Palaeozoic rise of the seed plants.

We take this opportunity to question particularly the hypothesis of DiMichele *et al.*, and ask whether the fossil evidence is compatible with the proposition that seeds arose directly from plants with a homosporous life cycle. In particular, since exines of the spores of vascular plants have a particular facility for surviving intact in fossil reproductive structures, we consider whether exine structure can give any clear evidence on this matter.

Exine structure and the heterospory–seed transition

Archaeopteris

Ever since Arnold (1939) first described mega- and microspores in sporangia associated with the Upper Devonian progymnosperm *Archaeop-*

teris, this genus has had a special status in debates concerning the origin of the seed. Arnold himself (1947), who at that time saw *Archaeopteris* simply as a late Devonian heterosporous fern, wrote that it indicated 'a trend in the direction of seed plant evolution, and it may constitute the necessary heterosporous stage in the derivation of the seed from the terminal sporangium of the Psilotales'. Certainly it is the most completely known of the heterosporous progymnosperms of the late Devonian. It also occupies an intermediate position, both structurally and stratigraphically, between the homosporous, megaphyllous plants of the Middle Devonian and the earliest seeds which appear more or less synchronously with it in the (last) Famennian stage of the Devonian (Pettitt and Beck 1968; Gillespie *et al.* 1981; Fairon-Demaret and Scheckler 1989).

Pettitt (1965) was able to confirm Arnold's suggestion by obtaining sporangia which contained either megaspores or microspores, in organic connection with *Archaeopteris* fertile branch systems. He showed that the ultrastructure of both megaspores and microspores was basically similar, both being bilayered (commonly splitting between the two exine layers to give a cavate exine). In both the megaspores and the microspores the inner layer is homogeneous (Pettitt 1966) while the outer layer is granular. The thickness of the megaspore exines is about 10 times that of the microspores, which is roughly in the same proportion as the diameter of the two spore sizes.

All seed plants differ from heterosporous free-sporing plants (such as *Archaeopteris*) in two fundamental respects. First, a single tetrad only is involved in forming the haploid gametophytic structure—that is, the female gametophyte, the embryo sac, (the megaspore, of our usage). Further, in the seed all members of the tetrad bar one abort, so as to leave a single gamete-producing haploid spore. Considerable interest therefore attaches to the number of tetrads in a megasporangium of any possible seed-precursor, and also to any evidence that abortion is occurring in some tetrad members. Material of *Archaeopteris latifolia* studied by Pettitt from the Upper Devonian of Pennsylvania shows a wide range of megaspore number per sporangium, and some evidence that this variation is in part due to a varied number of tetrads in the sporangium, but also to abortion of individual products of meiosis (Chaloner and Pettitt 1987). Some of the range of megaspore content in these sporangia is illustrated here, with 11 tetrads of megaspores (Fig. 8.1a), five tetrads (Fig. 8.1b), and only two tetrads (Fig. 8.1c) per sporangium. All these sporangia came from a transfer specimen of the same fertile branch of *A. latifolia*. They are offered here as tangible evidence that *Archaeopteris* shows great variability in megaspore number; further loss of a single tetrad would take it to the state seen in the Devonian seed *Archaeosperma*, in which a single tetrad occupies the whole of the megasporangium.

Bensonites

The Lower Carboniferous stauropteridalean sporangium, *Bensonites*, is of some interest here in representing a sporangium in which a single tetrad develops in the sporangium, with two megaspores reaching an appreciable size, while two (? abortive) members of the tetrad remain small [Fig. 8.1d, a dispersed tetrad of this species; the two aborted spores are seen at right angles to the two (?) functional megaspores set transversely in the figure]. While *Bensonites* considerably post-dates the earliest seeds, it is of some interest in being borne by a megaphyllous plant (a fern, *sensu lato*) and in representing a state intermediate between that in, say, *Selaginella* with a single megaspore tetrad in the sporangium, and a seed in which three members abort and one goes to maturity. Andrews (1961) saw in *Bensonites* a possible evolutionary stage towards seed formation, although accepting that chronologically it could only represent an analogue of what might have preceded the earliest seeds.

The ultrastructure of the *Bensonites* megaspore tetrad is of some interest in lending support to the similarity between the megaspores of the species and that of some extant gymnosperms (Hemsley 1990). The exine of the megaspore itself is completely homogeneous, but external to this an elaborate structure of sporopollenin-like material forms a 'tapetal membrane' of a cellular character (Fig. 8.2), comparable to that seen in *Ginkgo* and some cycads (Pettitt 1970). However, similar structures form around the megaspores of the heterosporous fern *Marsilea*, so that they may merely represent the residue of a

---

Fig. 8.1 Devonian and Lower Carboniterous megasporangia (a)–(c) Megasporangial contents of the Devonian progymnosperm *Archaeopteris*, showing variation in the number of tetrads in the sporangium. This material, prepared by Dr J.M. Pettitt, is from the Famennian of Pennsylvania; for locality details see Chaloner and Pettitt (1987, p. 47). Sample (a) contains the greatest number of megaspores—some 11 tetrads—seen in any sporangium. (This is greater than the highest number plotted in Chaloner and Pettitt 1987, fig. 5.) The sporangium in (b) contains fewer tetrads, with proportionately larger megaspores, while (c) contains only two tetrads. (d), The single tetrad from *Bensonites*, the sporangium of *Stauropteris burntislandica*, is of Lower Carboniferous age, and has two large (functional?) megaspores and two small (abortive?) members in the tetrad (photographed by transmitted light) (e) The large (functional) megaspore of the *Spermasporites* tetrad (borne in the Devonian seed, *Archaeosperma*), with the three markedly smaller (abortive) megaspores at the upper pole; shown at higher magnification in f (both SEMs).

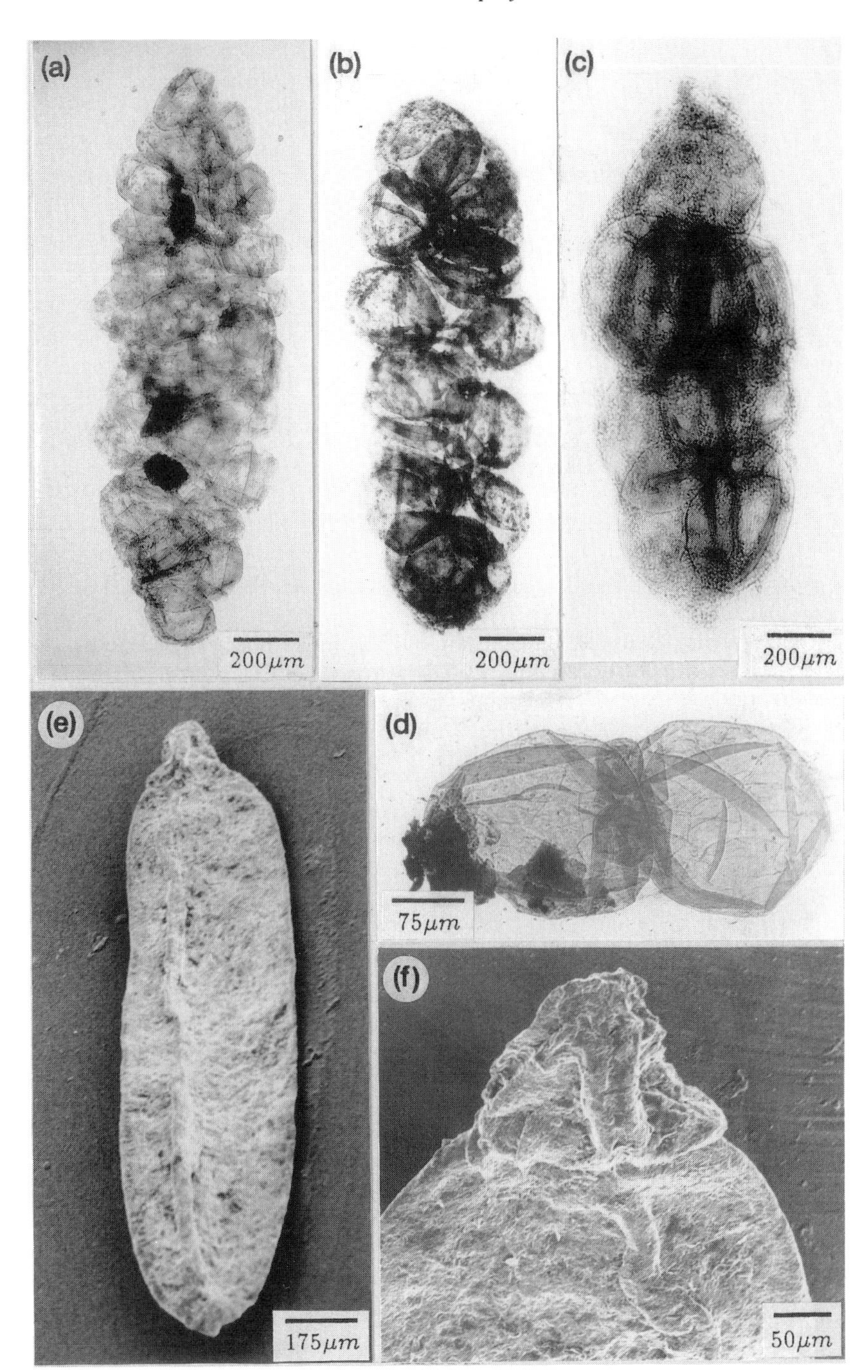
(a)
200μm
(b)
200μm
(c)
200μm
(e)
175μm
(d)
75μm
(f)
50μm

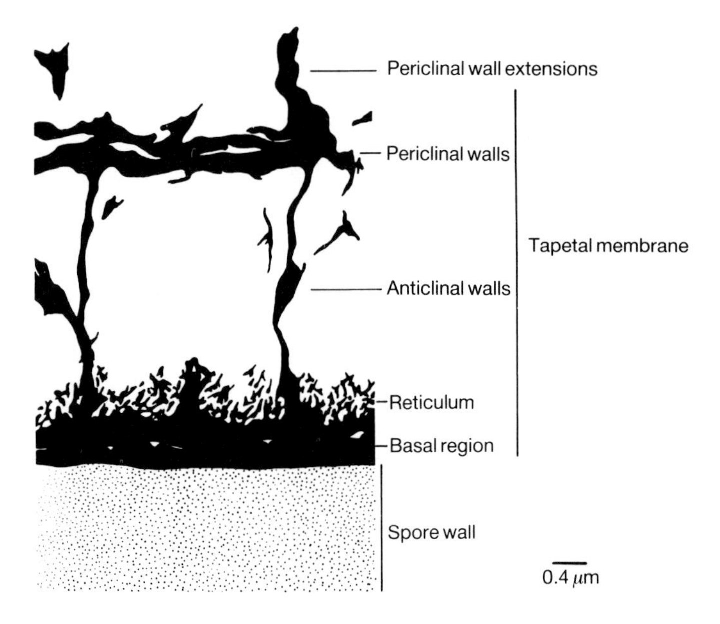

Fig. 8.2 A diagrammatic interpretation of the megaspore exine and adjoining structures in the sporangium of *Stauropteris*. The exine proper is homogeneous, but the material external to it has a chambered (?cellular) structure and resembles the megaspore membrane in some gymnosperms and the megaspore wall of *Marsilea*.

similar tapetal function rather than indicating any particular level of seed-like organization.

Archaeosperma

Much greater interest attaches to the exine structure of the megaspore of the Devonian seed *Archaeosperma*, which is one of the earliest seeds of which there is knowledge of the megaspore membrane. The use of the label 'seed' for this and other Palaeozoic seed-like bodies is, of course, presumptuous. Technically it may be argued that without knowledge of an embryo within it, this structure should be referred to as an ovule rather than as a seed. The finer shades of meaning of ovule/pre-ovule/seed are discussed in Jonker (1977), Gensel and Andrews (1984), and Chaloner and Pettitt (1987), and in a number of earlier works cited therein. For the purpose of this treatment, we follow the convention of most palaeobotanists in referring to detached and dispersed Palaeozoic ovules, probably borne by pteridosperms or cordaites or other gymnosperm groups, as seeds, although none of Carboniferous or earlier age have been found to contain an embryo.

The megaspore tetrad that forms the central feature of *Archaeosperma* was identified by Pettitt and Beck (1968) as indistinguishable from the dispersed megaspore tetrad described by Chaloner and Pettitt (1964) as *Cystosporites devonicus* now *Spermasporites devonicus* Hemsley). Material of the latter from the original Canadian type locality has recently been subjected to examination by transmission electron microscopy (TEM) (Hemsley 1990). The dispersed megaspore tetrads, believed to represent the contents of dispersed seeds of which these are the only part surviving bulk maceration, show an elongated functional megaspore, with three aborted spores at the apex (Fig. 8.1e,f). Sections of the fertile megaspore exine examined by TEM show it to be quite structureless (Fig. 8.3a).

The other 'seed megaspore' which offers immediate comparison with that contained in *Archaeosperma* is that of the lycopsid seed-like body *Lepidocarpon*, which contained megaspore tetrads assignable to the genus *Cystosporites*. These resemble the megaspore of *Archaeosperma* in having a single, large, elongated, 'functional' megaspore and three aborted spores. The site at which the three aborted spores were appressed to the functional spore in a Carboniferous *Cystosporites* is shown in Fig. 8.3b. However, the exine structure of *Lepidocarpon* megaspores is very different from that of the Devonian seed. Above a thin homogeneous basal lamina, there is a thick, loosely fibrous outer exine layer, which makes up more than 90 per cent of its thickness (Figs 8.3c,d and 8.4). This fundamental difference in exine organization between *Archaeosperma* and *Lepidocarpon* parallels other major differences in the symmetry and structure of these two seeds—the former radiospermic, and a presumed pteridosperm, while *Lepidocarpon* is simply an integumented megasporangium, of which the 'micropyle' is a slit between two marginal flaps of the sporophyll. Few would regard it as a seed, or its parent plant a gymnosperm, although functionally it must have behaved like one. Indeed, *Lepidocarpon* is the earliest seed-like structure in which an embryo has been demonstrated (Phillips 1979). There is an obvious possible correlation between the fibrous exine and the sustained growth of the megaspore membrane which might have accompanied embryo development, and the potential for transport of food material from the parent sporophyte to the female gametophyte through the megaspore exine.

Stamnostoma

An interesting comparison with the *Lepidocarpon* exine is offered by that of a pteridosperm seed, of which the megaspore and its ultrastructure are shown in Figs 8.3e,f and 8.5. This megaspore exine was obtained from a Lower Carboniferous radiospermic seed, cf.

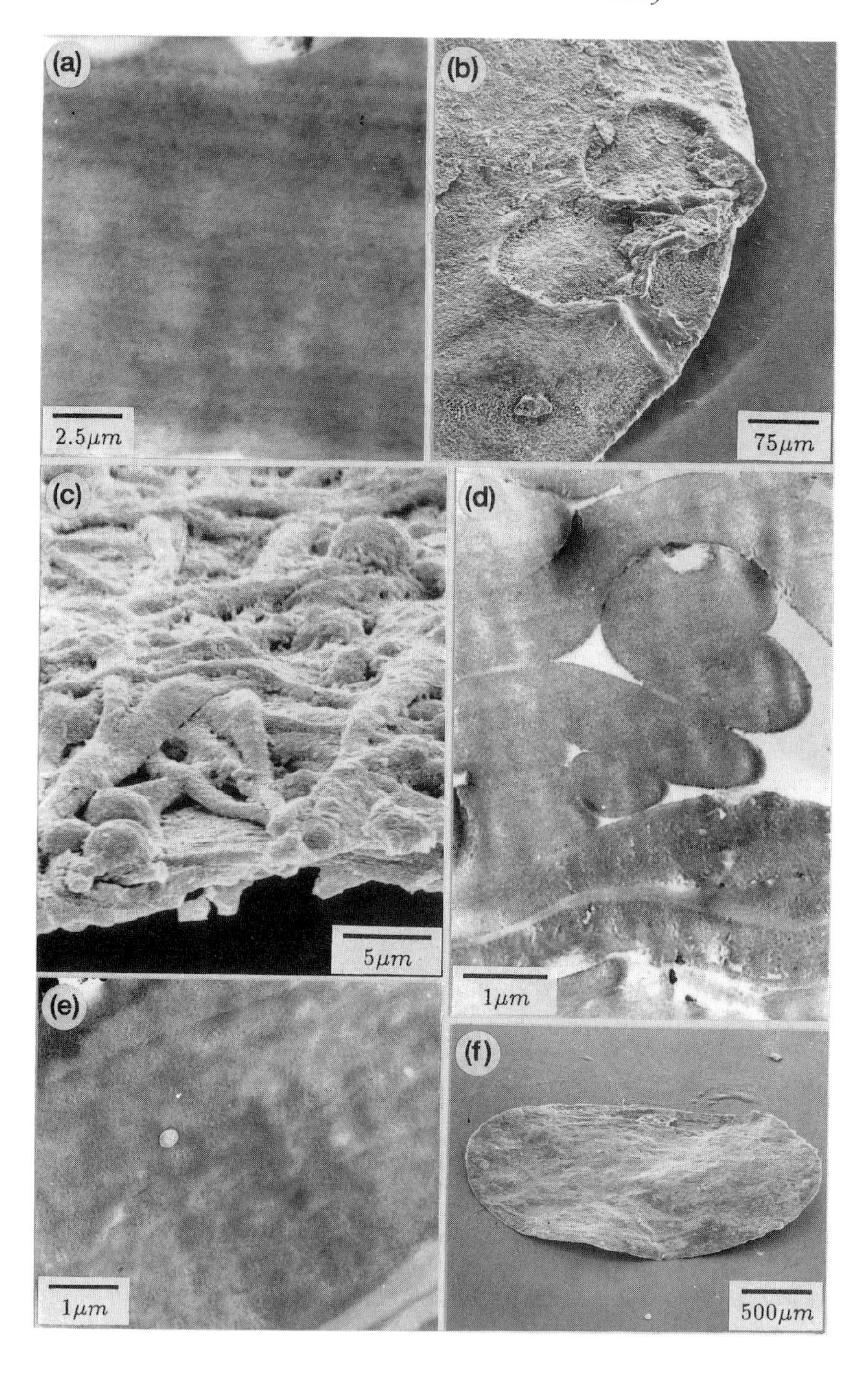
(a)
2.5μm
(b)
75μm
(c)
5μm
(d)
1μm
(e)
1μm
(f)
500μm

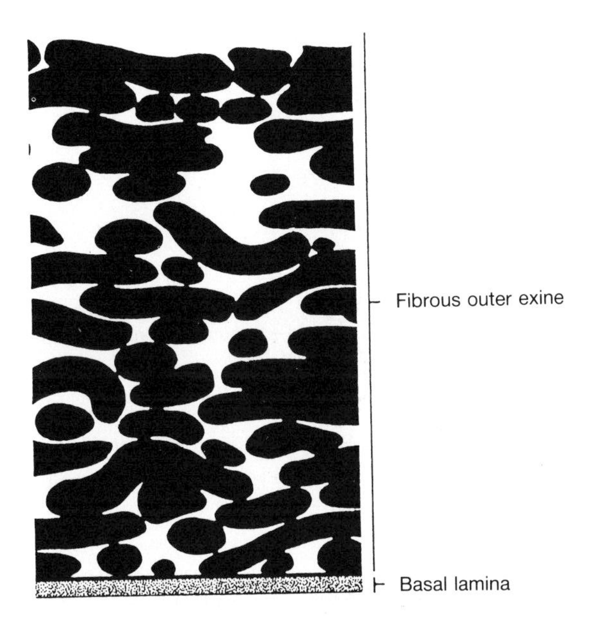

Fig. 8.4 A diagrammatic interpretation of the ultrastructure of the exine of the megaspore of *Lepidocarpon* (see Fig. 8.3d). The exine thickness ranges from around 6 μM in the centre of the functional spore to 12 μm or more at the ends.

Fig. 8.3 (a) TEM of a transverse ultra-thin section of the megaspore of *Spermasporites* (the megaspore of *Archaeosperma*) extending through the entire thickness of the exine, showing its homogenous nature; the external surface is above, the inner, below). The near-horizontal, lineations are knife marks. (b)–(d) *Cystosporites* sp., the seed megaspore of the Upper Carboniferous lycopod *Lepidocarpon*. (b) Detail of the apex of the single megaspore, where the three abortive spores have been appressed to it. (c) SEM of the outer surface of this megaspore, showing its fibrous character. (d) TEM of an ultra-thin section of the same exine, which consists of a thin basal lamina (seen double, from both sides of the spore lumen, below) surmounted by the fibrous exine. This is interpreted in Fig. 8.4. (e,f), The megaspore from inside the Lower Carboniferous seed cf. *Stamnostoma*, from Foulden, Berwickshire (Scott and Meyer-Berthaud 1985). (e) The ultrastucture of the exine under TEM—the thin electron-lucent basal structure is seen double, below, from both surfaces of the lumen, overlain by granular exine units. (f) The outer surface of the megaspore (SEM) gives no indication of abortive spores at the apex.

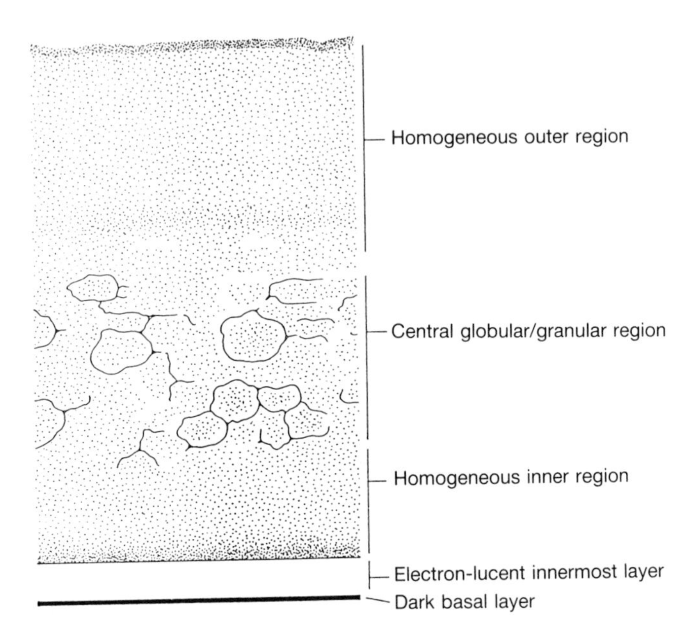

Fig. 8.5 An interpretation of the ultrastructure of the exine of the Lower Carboniferous seed, cf. *Stamnostoma* (see Fig. 8.3e). The thickness of this seed megaspore membrane is between 6 and 10 μm.

Stamnostoma huttonense (the specimen figured in Scott and Meyer-Berthaud 1985, Fig. 8.3g). In its ultrastructure, it is far closer to the simple homogeneous exine of *Archaeosperma* than to its contemporary Lepidocarp. Above a thin dark basal layer, and a somewhat thicker electron-lucent innermost layer (base of Fig. 8.5, and bottom right in Fig. 8.3e) the thick exine shows a somewhat granular middle layer sandwiched between entirely homogeneous regions. The megaspore itself, unlike that of either *Lepidocarpon* or *Archaeosperma* is a simple ellipsoidal sac, without any indication of polarity, or any sign of aborted members of the tetrad.

Discussion

A number of features of the megaspore exines shown by the seeds (*Archaeosperma* and cf. *Stamnostoma*) and seed-like bodies (*Bensonites*, *Lepidocarpon*) discussed above are relevant to the matter of seed origins. First, all show quite a robust exine which presumably enclosed the female gametophyte. Although they are of quite varied thickness and structure, it is particularly noteworthy that the megaspore exine of the early seed *Archaeosperma* (25–60 μm) is materially thicker than that of the free-sporing *Archaeopteris* megaspore. If these early seeds had arisen from homosporous stock, in which the endosporic gametophytes came to be retained by the parent sporophyte, the formation of such a robust exine is puzzling. Particularly so, since the ensuing history of seed plants is one of reduction in thickness of the megaspore membrane, from pteridosperms and cordaites through the modern conifers, ginkgos, and cycads to the angiosperms, in which no exine is detectable around the embryo sac.

Some light may be thrown on this possibility by the plot of the megaspore exine thickness against the maximum megaspore size, in Palaeozoic free-sporing heterosporous plants and in some seed- or seed-like plants (Fig. 8.6). The data are limited to some measurements that we have made ourselves, and a few derived from the literature. However, even in this small data set there is a broad correlation for the free-sporing plants between megaspore size and exine thickness (the continuous line encircling the relevant items in Fig. 8.6): the larger the megaspore, the thicker the exine. However, in those seed megaspores or equivalent structures (*Bensonites*, *Lepidocarpon*) that simple proportionality does not apply. The earliest seed megaspore of which we have data, *Spermasporites devonicus* (megaspore of *Archaeosperma*), lies at the top end of the free-sporing group (item 12 in Fig. 8.6), but the remainder of the seeds and seed-like bodies are displaced in the direction of a proportionately thinner exine (dotted outline, Fig. 8.6).

This is construed as possibly representing the thinning of the exine following its retention within a megasporangium/nucellus/integumentary structure. The thickness of the *Spermasporites* megaspore encourages this interpretation, rather than the possibility that these Palaeozoic seeds have evolved independently from homosporous sources. It seems more plausible that the early seed megaspore exines are indeed a relictual feature of a free-sporing and heterosporous ancestry, which lingered after its initial function in protecting the dispersed megaspore had been superceded by retention within nucellus and integument. There are many analogies to this in seed plant evolution—perhaps the retention of flagellated sperm in cycads and ginkgos, long after the evolution of the pollen tube, is a case in point.

Even if the megaspore membrane in early seeds is 'relictual', its varied structure in the cases described above strongly suggests two further features of the processes involved in seed evolution. First, that the lepidocarps and pteridosperms had entirely separate origins, as is suggested by all other aspects of their vegetative and reproductive structures. This striking parallelism of evolution of seed-like bodies with a single surviving functional megaspore enhances the probability,

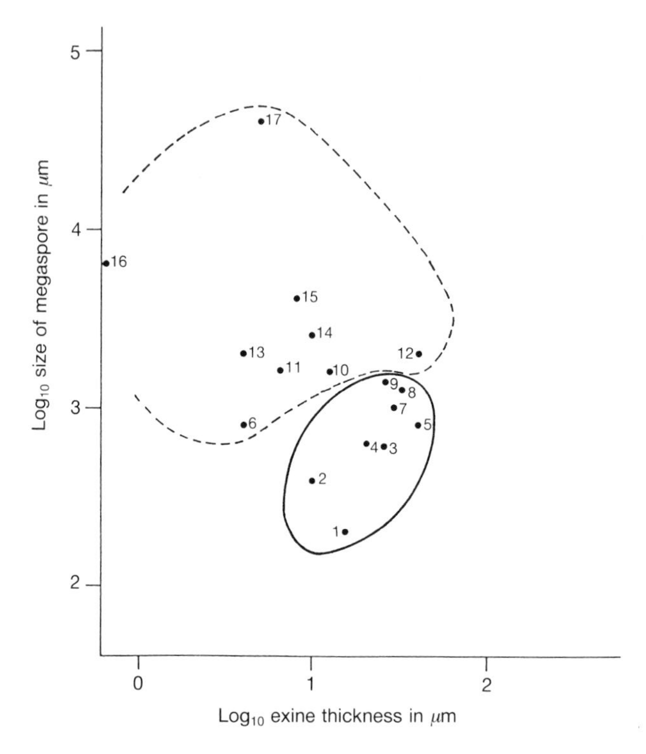

Fig. 8.6 A plot of megaspore exine thickness against maximum megaspore size, for various Palaeozoic plants—some free-sporing heterosporous, some from seeds or seed-like bodies. The numbers relate to the names of the plants given below, together with reference sources. Where no reference source is given, the measurements are the authors'. The free-sporing megaspores, mainly from Carboniferous lycopods (enclosed in a solid line), show a broad correlation between increasing size and increasing exine thickness. Those (within the dotted outline) which were enclosed in seeds or seed-like structures are displaced, relative to their maximum diameter, in the direction of a thinner exine. This is most marked for the Upper Carboniferous *Pachytesta* (16, 17), somewhat less so for the Lower Carboniferous *Stauropteris* (6) and *Stamnostoma* (11), and least for the megaspore membrane of the Devonian seed *Archaeosperma* (12). 1, *Archaeopteris* megaspore (Pettitt 1966); 2, *Triangulatisporites* sp.; 3, *Lagenoisporites nudus* (Taylor 1990); 4, *Lagenicula subpilosa* f. *major*; 5, *Valvisisporites auritus* (Gastaldo 1981); 6, *Stauropteris* megaspore; 7, *Setispora subpaleocristatus*; 8, *Lagenicula crassiaculeata*; 9, *Tuberculatisporites mamillarius*; 10, *Cystosporites verrucosus* (Brack-Hanes 1981); 11, cf. *Stamnostoma* megaspore membrane; 12, *Spermasporites devonicus*; 13, *Cystosporites bennholdii*; 14, *C. varius* (Taylor and Brack-Hanes 1976); 15, *C. giganteus*; 16, *Pachytesta berryvillensis* megaspore membrane (Taylor and Eggert 1969; Zimmerman and Taylor 1970); 17, *P. illinoensis* (Zimmerman and Taylor 1970; Stewart 1983).

referred to above, that even within the Carboniferous gymnosperms (excluding *Lepidocarpon*) 'the seed' may have arisen more than once. Secondly, the differences in megaspore membrane ultrastructure may also reflect differences in function which post-date the retention of the megaspore and reflect as-yet-unknown peculiarities of the reproductive biology of the plants in question. Collectively, the evidence from early seed-megaspore exines certainly gives no support for a derivation of any seed plant directly from the homosporous condition, and may be more plausibly seen as being derived from megaspores previously adapted to the process of free spore dispersal.

Acknowledgements

We are pleased to have this opportunity to record our thanks to Dr J.M. Pettitt for allowing us to study his *Archaeopteris* sporangia from Pennsylvania, to Dr A. Scott for giving us access to his Foulden seeds, and to the staff of the Electron Microscope Unit of RHBNC, Egham, for all their help and patience. One of us (A.R.H.) gratefully acknowledges the award of a NERC studentship, during the course of which some of this research was carried out.

References

Andrews, H.N. (1961). *Studies in Paleobotany*. Wiley, New York.

Andrews, H.N. (1963). Early seed plants. *Science* **142**, 925–31.

Arnold, C.A. (1939). Observations on fossil plants from the Devonian of Eastern North America. IV. Plant remains from the Catskill Delta deposits of Northern Pennsylvania and Southern New York. *Contributions from the Museum of Paleontology, University of Michigan* **5**, 271–313.

Arnold, C.A. (1947). *An introduction to paleobotany*. McGraw Hill, New York.

Beck, C.B. (1981). *Archaeopteris* and its role in vascular plant evolution. In *Paleobotany, paleoecology and evolution*, Vol. I, (ed. K.J. Niklas), pp. 193–230. Praeger, New York.

Brack-Hanes, S.D. (1981). On a lycopsid cone with winged spores. *Botanical Gazette* **142**, 294–304.

Chaloner, W.G. (1967). Spores and land-plant evolution. *Review of Palaeobotany and Palynology* **1**, 83–93.

Chaloner, W.G. (1989). A missing link for seeds. *Nature* **340**, 185.

Chaloner, W.G. and Pettitt, J.M. (1964). A seed megaspore from the Devonian of Canada. *Palaeontology* **7**, 29–36.

Chaloner, W.G. and Pettitt, J.M. (1987). The inevitable seed. *Bulletin de la Société Botanique de France 134, Actualités botaniques* **1987**, (2), 39–49.

Chaloner, W.G., Hill, A.J., and Lacey, W.S. (1977). First Devonian platyspermic seed and its implications in gymnosperm evolution. *Nature* 265, 233-5.

DiMichele, W.A., Davis, J.I., and Olmstead, R.G. (1989). Origins of heterospory and the seed habit: the role of heterochrony. *Taxon* **38**, (1), 1-11.

Doyle, J. (1953). Gynospore or megaspore—a restatement. *Annals of Botany* **17**, (67), 465-76.

Fairon-Demaret, M. and Scheckler, S. (1989). Typification and redescription of *Moresnetia zalesskyi* Stockmans 1948, an early seed plant from the late Devonian (Famennian) of Belgium. *Bulletin de l'Institut Royal des Sciences Naturelles de Belgique, Sciences de la Terre* **57**, 183-99.

Galtier, J. and Rowe, N.P. (1989). A primitive seed-like structure and its implications for early gymnosperm evolution. *Nature* **340**, 225-7.

Gastaldo, R.A. (1981). An ultrastructural and taxonomic study of *Valvisisporites auritus* (Zerndt) Bhardwaj, a lycopsid megaspore from the middle Pennsylvanian of Southern Illinois. *Micropaleontology* **27**, 84-93.

Gensel, P.G. and Andrews, H.N. (1984). *Plant life in the Devonian*. Praeger, New York.

Gifford, E.M. and Foster, A.S. (1989). *Morphology and evolution of vascular plants*, (4th edn). Freeman, New York.

Gillespie, W.H., Rothwell, G.W., and Scheckler, S.E. (1981). The earliest seeds. *Nature* **293**, 462-4.

Hemsley, A.R. (1990). The ultrastructure of the exine of the megaspores in two Palaeozoic seed-like structures. *Review of Palaeobotany and Palynology* **63**, 137-52.

Jonker, F.P. (1977). Spermatophytes and pre-spermatophytes, ovules and pre-ovules, pollen and pre-pollen - a comparison of pollen and seed evolution. *Advances in Pollen-spore Research* **2**, 6-13.

Pettitt, J.M. (1965). Two heterosporous plants from the Upper Devonian of North America. *Bulletin of the British Museum (Natural History) Geology* **10**, 83-92.

Pettitt, J.M. (1966). Exine structure in some fossil and recent spores and pollen as revealed by light and electron microscopy. *Bulletin of the British Museum (Natural History) Geology* **13**, 221-57.

Pettitt, J.M. (1970). Heterospory and the origin of the seed habit. *Biological Reviews* **45**, 401-15.

Pettitt, J.M. and Beck, C.B. (1968). *Archaeosperma arnoldii*—a cupulate seed from the Upper Devonian of North America. *Contributions from the Museum of Paleontology, University of Michigan* **10**, 139-54.

Phillips, T.L. (1979). Reproduction of heterosporous arborescent lycopods in the Mississippian–Pennsylvanian of Euramerica. *Review of Palaeobotany and Palynology* **27**, 239-89.

Rothwell, G.W. (1986). Classifying the earliest gymnosperms. In *Systematic and taxonomic approaches in palaeobotany*, Systematics Association Special

Volume 31, (ed. R.A. Spicer and B.A. Thomas), pp. 137–61. Oxford University Press.

Scott, A.C. and Meyer-Berthaud, B. (1985). Plants from the Dinantian of Foulden, Berwickshire, Scotland. *Transactions of the Royal Society of Edinburgh: Earth Sciences* **76**, 13–20.

Sporne, K.R. (1974). *The Morphology of Gymnosperms*, (2nd edn). Hutchinson, London.

Stewart, W.N. (1983). *Paleobotany and the evolution of plants*. Cambridge University Press.

Taylor, T.N. and Brack-Hanes, S.D. (1976). The structure and reproductive significance of the exine in fossil lycopod megaspores. *Scanning Electron Microscopy* **7**, 513–18.

Taylor, T.N. and Eggert, D.A. (1969). On the structure and relationships of a new Pennsylvanian species of the seed *Pachytesta*. *Palaeontology* **12**, 382–7.

Taylor, W.A. (1990). Comparative analysis of megaspore ultrastructure in Pennsylvanian lycophytes. *Review of Palaeobotany and Palynology* **62**, 65–78.

Tiffney, B.H. (1981). Diversity and major events in the evolution of land plants. In *Paleobotany, paleoecology and evolution*, Vol. 2, (ed. K.J. Niklas), pp. 193–230. Praeger, New York.

Thomson, R.B. (1927). Evolution of the seed habit in plants. *Transactions of the Royal Society of Canada* **21**, 229–72.

Zimmerman, R.P. and Taylor, T.N. (1970). The ultrastructure of Palaeozoic megaspore membranes. *Pollen et Spores* **12**, 451–68.

9. Diversification of early angiosperm pollen in a cladistic context

JAMES A. DOYLE
Department of Botany, University of California, Davis, USA

CAROL L. HOTTON
Department of Biological Sciences, State University of New York, Binghamton, USA

Abstract

Comparisons of early fossil angiosperm pollen types with cladograms of extant angiosperms permit firmer inferences on affinities of fossils and bear on competing hypotheses of relationships among taxa. The shortest trees root angiosperms in or near Magnoliales, implying that granular monosulcate pollen is primitive, but trees rooted among herbaceous magnoliids and monocots (paleoherbs), supported by molecular data, are almost as parsimonious. The five main clades of angiosperms are all represented by the Barremian–Aptian, confirming their status as early lines. Large, granular monosulcates from the Potomac Group (*Lethomasites*) may represent core Magnoliales. Reticulate monosulcates with spinulose muri and a sculptured sulcus (‘*Clavatipollenites*’) represent Laurales near but not necessarily in Chloranthaceae. Possible extinct relatives include tectate monosulcates from Northern Gondwana (*Tucanopollis*) and coarsely reticulate, secondarily non-columellar forms (*Brenneripollis*). A clade including Winteraceae and Illiciales is represented in Northern Gondwana by ulcerate tetrads and possibly zonasulculate monads (*Afropollis*, *Schrankipollis*). Eudicots (with tricolpate and derived pollen) have been associated with Chloranthaceae, but the earliest tricolpates from Gabon lack chloranthaceous features and have sculpture suggestive of monocots, favouring instead cladistically inferred links with paleoherbs. Typical *Liliacidites* grains, with finer sculpture at the ends, are clearly monocots, but monosulcates with

Pollen and Spores (ed. S. Blackmore and S. H. Barnes), Systematics Association Special Volume No. 44, pp. 169–195, Clarendon Press, Oxford, 1991.

finer poles (*Similipollis*) could be near the common ancestor of tricolpates and paleoherbs. Triassic reticulate monosulcates with coarser proximal sculpture and thick endexine (Crinopolles), if related to angiosperms, would favour trees rooted among paleoherbs.

Introduction

For over 20 years, palynology has been used to trace the Cretaceous diversification of angiosperms. The first of these studies emphasized evidence for character evolution, such as the sequential appearance of monosulcate, tricolpate, tricolporate, and triporate pollen (Doyle 1969; Muller 1970), rather than diversification of angiosperms into extant systematic groups. This was partly a reaction to the tendency of early palaeobotanists to misidentify Cretaceous leaves with low-rank modern taxa, which hampered recognition that most of the angiosperm radiation occurred in the Cretaceous. It also reflected a belief that fossil pollen should be more reliable for elucidating character evolution than relationships of groups, because of the small number of characters available (cf. Hennig 1966), especially in nondescript monosulcates and tricolpates of the sort that dominate mid-Cretaceous floras.

Several developments are promoting more integration of the pollen record with systematics and phylogeny of modern angiosperms. First, techniques for scanning electron microscopy (SEM) and transmission electron microscopy (TEM) of single grains have greatly increased the number of characters in Cretaceous pollen (Doyle *et al.* 1975; Walker and Walker 1984). Secondly, comparative data have accumulated on distribution of the same characters in modern primitive groups (e.g. Walker 1976*a*, *b*; Le Thomas 1980, 1981). These advances have led to proposed relationships between *Liliacidites* and monocots (Doyle 1973; Walker and Walker 1986), '*Clavatipollenites*' and Chloranthaceae (Walker 1976*b*; Muller 1981; Walker and Walker 1984), and ulcerate tetrads and Winteraceae (Walker *et al.* 1983). Third, cladistic analyses based on a broad spectrum of characters have provided more explicit and testable hypotheses on basal relationships among modern angiosperms and a more objective assessment of the systematic value of pollen characters (Donoghue and Doyle 1989). This paper continues these trends by comparing results of SEM and TEM studies on early angiosperm pollen and the cladistic analysis of Donoghue and Doyle (1989).

One way to view these comparisons, which tends to assume that results of cladistic analysis are valid independent of fossils, is as a source of evidence on the age and/or place of origin of extant clades.

In so far as character combinations in pollen faithfully define phylogenetic groups derived from the totality of characters, we can be more confident in using these characters to relate fossil pollen groups to modern clades. Unlike the method of 'matching' fossils with lower-level taxa, this approach may work even when the fossils have no exact modern analogues, whether they are intermediate morphotypes or 'side-lines' with specializations of their own (autapomorphies), as long as they share enough advances (synapomorphies) with higher-level groups. Of course, such comparisons provide only a 'best guess' of relationships; it is always possible that features shared with a modern clade arose convergently in some extinct group. However, this risk presumably decreases as the number of distinctive features increases.

Besides dating inferred clades, fossils may help test and modify results of cladistic analyses of modern plants. Although some have argued that fossils rarely or never overturn inferred relationships among modern organisms (Patterson 1981; Ax 1987), cladistic experiments involving addition and subtraction of fossil taxa show that fossils can markedly affect cladogram topology (Doyle and Donoghue 1987; Gauthier *et al.* 1988; Donoghue *et al.* 1989). In other cases, fossils strengthen one tree relative to others that are equally parsimonious on extant data, which is important because cladistic results are often unstable and sensitive to re-evaluation of characters. Furthermore, even when fossils do not affect cladogram topology, they may alter the inferred course of character evolution or resolve the sequence of origin of apomorphies of modern taxa. It may be impractical to include dispersed pollen and living taxa in a single global analysis (because pollen taxa have so few characters, their position may be highly unstable), but sometimes their effects can be assessed without this.

Finally, the fossil record may contribute relevant stratigraphic data. One use is as independent evidence on character polarity (Hennig 1966; Doyle 1969). Furthermore, alternative trees may be more or less consistent with the observed sequence of taxa (Fisher 1980; Doyle *et al.* 1982). Stratigraphic evidence has been rejected because of the spottiness of the fossil record (Schaeffer *et al.* 1972), but it must be more reliable when the fossil record is relatively dense (Eldredge and Novacek 1985), as it is in palynology, and it is testable in the sense that trends may or may not hold up with increased sampling.

Background

The analysis of Donoghue and Doyle (1989) polarized characters using outgroup comparison, building on cladistic analyses of seed plants

(Crane 1985; Doyle and Donoghue 1986), which concluded that angiosperms belong to an 'anthophyte' clade also including Bennettitales, *Pentoxylon*, and Gnetales, which is in turn nested among 'Mesozoic seed ferns' such as *Caytonia*. It concentrated on monosulcate groups ('magnoliids' and monocots), lumped or split depending on their presumed naturalness and phylogenetic importance; 'higher dicots' were represented by six tricolpate taxa considered 'placeholders' for more advanced groups. Pollen data were derived primarily from Walker (1976*a*, *b*); where more than one state occurs in a taxon and the basic state is uncertain, characters were scored as unknown. Outgroup comparison, especially with Bennettitales and *Pentoxylon*, implies that the first angiosperm pollen grains were monads, boat-shaped, monosulcate, medium-sized, granular, tectate, with no aperture sculpture, no supratectal spinules, and no endexine (based on developmental arguments that the 'endexine' of gymnosperms corresponds to the footlayer of angiosperms, and the angiosperm endexine is a new layer: Zavada 1984).

The many resulting most parsimonious trees (cf. Figs 9.2–9.4) 'root' the angiosperms next to or in Magnoliales in a strict sense, namely those taxa with granular exines; the remaining groups are united by columellae and endexine, except for reversals in Laurales (associated with a shift to inaperturate pollen) and Nymphaeales. The position of Magnoliales is, of course, influenced by the assumption that granular exines are primitive, but it also indicates that other characters are consistent, which is not the case in Laurales and Nymphaeales. The 'columellates' form an unresolved trichotomy: Laurales (including Chloranthaceae), Winteraceae and tricolpate Illiciales ('winteroids'), and an unexpected grouping of other tricolpates ('eudicots') and herbaceous magnoliids plus monocots ('paleoherbs'), united by palmate leaf venation and differentiated stamens. However, there are trees rooted in or near paleoherbs that are only one or two steps less parsimonious, which is significant because similar trees are obtained from rRNA sequence data (Zimmer *et al.* 1989).

Our primary conclusion, discussed briefly by Donoghue and Doyle (1989), is that all five major clades are recognizable in the pollen record by the Barremian or Aptian stage of the Early Cretaceous, strengthening the view that they represent early lines. We will treat these groups in sequence, using the program MacClade (Maddison and Maddison 1987) to clarify the position of fossils by plotting the distribution of crucial characters, and then explore the bearing of the fossil record on the rooting problem.

Magnoliales

The most distinctive pollen in Magnoliales is large, boat-shaped, granular, and monosulcate, as in *Degeneria*, some Magnoliaceae and, some Annonaceae. Walker (1976*b*), Le Thomas (1980, 1981), and Walker and Walker (1984) argued that this pollen type is primitive in angiosperms, and that Early Cretaceous reticulate monosulcates, often cited as the oldest known angiosperms, are already rather advanced (cf. also Doyle 1969; Muller 1970). Until recently, such pollen was not known from the Early Cretaceous, but Ward *et al.* (1989) described apparent examples from the probable early Aptian of the Potomac Group of eastern USA, as *Lethomasites fossulatus*. These grains (Fig. 9.1a,b) are large, boat-shaped, and glassy-appearing under light microscopy (LM); SEM and TEM show that their exine consists of a thick, smooth tectum perforated by small foveolae or fossulae, a layer of fine granules, and a very thin nexine, apparently footlayer. *Lethomasites* is enough like pollen of Bennettitales to raise some question concerning its angiospermous affinities; however, Bennettitales sectioned so far have a uniformly thick, darkly staining endexine, sometimes with fine laminations (Ward *et al.* 1989).

The character distributions plotted in Fig. 9.2 show that *Lethomasites* fits best in a core group of Magnoliales, consisting of Magnoliaceae, Annonaceae, *Degeneria*, and Myristicaceae. It is here that granular structure is combined with boat shape, lack of endexine, and large size ($> 50\ \mu m$), although significance of the size character is equivocal because the small pollen of Myristicaceae could be either primitive or reversed. The next-best position is with Nymphaeales, but these differ in having endexine (Walker 1976*b*).

Discovery of *Lethomasites* in the early Aptian does not prove that such pollen is primitive, since there are already diverse reticulate-columellar monosulcates by this time. However, it removes an obstacle to this view, since previously such pollen was unknown. Trees as in Fig. 9.2, where Magnoliales form a monophyletic basal clade, imply that the *Lethomasites* pollen type is primitive in being granular and lacking endexine but advanced in being large and boat-shaped. It would be preceded by medium-sized, rounder monosulcates, as in Himantandraceae and some Canellaceae, which would be very hard to separate from pollen of Bennettitales without TEM. However, there are other equally parsimonious trees rooted within Magnoliales (which are then paraphyletic), where the core Magnoliales form a basal clade and boat-shaped pollen is primitive (Fig. 9.3). Given the polarity of Donoghue and Doyle (1989), based on the medium-sized pollen of most Bennettitales, large pollen is still derived. However, this polarity assessment

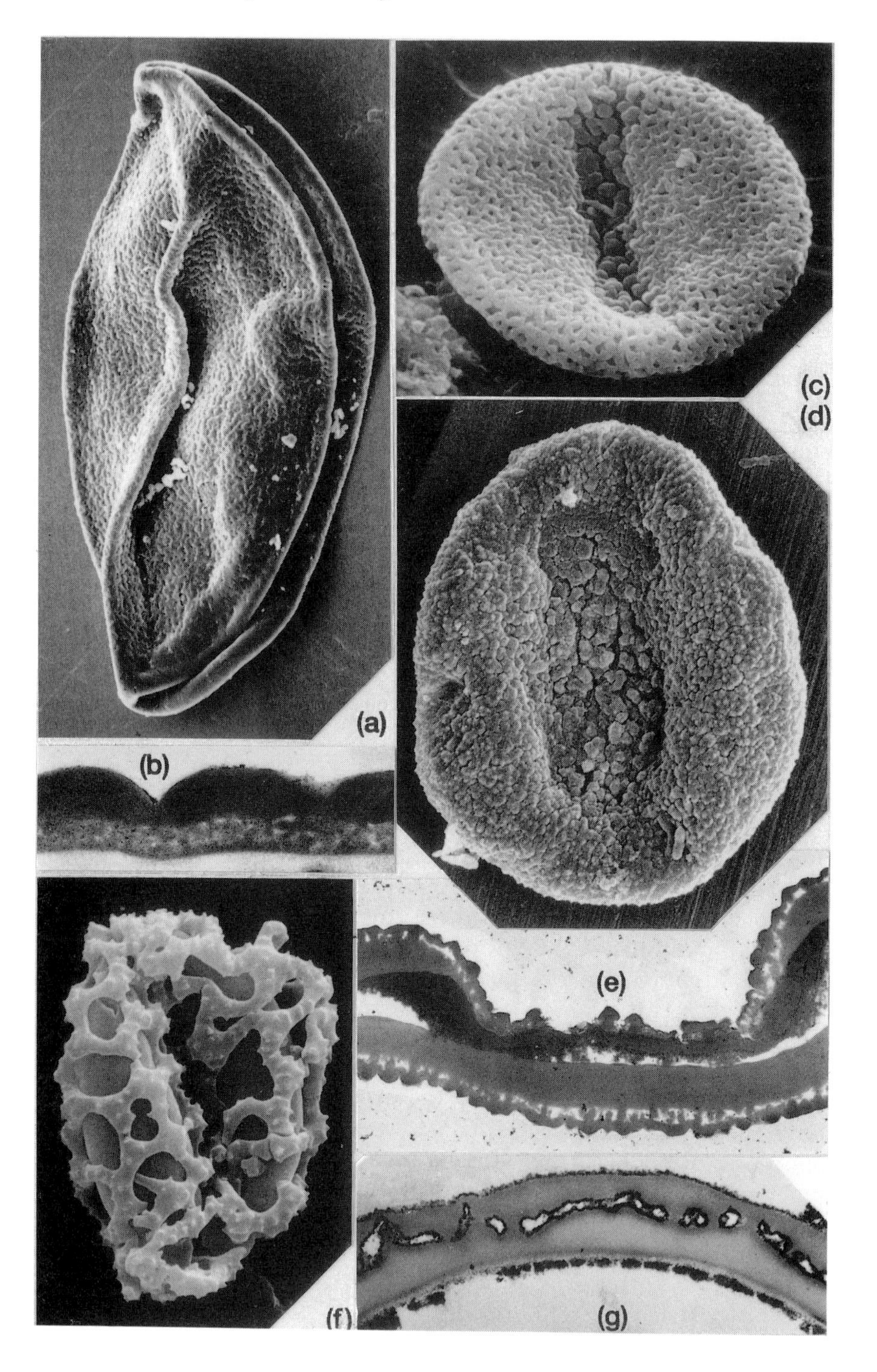
(a)
(b)
(c)
(d)
(e)
(f)
(g)

may have been premature, since some Bennettitales had large pollen (Ward *et al.* 1989). If size had not been polarized, *Lethomasites* could be primitive in all characters considered; it could represent either core Magnoliales or a 'stem-group' taxon below the common ancestor of all extant angiosperms. All these conclusions need revision if angiosperms are rooted among paleoherbs.

Laurales

Laurales appear to be represented from the Barremian on by reticulate monosulcates widely identified as *Clavatipollenites* Couper (1958). Use of this name was questioned by Hughes and McDougall (1987), since SEM studies show that the type sample of the type species *C. hughesii* contains several microsculpturally diverse reticulate monosulcates, and it is unclear which corresponds to the holotype described with LM. Grains here designated '*Clavatipollenites*' have supratectal spinules, a sculptured (verrucate) sulcus membrane, and a thick nexine composed mostly of footlayer, with endexine restricted to the sulcus (Doyle *et al.* 1975; Walker and Walker 1984; Chapman 1987; Fig. 9.1c). Couper (1958) compared '*Clavatipollenites*' with *Ascarina* in the Chloranthaceae, which are notable among magnoliids for having extremely simple flowers. Chloranthaceae have been variously interpreted as Piperales, reduced Laurales, or a link between these orders. However, Endress (1987) showed that Chloranthaceae are most like Laurales, especially Trimeniaceae and *Amborella*, and Donoghue and Doyle (1989) found that at least seven extra steps are needed to associate them with Piperales (Piperaceae, Saururaceae), which are nested in the paleoherbs, or to consider them a link between Laurales and Piperales.

Fig. 9.1 (a) *Lethomasites fossulatus* Ward *et al.*, single-grain preparation D12-770-27, Delaware City well D12, 770 ft, lower Zone I, early Aptian? (Ward *et al.* 1989), SEM, × 800; (b) same grain, TEM, exine section showing granular infratectal structure, × 20 000; (c) '*Clavatipollenites*' cf. *tenellis* Phillips and Felix, JAD sample 65-1, near Bladensburg, Maryland, lower Subzone II-B, middle Albian? (Doyle 1969), SEM, × 2400; (d) *Tucanopollis crisopolensis* (Regali *et al.*) Regali, single-grain preparation 35260-13, Elf-Aquitaine Pointe Indienne No. 1 well, Congo, 1399 m, Zone C-VI, Barremian?, SEM, × 2400; (e) same grain, TEM, section showing thin tectum, thick footlayer, and endexine below the sulcus, × 6000; (f) *Brenneripollis peroreticulatus* (Brenner) Juhász and Góczán, single-grain preparation 6921-27, United Clay Mine, Maryland, upper Zone I, Aptian? (Doyle *et al.* 1975), SEM, × 3000; (g) *Saururus cernuus* L., Louisiana, C. E. Davis 80, DAV, TEM, exine section, × 16 500.

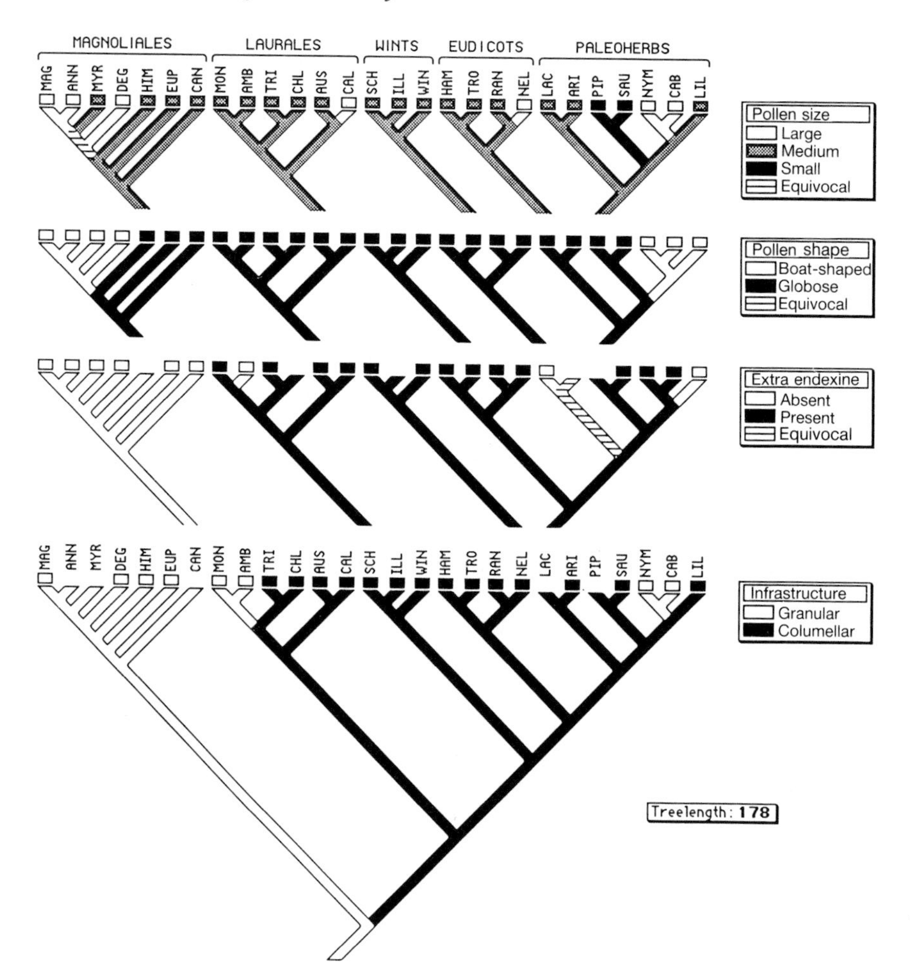

Fig. 9.2 Representative most parsimonious angiosperm cladogram of Donoghue and Doyle (1989), with the distribution of characters relevant to placement of *Lethomasites* indicated by shading. Boxes below names of taxa indicate character states; blanks indicate taxa scored as unknown, usually because both states occur and the basic state is uncertain. MAG, Magnoliaceae; ANN, Annonaceae; MYR, Myristicaceae; DEG, Degeneriaceae; HIM, Himantandraceae; EUP, Eupomatiaceae; CAN, Canellaceae; MON, Monimiaceae s. l., Gomortegaceae, Hernandiaceae, Lauraceae; AMB, Amborellaceae; TRI, Trimeniaceae; CHL, Chloranthaceae; AUS, Austrobaileyaceae; CAL, Calycanthaceae; SCH, Schisandraceae; ILL, Illiciaceae; WIN, Winteraceae; HAM, Hamamelidales; TRO, Trochodendrales; RAN, Ranunculidae; NEL, Nelumbonaceae; LAC, Lactoridaceae; ARI, Aristolochiaceae; PIP, Piperaceae; SAU, Saururaceae; NYM, Nymphaeaceae; CAB, Cabombaceae; LIL, Liliopsida (monocots); WINTS, winteroids; EUDICOTS, eudicots (taxa with tricolpate and derived pollen).

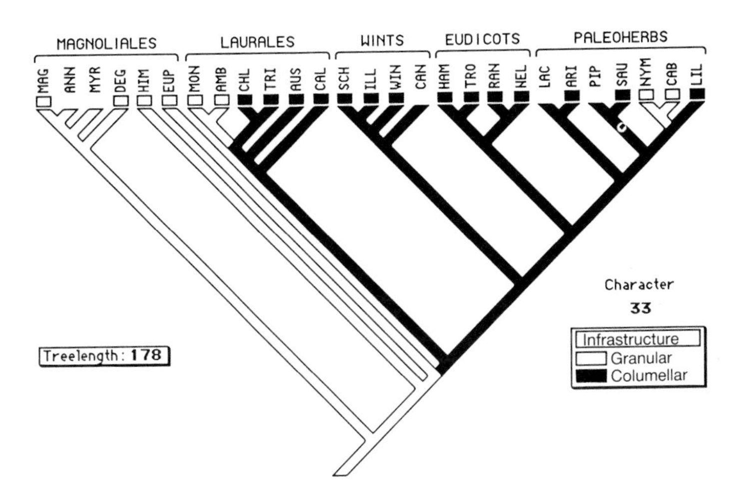

Fig. 9.3 Alternative most parsimonious cladogram of Donoghue and Doyle (1989) rooted within Magnoliales, with the distribution of granular v. columellar infratectal structure. See Fig 9.2 for explanation and abbreviations of taxa.

Walker and Walker (1984) adduced strong EM evidence for chloranthaceous affinities of '*Clavatipollenites*' and two related Albian forms, *Asteropollis asteroides* (polychotomosulcate) and *Stephanocolpites fredericksburgensis* (polycolpate). *Ascarina* matches '*Clavatipollenites*' in size, shape, spinules, sulcus sculpture, and thick nexine (although there is some disagreement over presence of non-apertural endexine in the two taxa: Chlonova and Surova 1988), whereas *Hedyosmum* and *Chloranthus* resemble *Asteropollis* and *S. fredericksburgensis*, respectively. However, although the apomorphic aperture conditions of *Asteropollis* and *S. fredericksburgensis* suggest that they are nested within Chloranthaceae, '*Clavatipollenites*' and *Ascarina* have the presumed ancestral pollen type for the family, so '*Clavatipollenites*' could represent a stem-group taxon more primitive than modern Chloranthaceae. This is consistent with the recent association of '*Clavatipollenites*' pollen (resembling '*C.*' *rotundus* Kemp and forms identified as '*Retimonocolpites*' *dividuus* Pierce) with fruits similar to those of Chloranthaceae but differing in having anatropous rather than orthotropous ovules (Friis *et al.* Chapter 10, this volume). Referring to Fig. 9.4, spinules go back to the common ancestor of Chloranthaceae, Trimeniaceae, *Amborella*, and higher Laurales, while aperture sculpture is a synapomorphy of the whole order. However, open reticulate sculpture and thick nexine (not included in the data matrix because it is an autapomorphy of Chloranthaceae) serve to link '*Clavatipollenites*' directly with Chloranthaceae.

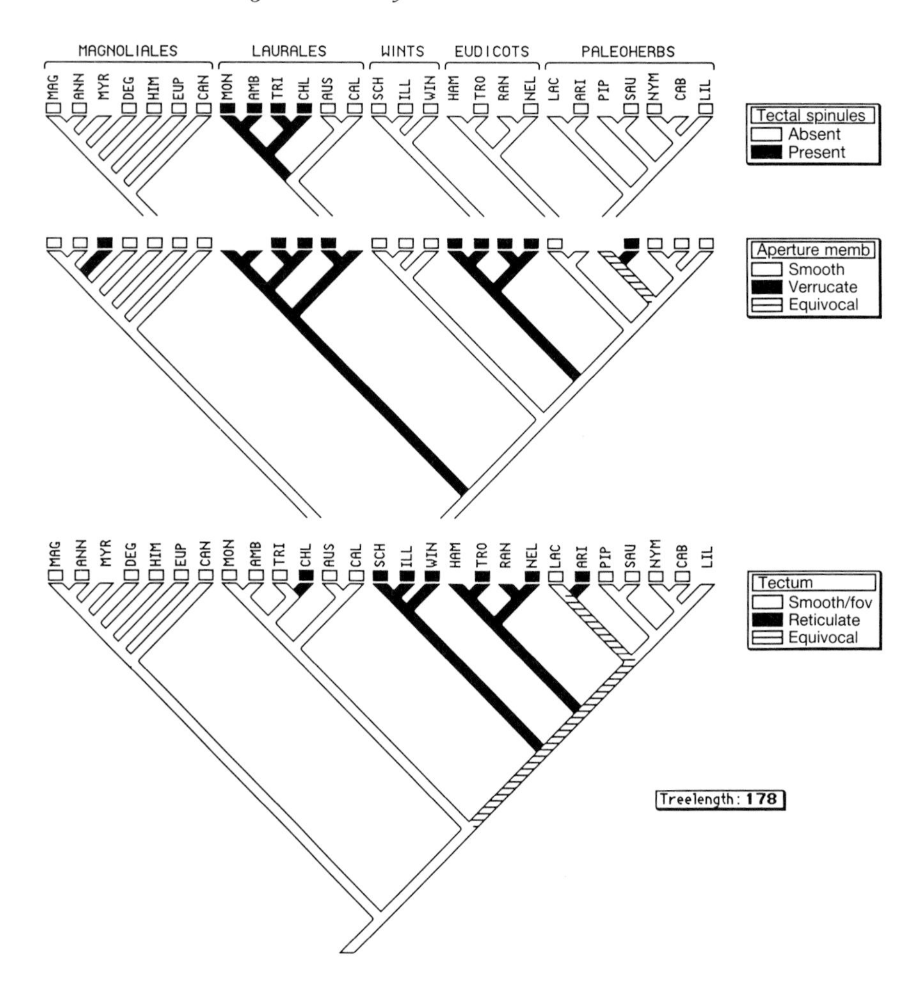

Fig. 9.4 Most parsimonious cladogram of Donoghue and Doyle (1989), with the distribution of characters relevant to placement of '*Clavatipollenites*', *Tucanopollis*, and *Brenneripollis*. See Fig. 9.2 for explanation and abbreviations of taxa.

'*Clavatipollenites*' also approaches some Myristicaceae, which have small, reticulate monosulcate pollen with supratectal spinules and sulcus sculpture (Walker and Walker 1979). This is not evident from Fig. 9.4, because Myristicaceae include both granular and columellar taxa and were therefore scored as unknown for exine structure (their position in Magnoliales supports the view that they are basically granular: Walker and Walker 1981). However, Myristicaceae differ from '*Clavatipollenites*' in their more elongate shape and lack of endexine. As noted by Walker and Walker (1984), certain other Early Cretaceous monosulcates are more like Myristicaceae.

A possible extinct lauralian group consists of tectate pollen with supratectal spinules and a sculptured, sometimes circular aperture, from the Barremian–Aptian of Northern Gondwana (Africa–South America). This group was first described from Brazil by Regali *et al.* (1974) as *Inaperturopollenites crisopolensis*, recognized as angiospermous by Doyle *et al.* (1977), and transferred to the new genus *Tucanopollis* by Regali (1989). Somewhat similar forms, but with a smoother tectum and less sculptured sulcus, were described from the Albian of Hungary by Góczán and Juhász (1984) as *Transitoripollis*. TEM shows that *Tucanopollis* resembles '*Clavatipollenites*' in having a thin tectum, thick footlayer, and endexine below the sulcus (Fig. 9.1d,e). Infratectal structure varies from columellar to granular, which was taken by Doyle *et al.* (1979) as evidence that granular structure is primitive. However, if *Tucanopollis* is associated with Chloranthaceae, which are nested among columellar groups (Fig. 9.2), its granular condition is probably a reversal. The best position for *Tucanopollis* is above the origin of the thick footlayer but below the origin of reticulate sculpture. Status of its imperforate tectum is equivocal, since Donoghue and Doyle (1989) lumped tectate perforate and imperforate as one state. However, the tectum varies from finely perforate in Calycanthaceae and some Trimeniaceae to almost reticulate in other Trimeniaceae and *Austrobaileya* (Endress and Honegger 1980; Sampson and Endress 1984). This suggests that perforate is basic in Laurales and both the imperforate condition of *Tucanopollis* and the reticulate condition of '*Clavatipollenites*' and Chloranthaceae are derived.

These considerations raise the possibility that *Tucanopollis* does not belong in Laurales at all, but rather near Piperales, which also have round, tectate pollen, sometimes with supratectal spinules and a sculptured sulcus (Walker 1976*a*; Chapman 1987; Bornstein 1989). However, pollen of Piperales is smaller, and TEM shows that although *Saururus* has a rather thick nexine, the tectum is thicker than in '*Clavatipollenites*' and *Tucanopollis*, and a thin endexine occurs all around the grain (Fig. 9.1g). Placement of *Tucanopollis* in Piperales would therefore imply that it converged with Chloranthaceae in the latter characters. On pollen characters alone, *Tucanopollis* might be considered a link between Piperales and Chloranthaceae. However, this must be weighed against the greater number of conflicting characters in the extant groups.

Another possible extinct lauralian group consists of coarsely reticulate Aptian–Albian monosulcates with spinulose muri but no columellae (Fig. 9.1f), studied with EM by Doyle *et al.* (1975), Walker and

Walker (1984), and Penny (1988). These were assigned to *Peromonolites* by Brenner (1963) but transferred to *Retimonocolpites* by Doyle *et al.* (1975) and finally to *Brenneripollis* by Juhász and Góczán (1985). Penny (1988) questioned use of *Brenneripollis* because the type species is described as having columellae, but the light micrographs do not convince us that it does, and in any case all these forms seem related on other characters. Under LM and SEM, *Brenneripollis* resembles pollen of the monocot *Aponogeton* (Thanikaimoni 1985), but *Aponogeton* has a thin nexine, whereas *Brenneripollis* has a thick footlayer and endexine below the sulcus, like '*Clavatipollenites*' and Chloranthaceae. The lack of columellae does not contradict lauralian affinities, since it is presumably an autapomorphic loss, quite unlike the original granular state. *Brenneripollis* lacks sculpture on the sulcus, which could be evidence that it belongs elsewhere. However, reduction in sulcus sculpture and number of columellae is also seen in the '*Clavatipollenites*' *rotundus* group, which is associated with chloranthoid fruits (Friis *et al.* Chapter 10, this volume). This suggests that *Brenneripollis* and '*C.*' *rotundus* may represent an extinct sister taxon of Chloranthaceae, derived from a common ancestor with typical '*Clavatipollenites*' pollen, which reduced its columellae while coarsening its exine sculpture.

Winteroids

The winteroid clade is clearly represented by tetrads from the late Aptian–early Albian of Israel (Walker *et al.* 1983) and the late Barremian–early Aptian of Gabon (Zone C-VII, formerly considered Aptian: Doyle *et al.* 1990*a*; Fig. 9.5a), based on the distinctively winteraceous tetrad condition and ulcerate, annulate aperture, with a ring of granulate sculpture and thick endexine around a central

Fig. 9.5 (a) *Walkeripollis gabonensis* Doyle *et al.*, ulcerate tetrad, single-grain preparation 2963–27, Elf-Aquitaine N'Toum No. 1 well, Gabon, 939–944 m, Zone C-VII, late Barremian–early Aptian? (Doyle *et al.* 1990*a*), SEM, × 1800; (b) *Afropollis operculatus* Doyle *et al.*, single-grain preparation 2963–16, N'Toum No. 1 well, 939–944 m, SEM, × 1600; (c) and (d) spiraperturate tricolpate relative, single-grain preparation 2968–15, N'Toum No. 1 well, 1629–1637 m, Zone C-VII, late Barremian–early Aptian? (Doyle *et al.* 1977), SEM, two sides of same grain, × 2200; (e) *Liliacidites* sp., two cohering grains, slide 71–8–ld, Trent's Reach, Virginia, lower Zone I, early Aptian? (Doyle and Robbins 1977), LM, × 1000; (f) and (g) *Similipollis* sp., with extended sulcus, single-grain preparation 2963–12, N'Toum No. 1 well, 939–944 m, Zone C-VII, late Barremian–early Aptian?, LM, distal and proximal views, × 1000; (h) same grain, SEM, proximal view, × 2200.

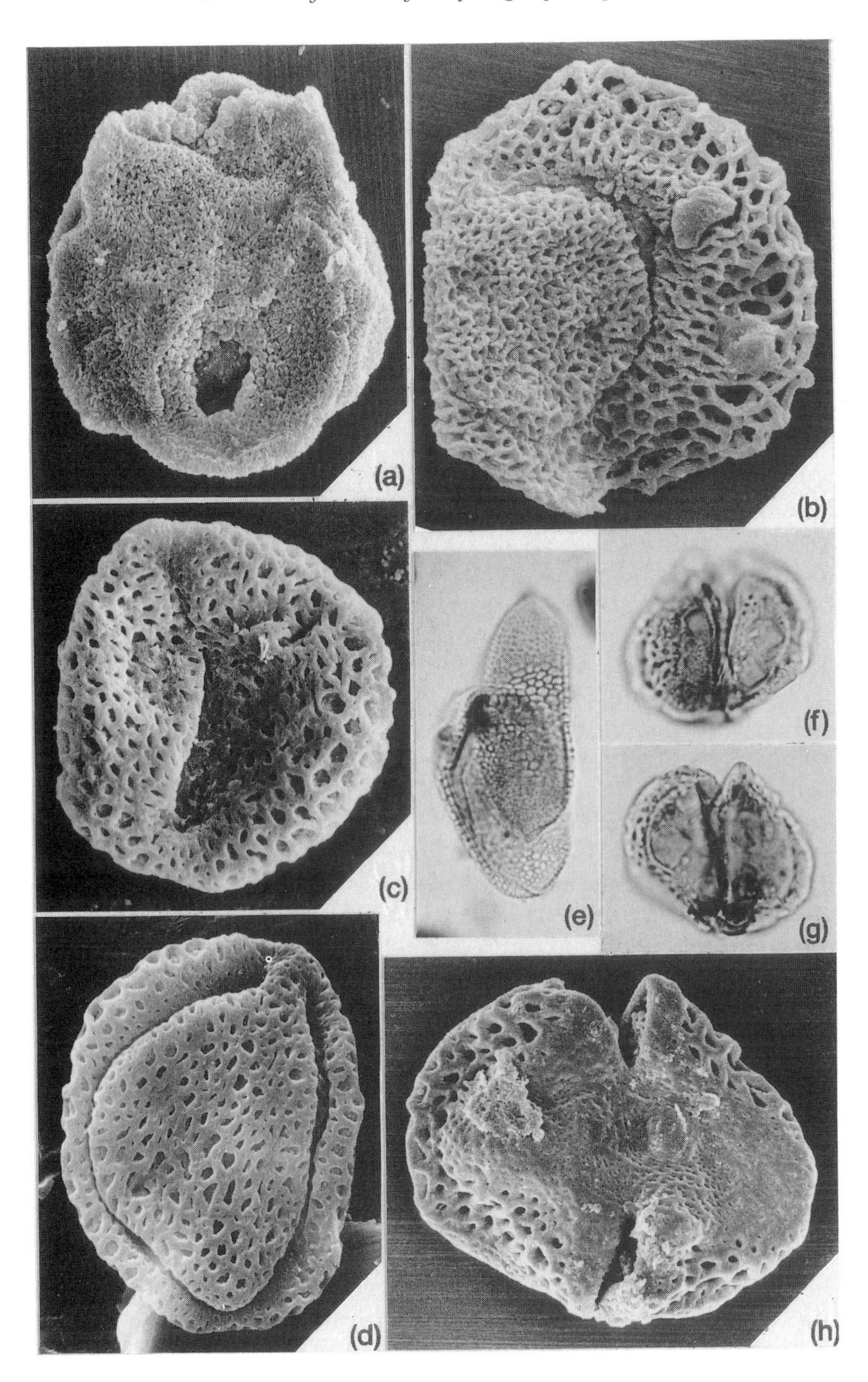
(a)
(b)
(c)
(d)
(e)
(f)
(g)
(h)

pore (*Walkeripollis* Doyle *et al.*). Ironically, these grains lack the one pollen character that unites Winteraceae and Illiciales in the analysis of Donoghue and Doyle (1989), coarsely reticulate sculpture, since they are foveolate. However, in addition to their winteraceous characters, they do have another exine feature of modern winteroids that Donoghue and Doyle omitted: relatively tall muri and short columellae.

The older tetrads (*Walkeripollis gabonensis*) are significant in having transversely ridged or segmented muri, a feature shared with two Cretaceous monad groups with no close modern analogues: *Afropollis* Doyle *et al.* (1982) from the late Barremian to Cenomanian throughout Northern Gondwana, which is spheroidal and distally or equatorially zonasulculate or inaperturate (Fig. 9.5b), and *Schrankipollis* Doyle *et al.* (1990*a*) from the Aptian of Egypt and the Potomac Group, which is elliptical and zonasulculate ('*Retimonocolpites*' *mawhoubensis* Schrank 1983; '*Schizosporis*' *microreticulatus* Brenner 1963). Both genera also resemble winteroids in having tall muri and short columellae. According to a cladistic analysis of both the fossils and modern winteroid pollen types (Doyle *et al.* 1990*b*; Fig. 9.6), *Schrankipollis* and *Afropollis* are the sister group of *Walkeripollis*, Winteraceae, and Illiciales, derived from a finely sculptured common ancestor with either an ulcerate or a zonasulculate aperture.

A surprising implication of Fig. 9.6 is that the monads of Illiciales are derived from ulcerate tetrads (by separation). Illiciales have three colpi, often fused at the distal pole, plus three alternating shorter colpi in most Schisandraceae. Derivation from tetrads explains a previously puzzling feature, thinning of the exine at the proximal pole in *Schisandra*. Since Recent data would imply that the common ancestor of Winteraceae and Illiciales had monads, this illustrates the principle that fossils may affect ideas on character evolution even when branching relationships are unaltered (Doyle and Donoghue 1987). These results are also interesting biogeographically: today Winteraceae are austral temperate to upland tropical and Illiciales are Laurasian, but Early Cretaceous members of the clade are centred on the palaeoequator in Northern Gondwana, suggesting that the modern taxa are temperate offshoots of an originally tropical group.

These results may also help resolve the trichotomy at the base of the columellate clade. Fig. 9.6 implies that the winteroid clade was originally foveolate, and we have noted that the basic condition in Laurales may be similar, as in *Austrobaileya*. In both lines, pollen would be medium-sized and almost round. Judging from extant members, the sulcus would be sculptured in Laurales but non-sculptured in winteroids. However, the annulus of *Walkeripollis gabonensis* is more or less verrucate. Since apertural sculpture is lacking or seems derived

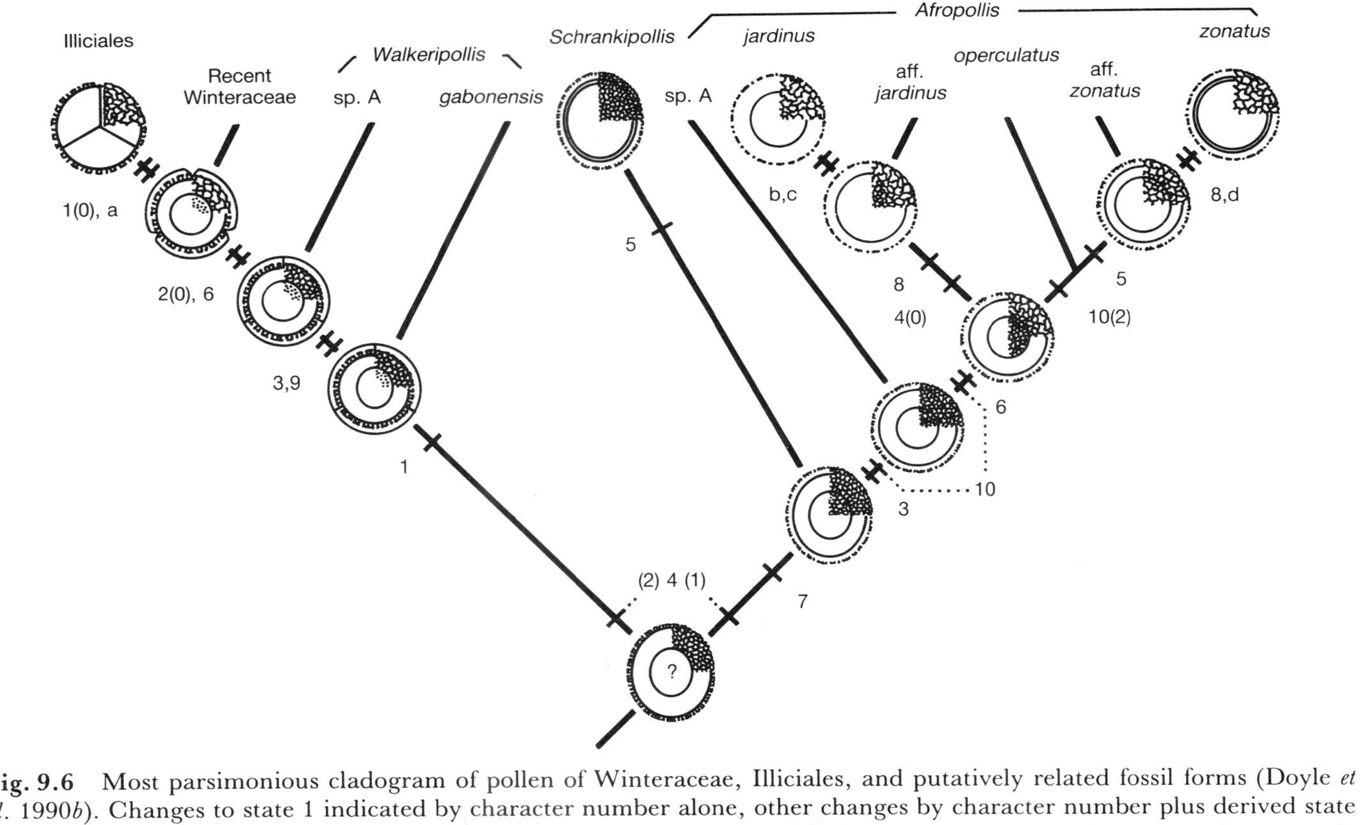

Fig. 9.6 Most parsimonious cladogram of pollen of Winteraceae, Illiciales, and putatively related fossil forms (Doyle *et al.* 1990*b*). Changes to state 1 indicated by character number alone, other changes by character number plus derived state in parentheses, ambiguous character placements by dotted lines. 1, monads–tetrads; 2, tetrads acalymmate–calymmate; 3, symmetry bilateral–radial; 4, pollen sulcoidate–zonate–ulcerate (unordered); 5, aperture distal–(sub)equatorial; 6, reticulum fine–coarse; 7, reticulum attached–detached; 8, columellae present–absent; 9, muri segmented–smooth; 10, endexine thin–medium–thick (ordered); a, tricolpate aperture condition; b, reduction of aperture; c, reduction of nexine; d, equatorial zonasulculus.

in other clades (cf. Fig. 9.4 and below), it could therefore be a synapomorphy uniting Laurales and winteroids, as the sister group of eudicots and paleoherbs. The first columellates might also have had foveolate sculpture but smooth muri and a smooth sulcus. Early Cretaceous monosulcates show many combinations of these features, and detailed study of these forms could test this hypothesis. Canellaceae, which vary from granular to reticulate (Wilson 1964; Walker 1976*b*), could also be informative, since they are basal members of the winteroid clade rather than Magnoliales in some most parsimonious trees of Donoghue and Doyle (1989; Fig. 9.3).

Eudicots

Most trees of Donoghue and Doyle (1989; Figs 9.2–9.4) imply that tricolpate pollen arose twice: in Illiciales and in hamamelids (Trochodendrales, Hamamelidales) plus ranunculids (*Nelumbo*, Ranunculidae). Other trees split hamamelids and ranunculids and associate the latter with paleoherbs, implying that tricolpate pollen arose three times (or arose twice and was lost in paleoherbs, a less plausible scenario on functional grounds). These results are supported by the observation of Huynh (1976) that the three colpi in Illiciales (or three fused colpi in hexacolpate grains) are oriented according to Garside's rule, not Fischer's rule, as in most dicots. Donoghue and Doyle (1989) assumed that the ranunculid–hamamelid clade also includes all dicot subclasses other than Magnoliidae (*sensu* Takhtajan 1980). They called this clade 'tricolpates', but most members have more advanced pollen, and Illiciales are also tricolpate. Others have used the term 'non-magnoliid dicots' (e.g. Walker and Doyle 1975; Crane 1989), but this defines a putatively monophyletic group by its non-membership in a non-monophyletic (paraphyletic) group. We therefore propose the term 'eudicotyledons' or 'eudicots', which recognizes that this group conforms to the concept of a dicot held by most non-specialists and emphasizes its proposed monophyletic status. Tricolpate pollen appears near the Barremian–Aptian boundary in Northern Gondwana, although most modern hamamelids and ranunculids are Laurasian. Most Early Cretaceous tricolpates clearly represent eudicots, since they lack illicialian features (tall muri, fused colpi with a distinctive median ridge) and have been isolated from Albian flowers referable to hamamelids (Crane *et al.* 1986; Friis *et al.* Chapter 10, this volume).

The oldest tricolpates from the late Barremian–early Aptian of Gabon (Zone C-VII, Doyle *et al.* 1977; Doyle and Ward, in preparation) may provide hints on the origin and relationships of eudicots. Within one species as defined on other characters, grains vary from almost radially symmetrical, to tricolpate with oblique, unevenly spaced colpi, to spiraperturate, with a single sinuous furrow resembling three colpi joined at the ends (Fig. 9.5c, d). The aperture variations suggest that tricolpate pollen was derived not via trichotomosulcate or inaperturate intermediates (Straka 1963; Wilson 1964; Muller 1970; Walker 1974), but by an unsuspected process of extension, twisting, and fragmentation of a distal sulcus. There is no clear vestige of this process in extant eudicots: Berberidaceae *s. str.* have spiraperturate pollen (Furness 1985), but they have granular exines and appear to be derived within ranunculids (Meacham 1980; Loconte and Estes 1989). Aperture irregularities in *Nelumbo* (Kuprianova 1979) could be more relevant. However, this evidence does further weaken any relationship with Illiciales, where the Garside's rule arrangement of the colpi suggests derivation from trichotomosulcate pollen, in which the three arms of the aperture are also oriented according to Garside's rule (Walker 1974; Huynh 1976; Liu and Yang 1989; Doyle *et al.* 1990*b*).

Other relevant characters of the Gabon tricolpates concern exine sculpture. Walker and Walker (1984) linked tricolpate Hamamelidae with Chloranthaceae and Piperales because they often have supratectal spinules and sculptured apertures, as do many ranunculids. According to Fig. 9.4, spinules were originally lacking in Laurales and eudicots [muri are smooth in *Nelumbo* and Trochodendrales (Endress 1986), and Hamamelidales and Ranunculales have some members with spinules and some without], but sculptured apertures are basic in both groups. However, the Gabon tricolpates have neither supratectal spinules nor apertural sculpture. Assuming that these forms are related to and more primitive than modern tricolpates, a view supported by intermediates in the later Aptian and Albian (Doyle and Ward, in preparation), they further weaken any relationship of eudicots with Laurales and by default strengthen ties with paleoherbs. Sculptured colpi become common in the Albian; even if they are not basic in eudicots, they could unite a subgroup including modern hamamelids and ranunculids.

Besides lacking lauralian advances, the Gabon tricolpates show two more positive similarities with paleoherbs: their sculpture consists of intermixed large and small lumina (heterobrochate) and varies from coarser at the equator to finer at the poles (graded). Similar sculpture occurs in many Albian tricolpates (*Rousea* spp.) and modern Trochodendrales (Endress 1986), but it is found elsewhere only in the largest paleoherb group, the monocots (Doyle 1973; Walker and Walker 1986). However, evaluation of its significance depends on relationships among paleoherbs.

Paleoherbs

Paleoherbs are most clearly represented in the Early Cretaceous pollen record by monocot-like, boat-shaped monosulcates with sculpture grading from fine at the ends of the grain to coarse elsewhere (*Liliacidites*), which extend to the early Aptian in the Potomac Group (Doyle 1973; Ward *et al.* 1989; Fig. 9.5e). Other monocot-like features include heterobrochate lumina, smooth muri, and a thin nexine consisting of footlayer (Walker and Walker 1986). This may be the basic pollen type in monocots, since it occurs in several potentially primitive groups (*Butomus*, Liliales, Araceae: Thanikaimoni 1969; Doyle 1973). However, other paleoherbs are different. Piperales have very small, round, tectate, spinulose pollen with a sculptured sulcus, noted above as converging with Chloranthaceae. *Lactoris* has tetrads of tectate grains that might be derived from the same type. In Aristolochiaceae, the primitive genus *Saruma* has reticulate monosulcate pollen with smooth muri, rather suggestive of some Barremian–Aptian forms, but the reticulum is uniform (Long Huo, personal communication). Nymphaeales have large, boat-shaped, granular monosulcate or derived pollen, like pollen of Magnoliales but with endexine (Walker 1976*b*).

Given the trees in Figs 9.2–9.4, graded reticulate sculpture is unlikely to be homologous in eudicots and monocots. Donoghue and Doyle (1989) scored monocots as unknown for sculpture, so some modification is needed to address this question. If the sculpture character is redefined as having three states, smooth/foveolate, even reticulate, and graded reticulate, and eudicots and monocots are scored as graded reticulate, it is two steps less parsimonious to assume that graded reticulate is homologous in eudicots and monocots and was lost in other paleoherbs than it is to assume that graded reticulate arose twice (Fig. 9.7a,b). However, relationships among paleoherbs are poorly resolved; other equally parsimonious trees link Aristolochiaceae, *Lactoris*, and Piperales, in which case the differential between the two scenarios shrinks to one step (Fig. 9.7c,d). Addition or reinterpretation of characters in modern paleoherbs or discovery of fossil paleoherbs with new character combinations might shift the balance further.

A Barremian–Albian pollen group that deserves closer attention in this context is *Similipollis* Góczán and Juhász (1984), in which we include grains previously designated *Liliacidites* but with finer sculpture at the sulcus margins and proximal pole rather than the ends of the grain (Doyle 1973; Doyle *et al.* 1977; Walker and Walker 1986). We do not know any examples of this sculpture gradient relative to the polar axis in extant monocots, but it does correspond to the gradient in *Rousea*-type tricolpates. Interestingly, some *Similipollis* grains in the late Barremian–early Aptian of Gabon show extension of the ends

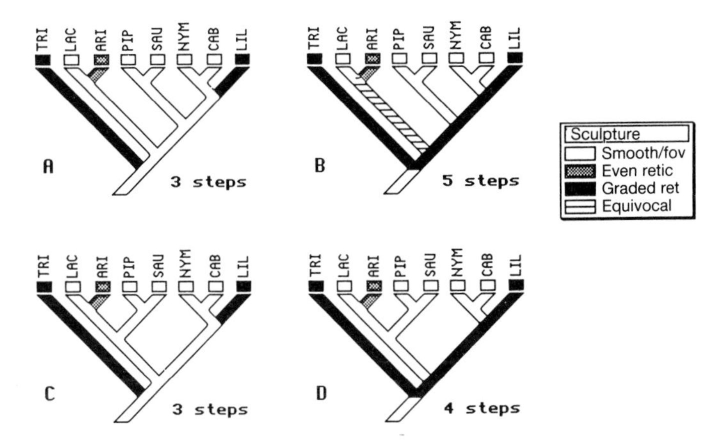

Fig. 9.7 Evaluation of homology of graded sculpture in tricolpates and monocots: alternative mappings of character-state changes given two equally parsimonious arrangements of paleoherbs found by Donoghue and Doyle (1989), with numbers of steps in the sculpture character. Abbreviations as in Fig. 9.2, except TRI, tricolpates.

of the sulcus (Fig. 9.5f–h), as hypothesized above in the origin of tricolpates. *Similipollis* is also interesting biogeographically: whereas tricolpates apparently originated in Gondwana and typical *Liliacidites* is concentrated in Laurasia (cf. Walker and Walker 1986), *Similipollis* occurs in the Barremian–Aptian of both areas (Doyle *et al.* 1977).

Crinopolles, paleoherbs, and the rooting problem

So far, we have seen no clear stratigraphic evidence on which of the five major angiosperm clades is most primitive. Late Barremian winteroids and eudicots are only slightly younger than Laurales, possible paleoherbs, and diverse unassignable columellates. Magnoliales are not recognized before the Aptian, but this is hardly evidence against their postulated basal status: smooth monosulcates are abundant throughout the Mesozoic, and TEM is needed to sort them into angiosperms and Bennettitales.

Recently, however, there have been pre-Barremian reports that may change this situation. Specifically, these may favour trees rooted in or near paleoherbs, which are supported by molecular data and only one or two steps less parsimonious than trees rooted near Magnoliales in terms of morphology (Donoghue and Doyle 1989; Zimmer *et al.* 1989). Trevisan (1988, personal communication) reported monosulcates from the Valanginian of Italy with graded reticulate sculpture.

Potentially even more significant are Late Triassic reticulate-columellar pollen types described by Cornet (1989) as the Crinopolles group. The reticulum grades from coarse proximally to fine distally, reminiscent of *Liliacidites* and *Similipollis* but not exactly like either. Most Crinopolles are monosulcate and pontoperculate (resulting in two parallel furrows), but some have additional lateral furrows suggestive of the 'striations' of Gnetales, which Cornet took as evidence for a relationship between angiosperms and Gnetales. TEM reveals a thick endexine, which Cornet considered angiosperm-like because it is non-laminated.

We have confirmed these features in Crinopolles grains from the Late Triassic Chinle Formation of Arizona (Fig. 9.8). However, we are not convinced by all of Cornet's arguments. First, endexine laminations are not always preserved in fossil pollen (Doyle 1978), and they seem especially inconspicuous in Bennettitales (Taylor 1973; Ward *et al.* 1989). Secondly, the uniform thickness of the endexine is not angiosperm-like: in monosulcate angiosperms, endexine is either thin or lacking except at the aperture, and in this regard Crinopolles are gymnosperm-like. This does not rule out a relationship between Crinopolles and angiosperms, but it does suggest that if the two groups are related, Crinopolles are more primitive than extant angiosperms, which would make them more rather than less significant. Specifically, they would imply that columellae originated before the endexine was lost (or transformed into footlayer), opposite the order of events postulated by Walker and Walker (1984) based on granular Magnoliales. Conversely, their monocot-like sculpture might favour a primitive status for paleoherbs. Independent evidence on the plants producing Crinopolles pollen is therefore of high priority for understanding the origin of angiosperms.

Conclusions

The points made in this paper are summarized in Fig. 9.9, with fossil pollen types placed on a cladogram of extant groups, showing the two alternative rootings. If the ancestors of angiosperms had monosulcate pollen like that of Bennettitales, the basic angiosperm pollen type would probably be granular, round, and medium-sized. This gave rise, on the one hand, to the larger, boat-shaped *Lethomasites* type, representing core Magnoliales, and, on the other, to foveolate, columellar monosulcates. Fig. 9.9 incorporates our speculation that the latter split into two lines. One, with a verrucate sulcus, gave rise to Laurales, represented by '*Clavatipollenites*' and possibly *Tucanopollis* and *Brenneripollis*, and winteroids, represented by *Afropollis*, *Schranki-*

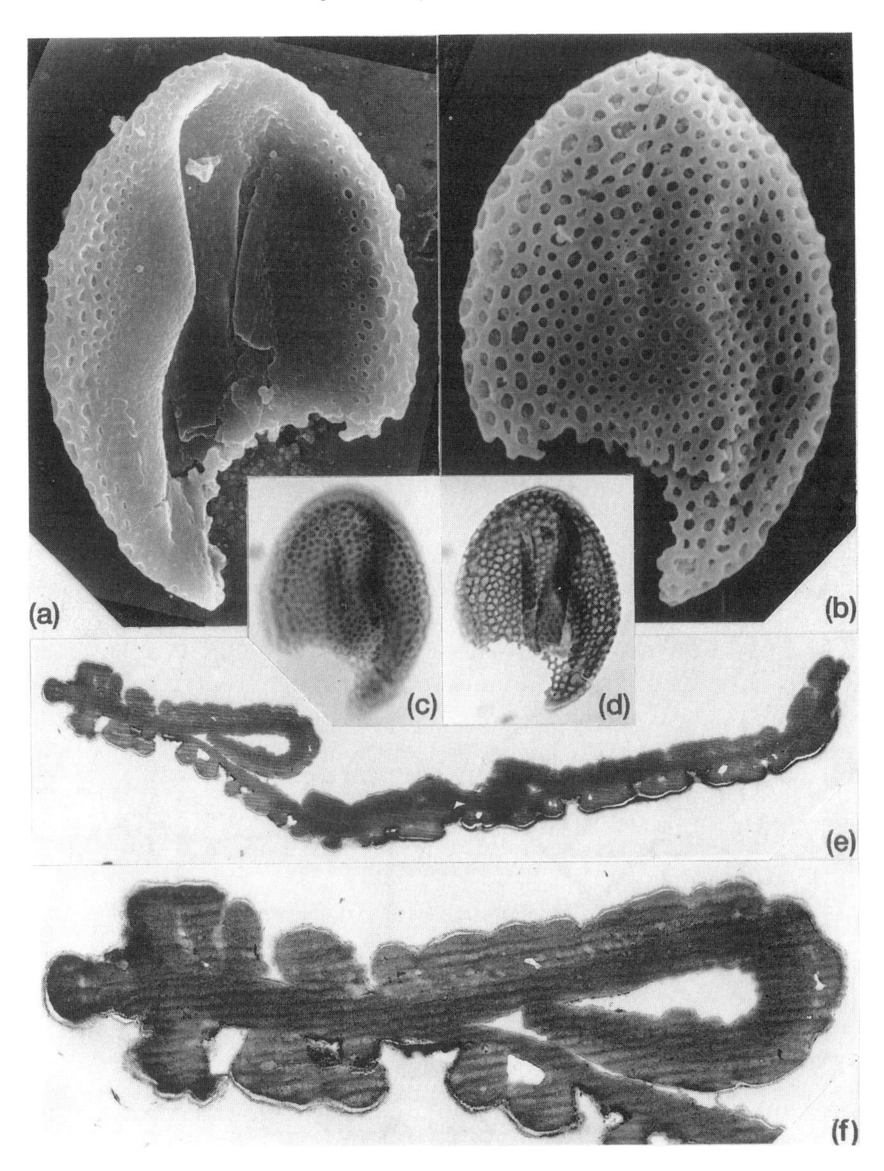

Fig. 9.8 *Monocrinopollis microreticulatus* Cornet, single-grain preparation CH277-5, Petrified Forest Member, Chinle Formation, Arizona (University of Michigan Museum of Paleontology No. 23021), Late Triassic. (a) SEM, distal view, showing pontoperculate (double) sulcus, × 2500; (b), proximal view, showing variation in size of lumina, × 2500; (c) and (d), LM, proximal view, × 1000; (e), TEM, section of whole grain, × 2500; (f), exine section, showing thick, darkly staining endexine and ectexine thinning toward the sulcus, × 16 000.

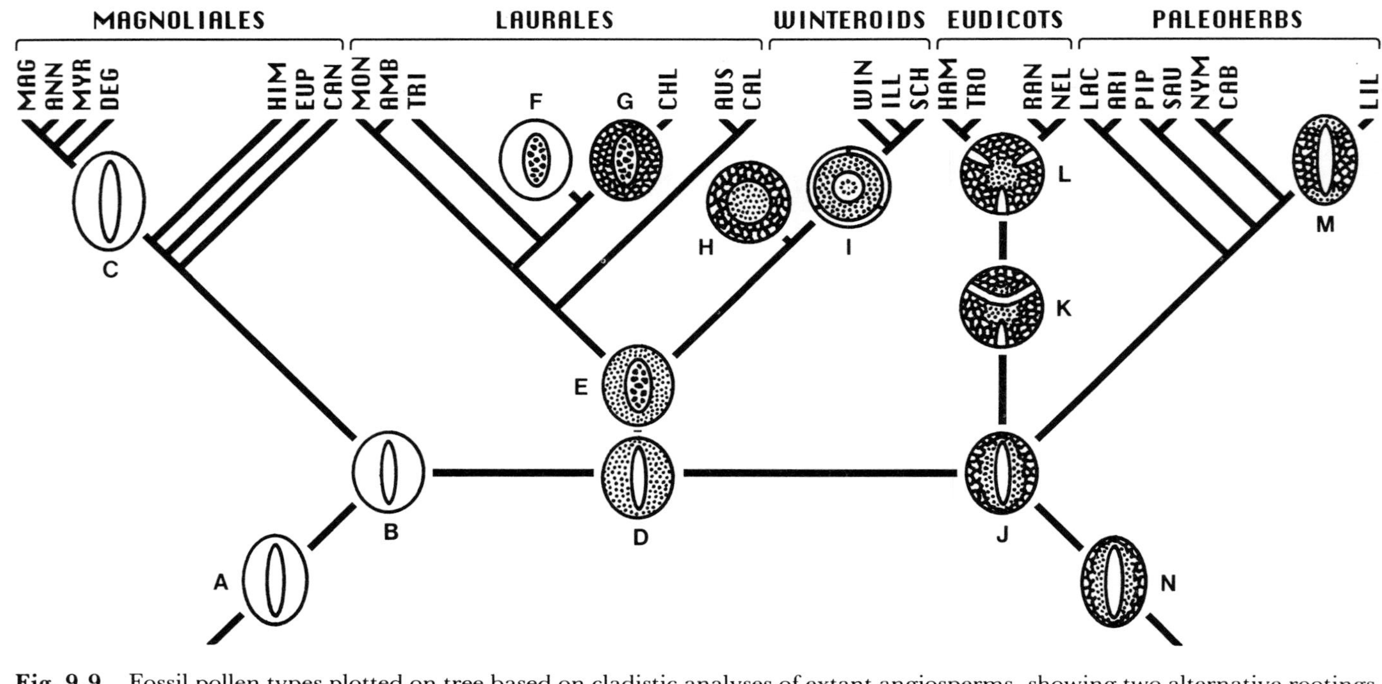

Fig. 9.9 Fossil pollen types plotted on tree based on cladistic analyses of extant angiosperms, showing two alternative rootings. See Fig. 9.2 for abbreviations of taxa. A, bennettitalian-type monosulcate; B, hypothetical medium-sized, round, granular monosulcate; C, *Lethomasites*; D, hypothetical foveolate monosulcate with unsculptured sulcus; E, hypothetical foveolate monosulcate with sculptured sulcus; F, *Tucanopollis*; G, '*Clavatipollenites*'; H, *Afropollis*; I, *Walkeripollis*; J, *Similipollis*; K, Gabon tricolpate relative; L, typical Albian tricolpate; M, *Liliacidites*; N, Crinopolles group.

pollis, and *Walkeripollis* tetrads. The other, with graded reticulate sculpture of the *Similipollis* type, gave rise to paleoherbs, represented by *Liliacidites*, and tricolpate eudicots, via spiraperturate intermediates. If, however, the ancestors of angiosperms had pollen of the Crinopolles type, it may be more parsimonious to rotate and reroot the tree so that reticulate paleoherbs and eudicots are basal, and Magnoliales are a specialized line that became secondarily granular and converged with Bennettitales. We are hopeful that EM studies of Early Cretaceous, Jurassic, and Triassic pollen will help test these hypotheses.

Acknowledgements

We wish to thank Jerome Ward for use of unpublished data and Peter Crane and Michael Donoghue for useful discussions. This work was supported in part by NSF grant BSR-8415772.

References

Ax, P. (1987). *The phylogenetic system: the systematization of organisms on the basis of their phylogenesis*. Wiley, New York.

Bornstein, A.J. (1989). Taxonomic studies in the Piperaceae—I. The pedicellate pipers of Mexico and Central America (*Piper* subg. *Arctottonia*). *Journal of the Arnold Arboretum* **70**, 1-55.

Brenner, G.J. (1963). The spores and pollen of the Potomac Group of Maryland. *Maryland Department of Geology, Mines and Water Resources Bulletin* **27**, 1-215.

Chapman, J. L. (1987). Comparison of Chloranthaceae pollen with the Cretaceous "*Clavatipollenites* complex". Taxonomic implications for palaeopalynology. *Pollen et Spores* **24**, 249–72.

Chlonova, A. F. and Surova, T. D. (1988). Pollen wall ultrastructure of *Clavatipollenites incisus* Chlonova and two modern species of *Ascarina* (Chloranthaceae). *Pollen et Spores* **30**, 29–44.

Cornet, B. (1989). Late Triassic angiosperm-like pollen from the Richmond rift basin of Virginia, U.S.A. *Palaeontographica Abteilung B* **213**, 37–87.

Couper, R.A. (1958). British Mesozoic microspores and pollen grains. *Palaeontographica Abteilung B* **103**, 75–179.

Crane, P.R. (1985). Phylogenetic analysis of seed plants and the origin of angiosperms. *Annals of the Missouri Botanical Garden* **72**, 716–93.

Crane, P.R. (1989). Paleobotanical evidence on the early radiation of non-magnoliid dicotyledons. *Plant Systematics and Evolution* **162**, 165–91.

Crane, P.R., Friis, E.M., and Pedersen, K.R. (1986). Lower Cretaceous angiosperm flowers: fossil evidence on early radiation of dicotyledons. *Science* **232**, 852–4.

Donoghue, M.J. and Doyle, J.A. (1989). Phylogenetic analysis of angiosperms and the relationships of Hamamelidae. In *Evolution, systematics, and fossil history of the Hamamelidae*, (ed. P.R. Crane and S. Blackmore), pp. 17–45. Clarendon Press, Oxford.

Donoghue, M.J., Doyle, J.A., Gauthier. J., Kluge, A.G., and Rowe, T. (1989). The importance of fossils in phylogeny reconstruction. *Annual Review of Ecology and Systematics* **20**, 431–60.

Doyle, J.A. (1969). Cretaceous angiosperm pollen of the Atlantic Coastal Plain and its evolutionary significance. *Journal of the Arnold Arboretum* **50**, 1–35.

Doyle, J.A. (1973). Fossil evidence on early evolution of the monocotyledons. *Quarterly Review of Biology* **48**, 399–413.

Doyle, J.A. (1978). Potentials and limitations of exine structure in studies of early angiosperm evolution. *Courier Forschungsinstitut Senckenberg* **30**, 54–61.

Doyle, J.A. and Donoghue, M.J. (1986). Seed plant phylogeny and the origin of angiosperms: an experimental cladistic approach. *Botanical Review* **52**, 321–431.

Doyle, J.A. and Donoghue, M.J. (1987). The importance of fossils in elucidating seed plant phylogeny and macroevolution. *Review of Palaeobotany and Palynology* **50**, 63–95.

Doyle J.A. and Robbins, E.I. (1977). Angiosperm pollen zonation of the continental Cretaceous of the Atlantic Coastal Plain and its application to deep wells in the Salisbury Embayment. *Palynology* **1**, 43–78.

Doyle, J.A., Van Campo, M., and Lugardon, B. (1975). Observations on exine structure of *Eucommiidites* and Lower Cretaceous angiosperm pollen. *Pollen et Spores* **17**, 429–86.

Doyle, J.A., Biens, P., Doerenkamp, A., and Jardiné, S. (1977). Angiosperm pollen from the pre-Albian Cretaceous of Equatorial Africa. *Bulletin des Centres de Recherches Exploration-Production Elf-Aquitaine* **1**, 451–73.

Doyle, J.A., Jardiné, S., and Doerenkamp, A. (1979). Noncolumellar monosulcate angiosperm pollen from the Lower Cretaceous of Africa. *Botanical Society of America Miscellaneous Series Publication* **157**, 32.

Doyle, J.A., Jardiné, S., and Doerenkamp, A. (1982). *Afropollis*, a new genus of early angiosperm pollen, with notes on the Cretaceous palynostratigraphy and paleoenvironments of Northern Gondwana. *Bulletin des Centres de Recherches Exploration-Production Elf-Aquitaine* **6**, 39–117.

Doyle, J.A., Hotton, C.L., and Ward, J.V. (1990*a*). Early Cretaceous tetrads, zonasulculate pollen, and Winteraceae. I. Taxonomy, morphology, and ultrastructure. *American Journal of Botany* **77**, 1544–57.

Doyle, J.A., Hotton, C.L., and Ward, J.V. (1990*b*). Early Cretaceous tetrads, zonasulculate pollen, and Winteraceae. II. Cladistic analysis and implications. *American Journal of Botany* **77**, 1558–68.

Eldredge, N. and Novacek, M.J. (1985). Systematics and paleobiology. *Paleobiology* **11**, 65–74.

Endress, P.K. (1986). Floral structure, systematics, and phylogeny in Trochodendrales. *Annals of the Missouri Botanical Garden* **73**, 297–324.
Endress, P.K. (1987). The Chloranthaceae: reproductive structures and phylogenetic position. *Botanische Jahrbücher für Systematik* **109**, 153–226.
Endress, P.K. and Honegger, R. (1980). The pollen of Austrobaileyaceae and its phylogenetic significance. *Grana* **19**, 177–82.
Fisher, D.C. (1980). The role of stratigraphic data in phylogenetic inference. *Geological Society of America Abstracts with Programs* **12**, 426.
Friis, E.M., Crane, P.R., and Pedersen, K.R. (this volume). Stamen diversity and *in situ* pollen of Cretaceous angiosperms.
Furness, C.A. (1985). A review of spiraperturate pollen. *Pollen et Spores* **27**, 307–19.
Gauthier, J., Kluge, A.G., and Rowe, T. (1988). Amniote phylogeny and the importance of fossils. *Cladistics* **4**, 105–209.
Góczán, F. and Juhász, M. (1984). Monosulcate pollen grains of angiosperms from Hungarian Albian sediments I. *Acta Botanica Hungarica* **30**, 289–319.
Hennig, W. (1966). *Phylogenetic systematics.* University of Illinois Press, Urbana.
Hughes, N.F. and McDougall, A.B. (1987). Records of angiospermid pollen entry into the English Early Cretaceous succession. *Review of Palaeobotany and Palynology* **50**, 255–72.
Huynh, K.-L. (1976). L'arrangement du pollen du genre *Schisandra* (Schisandraceae) et sa signification phylogénique chez les Angiospermes. *Beiträge zur Biologie der Pflanzen* **52**, 227–53.
Juhász, M. and Góczán, F. (1985). Comparative study of Albian monosulcate angiosperm pollen grains. *Acta Biologica Szeged* **31**, 147–72.
Kuprianova, L.A. (1979). On the possibility of the development of tricolpate pollen from monosulcate. *Grana* **18**, 1–4.
Le Thomas, A. (1980). Ultrastructural characters of the pollen grains of African Annonaceae and their significance for the phylogeny of primitive angiosperms. *Pollen et Spores* **22**, 267–342.
Le Thomas, A. (1981). Ultrastructural characters of the pollen grains of African Annonaceae and their significance for the phylogeny of primitive angiosperms. *Pollen et Spores* **23**, 5–36.
Liu H. and Yang C.-S. (1989). Pollen morphology of Illiciaceae and its significance in systematics. *Chinese Journal of Botany* **1**, 104–15.
Loconte, H. and Estes, J.R. (1989). Phylogenetic systematics of Berberidaceae and Ranunculales (Magnoliidae). *Systematic Botany* **14**, 565–79.
Maddison, W.P. and Maddison, D.R. (1987). MacClade, version 2.1. A phylogenetics computer program distributed by the authors.
Meacham, C.A. (1980). Phylogeny of the Berberidaceae with an evaluation of classifications. *Systematic Botany* **5**, 149–72.
Muller, J. (1970). Palynological evidence on early differentiation of angiosperms. *Biological Reviews of the Cambridge Philosophical Society* **45**, 417–50.

Muller, J. (1981). Fossil pollen records of extant angiosperms. *Botanical Review* **47**, 1-142.

Patterson, C. (1981). Significance of fossils in determining evolutionary relationships. *Annual Review of Ecology and Systematics* **12**, 195-223.

Penny, J. H. J. (1988). Early Cretaceous acolumellate semitectate pollen from Egypt. *Palaeontology* **31**, 373-418.

Regali, M. S. P. (1989). *Tucanopollis*, um gênero novo das angiospermas primitivas. *Boletim de Geociências da Petrobrás* **3**, 395-402.

Regali, M. S. P., Uesugui, N., and Santos, A. S. (1974). Palinologia dos sedimentos meso-cenozóicos do Brasil. *Boletim Técnico da Petrobrás* **17**, 117-91, 263-301.

Sampson, F. B. and Endress, P. K. (1984). Pollen morphology of the Trimeniaceae. *Grana* **23**, 129-37.

Schaeffer, B., Hecht, M. K., and Eldredge, N. (1972). Phylogeny and paleontology. *Evolutionary Biology* **6**, 31-46.

Schrank, E. (1983). Scanning electron and light microscopic investigations of angiosperm pollen from the Lower Cretaceous of Egypt. *Pollen et Spores* **25**, 213-42.

Straka, H. (1963). Über die mögliche phylogenetische Bedeutung der Pollenmorphologie der madagassischen *Bubbia perrieri* R. Cap. (Winteraceae). *Grana Palynologica* **4**, 355-60.

Takhtajan, A. L. (1980). Outline of the classification of flowering plants (Magnoliophyta). *Botanical Review* **46**, 225-359.

Taylor, T. N. (1973). A consideration of the morphology, ultrastructure and multicellular microgametophyte of *Cycadeoidea dacotensis* pollen. *Review of Palaeobotany and Palynology* **16**, 157-64.

Thanikaimoni, G. (1969). Esquisse palynologique des Aracées. *Travaux de la Section Scientifique et Technique, Institut Français de Pondichéry* **5**, (5), 1-31.

Thanikaimoni, G. (1985). Palynology and phylogeny. *Bibliotheca Botanica* **137**, 11-14.

Trevisan, L. (1988). Angiospermous pollen (monosulcate-trichotomosulcate phase) from the very early Lower Cretaceous of Southern Tuscany (Italy): some aspects. *Seventh International Palynological Congress, Brisbane, Abstracts*, p. 165.

Walker, J. W. (1974). Aperture evolution in the pollen of primitive angiosperms. *American Journal of Botany* **61**, 1112-37.

Walker, J. W. (1976*a*). Comparative pollen morphology and phylogeny of the ranalean complex. In *Origin and early evolution of angiosperms*, (ed. C. B. Beck), pp. 241-99. Columbia University Press, New York.

Walker, J. W. (1976*b*). Evolutionary significance of the exine in the pollen of primitive angiosperms. In *The evolutionary significance of the exine*, (ed. I. K. Ferguson and J. Muller), pp. 251-308. Academic Press, London.

Walker, J. W. and Doyle, J. A. (1975). The bases of angiosperm phylogeny: palynology. *Annals of the Missouri Botanical Garden* **62**, 664-723.

Walker, J. W. and Walker, A. G. (1979). Comparative pollen morphology of the American myristicaceous genera *Compsoneura* and *Virola*. *Annals of the Missouri Botanical Garden* **66**, 731–55.

Walker, J. W. and Walker, A. G. (1981). Comparative pollen morphology of the Madagascan genera of Myristicaceae (*Mauloutchia*, *Brochoneura*, and *Haematodendron*). *Grana* **20**, 1–17.

Walker, J. W. and Walker, A. G. (1984). Ultrastructure of Lower Cretaceous angiosperm pollen and the origin and early evolution of flowering plants. *Annals of the Missouri Botanical Garden* **71**, 464–521.

Walker, J. W. and Walker, A. G. (1986). Ultrastructure of Early Cretaceous angiosperm pollen and its evolutionary implications. In *Pollen and spores: form and function*, Linnean Society Symposium Series 12, (ed. S. Blackmore and I. K. Ferguson), pp. 203–17. Academic Press, London.

Walker, J. W., Brenner, G. J., and Walker, A. G. (1983). Winteraceous pollen in the Lower Cretaceous of Israel: early evidence of a magnolialean angiosperm family. *Science* **220**, 1273–5.

Ward, J. V., Doyle, J. A., and Hotton, C. L. (1989). Probable granular magnoliid angiosperm pollen from the Early Cretaceous. *Pollen et Spores* **33**, 101–20.

Wilson, T. K. (1964). Comparative morphology of the Canellaceae. III. Pollen. *Botanical Gazette* **125**, 192–7.

Zavada, M. (1984). Angiosperm origins and evolution based on dispersed fossil pollen ultrastructure. *Annals of the Missouri Botanical Garden* **71**, 444–63.

Zimmer, E. A., Hamby, R. K., Arnold, M. L., LeBlanc, D. A., and Theriot, E. C. (1989). Ribosomal RNA phylogenies and flowering plant evolution. In *The hierarchy of life*, (ed. B. Fernholm, K. Bremer, and H. Jörnvall), pp. 205–14. Elsevier, Amsterdam.

10. Stamen diversity and *in situ* pollen of Cretaceous angiosperms

ELSE MARIE FRIIS
Department of Palaeobotany, Swedish Museum of Natural History, Stockholm, Sweden

PETER R. CRANE
Department of Geology, Field Museum of Natural History, Chicago, USA

KAJ RAUNSGAARD PEDERSEN
Department of Geology, University of Aarhus, Århus, Denmark

Abstract

Stratigraphic and systematic analyses of dispersed fossil pollen grains have played a major role in developing current ideas on the timing and pattern of early angiosperm evolution. In most cases other parts of the plants that produced these grains are unknown but recent studies have recovered diverse floral remains from many Cretaceous localities, and these frequently have pollen preserved within anthers or on stigmatic surfaces. In the mid-Cretaceous (Albian–Cenomanian), probable magnoliid and 'lower' hamamelidid stamens are particularly diverse and usually contain small, monocolpate or tricolpate-tricolporate pollen, generally with a relatively thick endexine and a prominent footlayer. Typically these stamens show only weak differentiation between anther and filament, have a massive connective with a pronounced apical expansion, and exhibit well-developed valvate dehiscence. All of the hamamelidid stamens for which other floral organs are known occur in small, unisexual flowers. In the later Cretaceous, there is a greater variety of stamens, which occur in a greater range of flowers and contain more diverse pollen types. Most of the later Cretaceous stamens contain very small to small, tricolpate or tricolporate pollen, often with a foveolate or psilate tectum and a thinner endexine than in the mid-Cretaceous

Pollen and Spores (ed. S. Blackmore and S. H. Barnes), Systematics Association Special Volume No. 44, pp. 197–224, Clarendon Press, Oxford, 1991.

taxa. Typically these stamens do not have a massive connective or a pronounced apical extension, the anther and filament are often clearly differentiated, and they occur in bisexual, probable rosid and dilleniid flowers with a distinct calyx and corolla.

Introduction

Studies of dispersed fossil pollen have been highly influential in formulating current ideas of early angiosperm evolution. The stratigraphic distribution of columellate and triaperturate pollen grains have been used to assess the timing of angiosperm origin and diversification (Brenner 1963; Doyle 1969; Muller 1970; Doyle and Hickey 1976), and structural comparisons between fossil and Recent grains have been used to infer the systematic affinities of mid-Cretaceous taxa (e.g. Couper 1958; Walker *et al*, 1983; Walker and Walker 1984). Although this approach has yielded important insights into many aspects of early angiosperm evolution, studies of well-preserved fossil flowers with pollen *in situ* provide a greater range of characters from which to infer both the relationships and pollination biology of Cretaceous taxa.

The fossil record of early angiosperm reproductive structures has been traditionally regarded as poor (Hughes 1976), but recent studies have recognized many Albian to Maastrichtian floras that contain diverse and well-preserved assemblages of angiosperm flowers, dispersed stamens, fruits, and seeds. In Europe, a great variety of flowers have been recovered from the Santonian–Campanian of Åsen in Scania, southern Sweden (Friis and Skarby 1982; Friis 1983, 1984, 1985*a, b*, 1990; Friis *et al.* 1988; Crane *et al.* 1989; Friis and Crane 1989) and recent field work in central Portugal has also identified numerous localities at which angiosperm reproductive structures are well preserved (work in progress). Similar floras occur widely elsewhere in western and central Europe, but the fossil flowers in these floras remain to be studied in detail (Knobloch and Mai 1986). In North America, well-preserved material of Campanian age has been recovered from the Neuse River locality in North Carolina (Friis 1984, 1988; Friis *et al.* 1988) and a variety of angiosperm reproductive structures have been recovered from floras ranging in age from Aptian to Cenomanian in the Potomac Group of Maryland and Virginia (Friis *et al.* 1988; Crane *et al.* 1989; Drinnan *et al.* 1990, 1991; Pedersen *et al.* 1991, work in progress).

In this chapter we focus on the early fossil record of angiosperm stamens. We discuss the diversity in stamen morphology and *in situ* pollen that is known currently from Cretaceous localities. Pollen preserved on stigmatic surfaces is also considered briefly because it provides evidence of the systematic affinities of important dispersed pollen types. The range of stamen structures among Cretaceous Magnoliidae, Hamamelididae, and Rosidae is reviewed, and possible patterns in the evolution of stamen and pollen structure are discussed based on these data.

Stamen diversity and *in situ* pollen

Although considerable morphological diversity is present among mid-Cretaceous stamens containing angiosperm pollen, they all show the same general organization with a fertile part (anther) consisting of two elongated thecae connected by a sterile tissue (connective), and borne on a basal sterile part (filament). Anthers are typically tetrasporangiate with each theca consisting of two microsporangia (pollen sacs), more rarely they are bisporangiate with only a single pollen sac per theca. These features are also characteristic for stamens of extant angiosperms and their presence in all the fossil specimens indicates that the characteristic angiospermous stamen organization was fully developed at least by the late Early Cretaceous. A tetrasporangiate and synangiate organization of the microsporangia, similar to that of the angiosperm anther, is uncommon among other seed plants. Tetrasporangiate synangia do occur in *Caytonanthus* (Caytoniales), but these are borne on a branched microsporophyll and differ also in their distinct quasisaccate *Vitreisporites* pollen. In Gnetales the microsporangia do not occur regularly in groups of four and they are borne apically on the stalk. Some angiosperms have an similar apical arrangement (e.g. *Euphorbia* species), but the typical position of the thecae in angiosperm stamens is lateral.

Most of the angiosperm stamens currently known from the mid-Cretaceous (Albian–Cenomanian) are characterized by poor differentiation between anther and filament, a massive connective, and well-developed valvate dehiscence. Recent surveys of stamen morphology and anatomy in extant taxa (Endress and Hufford 1989; Hufford and Endress 1989; Endress and Stumpf 1991) show that this combination of features is restricted to the subclasses Magnoliidae and Hamamelididae.

Magnoliidae

One of the characteristic components of mid-Cretaceous fossil floras are leaves and pollen grains similar to those of extant Magnoliidae (Němejc and Kvaček 1975; Doyle and Hickey 1976; Upchurch 1984; Crabtree 1987; Kvaček 1990; Upchurch and Dilcher 1990), and a corresponding range of magnoliid reproductive structures, including a few with pollen, have now been recognized (e.g. Crane and Dilcher 1984; Dilcher and

Crane 1984; Crane *et al.* 1989; Drinnan *et al.* 1990*a*; Pedersen *et al.* 1990.

Chloranthoid stamens and pollen. Dispersed pollen assigned to the fossil genus *Clavatipollenites* has long been compared to pollen of extant Chloranthaceae and particularly the extant genus *Ascarina* (Couper 1958, 1960). More recently, detailed scanning electron microscopy (SEM) and transmission electron microscopy (TEM) studies have documented further similarities between fossil and recent chloranthoid grains (Walker and Walker 1984; Chlonova and Surova 1988). Other fossil reticulate pollen grains of this type have been compared to pollen of extant *Chloranthus, Hedyosmum*, and Myristicaceae (Walker and Walker 1984).

The only unequivocal macrofossil evidence of plants that produced *Clavatipollenites* pollen is provided by small fruits (*Couperites mauldinensis*) with grains referable to this genus attached to the stigmatic surface (Pedersen *et al.* 1991). The fruits are unicarpellate, unilocular, and contain a single, pendulous, anatropous, and probably bitegmic ovule. A sessile, crest-like stigma, frequently with attached *Clavatipollenites* pollen, is present at the apex of the fruit (Fig. 10.1a). Pollen grains are broadly ellipsoidal, 22–25 μm in diameter, monocolpate, semitectate, reticulate to foveolate (Fig. 10.1b). The colpus has a distinctly granular-verrucate sculpture. Exine thickness is about 1 μm. The tectum is thick, columellae are of moderate length and often expanded in the upper part. The footlayer is very thick (about half the exine thickness) and uniform. The endexine is granular and very thin, except under the colpus where it is very thick and layered (Fig. 10.1c).

In the mid-Cretaceous, a single specimen of a chloranthoid androecium is known from the West Brothers locality (late Albian) of the Potomac Group (Crane *et al.* 1989). The specimen consists of three small stamens that are fused at the base (Fig. 10.8a). Each stamen consists of a stout, swollen filament that grades into the basifixed anther

Fig. 10.1 Chloranthoid reproductive organs with pollen from the Cretaceous. A–C, *Couperites mauldinensis* Pedersen *et al.* from the early Cenomanian of Mauldin Mountain, Maryland. (a) Fruit, SEM, × 75. (b) Pollen grain attached to stigmatic surface, SEM, × 2200. (c) Pollen wall ultrastructure showing thick footlayer and thickened endexine under the aperture, TEM, × 4200. (d)–(f) *Chloranthistemon endressii* Crane, Friis, and Pedersen from the Santonian–Campanian of Åsen, southern Sweden. (d) Dorsal (?) view of median androecial lobe showing glands of connective tissue and recurved valves of anther, SEM, × 80. (e) Pollen grain attached to the inner surface of pollen sac, SEM, × 3500. (f), Pollen wall ultrastructure showing thick footlayer, TEM, × 4200.

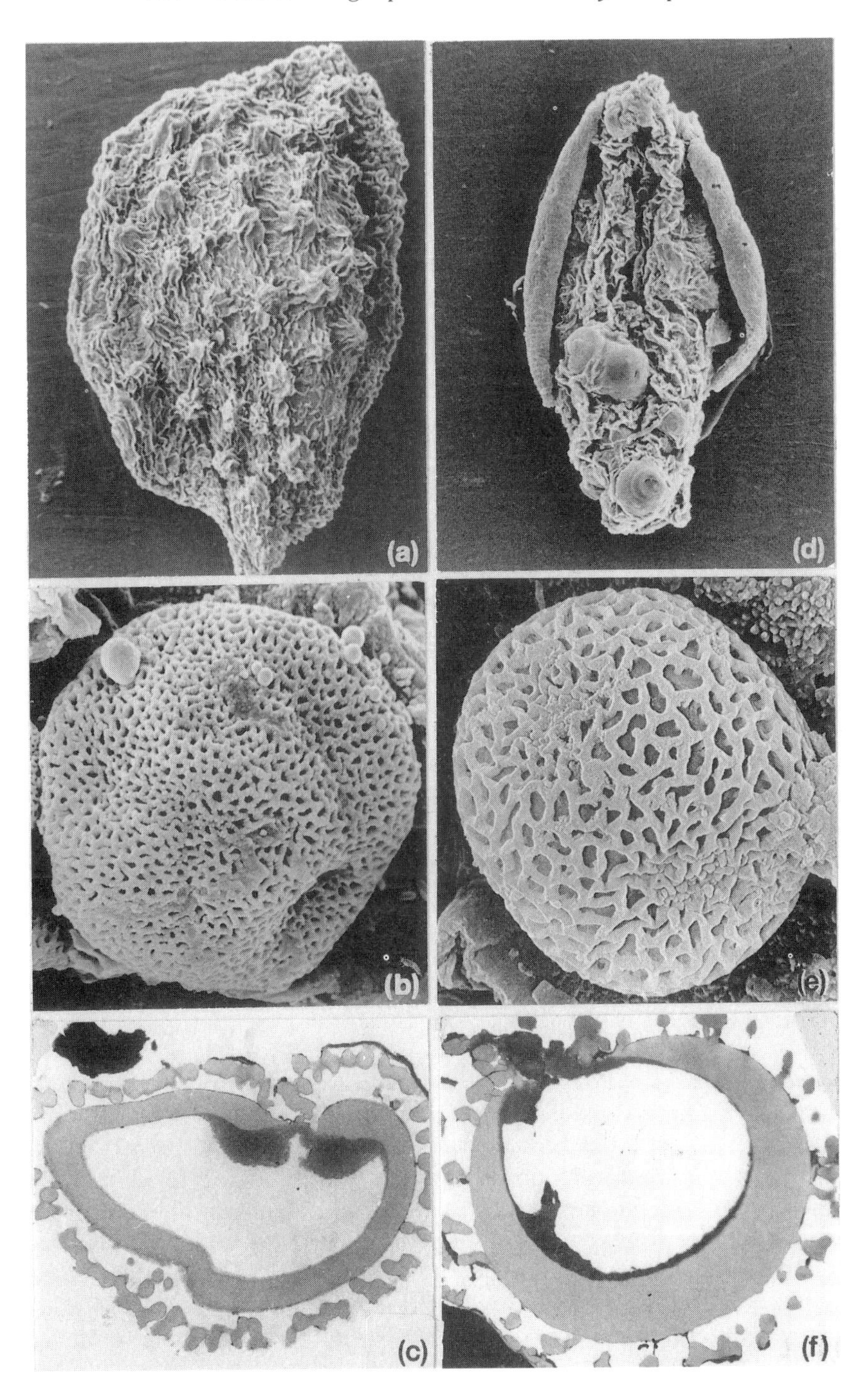
(a)
(d)
(b)
(e)
(c)
(f)

without a joint. Anthers are tetrasporangiate with two pairs of laterally arranged pollen sacs. Pollen sacs are of equal length and diameter. The connective is thick and extends into a short but prominent projection beyond the pollen sacs. The anthers show well-developed valvate dehiscence with each theca opening by two valves that split between the pollen sacs. The line of dehiscence apparently bifurcates into transverse slits at the base and apex of the pollen sacs, and the valves are strongly reflexed to expose their inner surface and pollen. Details of pollen grains are poorly known because they are partially covered, or embedded in a pollenkitt-like substance. Transmission electron miscroscopy is precluded by the availability of only one specimen. Pollen is prolate, about 10 μm long, with three or more colpi parallel to the polar axis of the grain, but the exact number of colpi, and their orientation is uncertain.

Chloranthistemon endressii from the Late Cretaceous (Santonian–Campanian) of southern Sweden is a distinctive, dorsiventrally flattened, three-lobed androecium with two pairs of pollen sacs on the central lobe, and one pair each on the two lateral lobes (Crane *et al.* 1989). Pollen sacs are borne laterally on massive connectives that form the androecial lobes and have a prominent extension beyond the apex of the pollen sacs (Figs 10.1d, 10.8b). Bulging glands on the connective surface correspond to large internal cavities that may have had a secretory function (Fig. 10.1d). Anther dehiscence is strongly valvate and occurs along a longitudinal split between the two pollen sacs that bifurcates at the base of the theca (Fig. 10.8b). The two valves of each theca curl back to expose their inner surfaces, and the pollen which is partially covered by a dense substance that may be residual pollenkitt. *In situ* pollen grains are spheroidal, 12–14 μm, with irregular apertures semitectate, reticulate (Fig. 10.1e). Exine thickness is about 1 μm. The tectum is thick; columellae are of moderate length. The footlayer is very thick and uniform. The endexine is granular and thin, except under the apertures where it is very thick (Fig. 10.1f).

Lauraceous stamens. Pollen of extant Lauraceae and related families has a predominantly cellulosic wall that contains relatively little sporopollenin. Because such grains rarely survive normal preservation processes, or preparation procedures used for isolating fossil pollen, these families are poorly represented in the palynological record. In contrast, leaves similar to those of extant Laurales are widespread in Cretaceous macrofloras (Němejc and Kvaček 1975; Upchurch 1984; Crabtree 1987; Kvaček 1990; Upchurch and Dilcher 1990) and unequivocal inflorescences and flowers of Lauraceae (*Mauldinia mirabilis*) are now known from the Mauldin Mountain locality (earliest

Cenomanian) of the Potomac Group (Drinnan *et al.* 1990). The androecium in *Mauldinia* consists of nine fertile stamens in three whorls and a fourth whorl of staminode-like structures. Stamens have a long, frequently flattened, filament that grades into the basifixed anthers. Anthers are bisporangiate and consists of two elongated pollen sacs separated by a thick connective. Each pollen sac dehisces by an elliptical flap-like valve with a distal hinge (Fig. 10.8c). Dehiscence of the two outer whorls of stamens is introrse, while that of the inner whorl of stamens is extrorse. Above the pollen sacs the connective forms a dorsiventrally flattened apical expansion, frequently with numerous prominent stomatal pores in its surface. Stamens opposite the outer tepals appear to have a pair of prominent staminal appendages attached close to the base of the filament. Each appendage has an expanded clavate–sagittate head. Pollen is not preserved.

Other magnoliid stamens and pollen. In addition to mid-Cretaceous pollen comparable to that of extant Chloranthaceae, several dispersed stamens from Potomac Group assemblages containing monocolpate pollen may represent non-chloranthoid magnoliids. One of these is distinguished by its rather large size compared to other Cretaceous stamens (Fig. 10.2a). Although the proximal part is missing, the stamen is clearly laminar and strongly flattened dorsiventrally. The anther is tetrasporangiate with the two pairs of pollen sacs arranged close to the margin. The pollen sacs are long and narrow and embedded in the extensive connective tissue. They are visible only on one surface (ventral?) while the other surface (dorsal?) is completely smooth and composed entirely of connective tissue. The connective tissue is considerably expanded above the level of the pollen sacs and grades without a joint into an elongated triangular protrusion that is longer than the pollen sacs. Dehiscence apparently took place between the pollen sacs, but the tissue over the pollen sacs is abraded and the mode of dehiscence unknown. This stamen conforms to the morphology of stamens of some modern Magnoliales, particularly Magnoliaceae, but its systematic relationship remains to be established. *In situ* pollen grains are boat-shaped, 20–30 μm long, monocolpate, semitectate, and reticulate, with a relatively coarse reticulum and distinct columellae (Fig. 10.2b). Along the margins of colpus the tectum is smooth (Fig. 10.2c). Pollen wall ultrastructure is unknown.

Hamamelididae

In addition to Magnoliidae, two other characteristic components of mid-Cretaceous macrofloras are fossil trochodendrophyll and platanoid leaves (Doyle and Hickey 1976; Upchurch 1984; Crabtree 1987; Crane

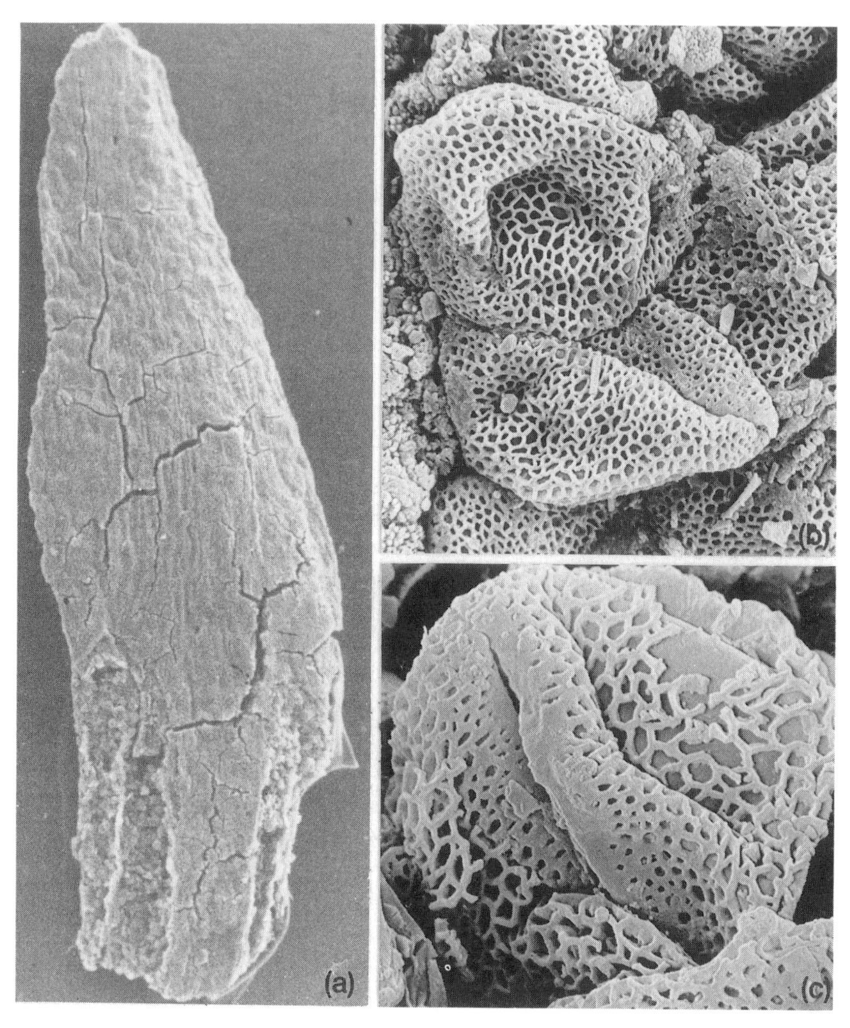

Fig. 10.2 Stamen and pollen of probably a magnoliid plant from the early Cenomanian of Mauldin Mountain, Maryland. (a) Stamen showing large apical extension of connective and four pollen sacs, SEM, × 55. (b) Pollen from stamen in (a) showing monocolpate aperture and reticulate surface, SEM, × 1500. (c) Pollen from fragmented stamen showing details of reticulum and aperture, SEM, × 3000.

1989) broadly comparable to those of extant 'lower' Hamamelididae (e.g. *Trochodendron, Tetracentron, Cercidiphyllum, Euptelea, Platanus*, and Hamamelidaceae). Taxa at this grade of angiosperm evolution are also well represented among reproductive structures known from the mid-Cretaceous, which include several kinds of stamen with *in situ* pollen.

Trochodendroid–buxoid stamens and pollen. The most complete evidence of mid-Cretaceous angiosperms related to extant Trochodendrales and Buxales is the fossil genus *Spanomera* from the Potomac Group that comprises two species with unisexual staminate and pistillate flowers (Drinnan *et al.* 1990). In one species, *S. mauldinensis*, both pistilate and staminate flowers are present and preserved together in a diclinous inflorescence. The flowers are small with few organs per whorl (4–5) and an undifferrentiated perianth.

Stamens of *Spanomera mauldinensis* from the Mauldin Mountain locality, Maryland (earliest Cenomanian), have a short, stout filament and dorsifixed, tetrasporangiate anther (Fig. 10.3a). Pollen sacs are arranged dorsiventrally and are of unequal size. Sterile tissue of the connective is mainly restricted to the dorsal side of the anther and extends apically into a short conical tip (Fig. 10.3a). Dehiscence of anthers is by simple longitudinal slits. *In situ* pollen grains are prolate, 18–23 μm in polar length tricolpate, semitectate, and reticulate to striate (Fig. 10.3b). The colpus has a verrucate sculpture. Exine thickness is about 1.5 μm. The tectum is very thick (muri about 1 μm high); columellae are short and simple. The footlayer is moderately thick and uniform. The endexine is granular, moderately thick; thicker and layered under the colpus (Fig. 10.3c).

Stamens of *Spanomera marylandensis* are very similar to those of *S. mauldinensis* in general morphology and also in distribution of connective tissue, but differ in their smaller size (Fig. 10.3d). *In situ* pollen grains are prolate, 16–19 μm in polar length, tricolpate, semitectate, reticulate (Fig. 10.3e). The colpus has irregular verrucate sculpture. Exine thickness is about 3 μm. The tectum is very thick (muri 1–2 μm high); columellae are short and simple. The footlayer is very thick (about half the exine thickness) and uniform. The endexine is granular and relatively thick, except under the colpus where it is thicker and layered (Fig. 10.3f).

Platanoid stamens and pollen. The characteristic ball-like staminate and pistillate inflorescences of *Platanus*-like plants and closely related forms are especially common during the mid-Cretaceous. They are characterized by their small, unisexual flowers with an undifferentiated perianth. Androecium and gynoecium are pentamerous, stamens and carpels are

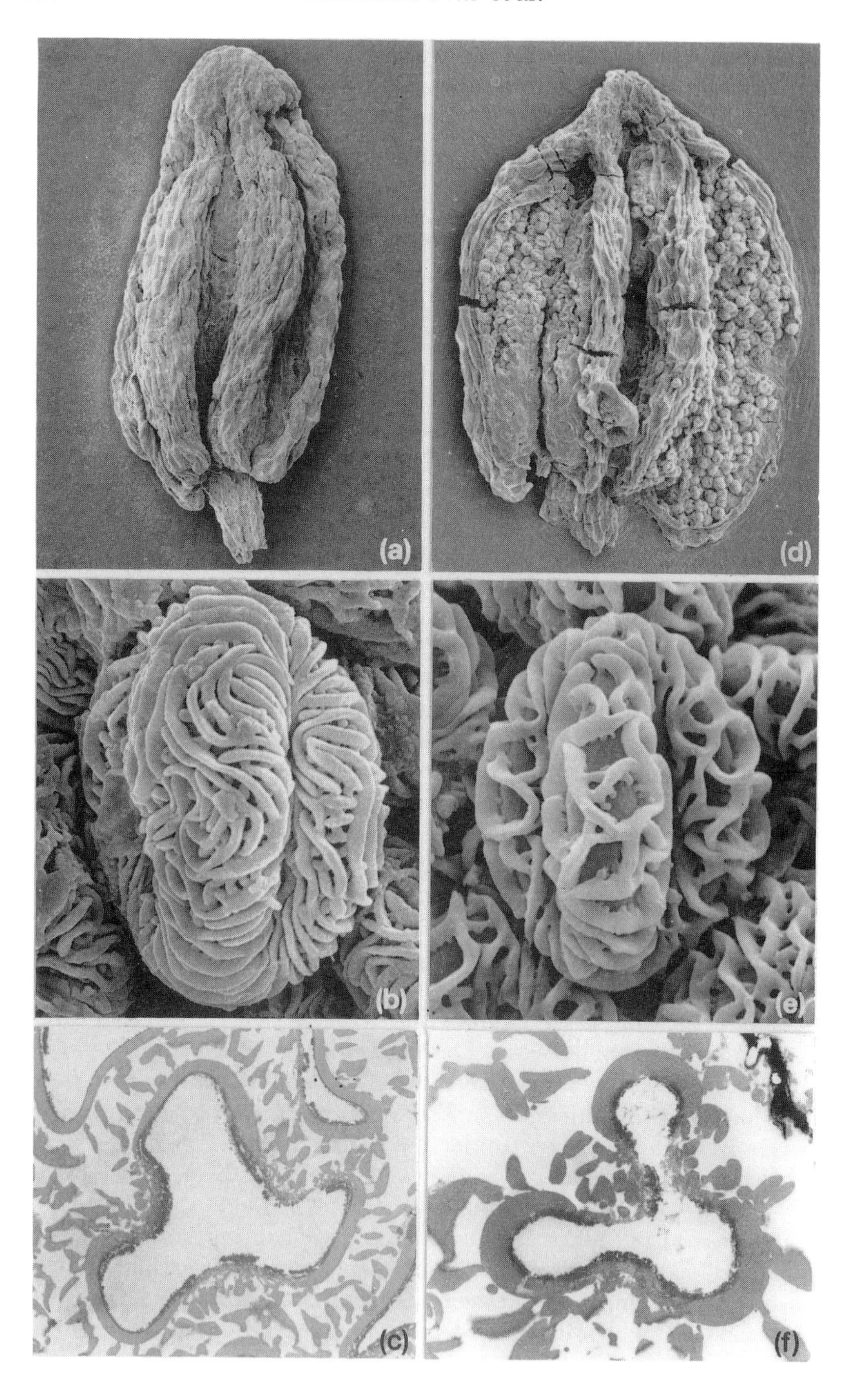
(a)
(d)
(b)
(e)
(c)
(f)

superficially similar in form and both have a swollen apical portion. Four different kinds of staminate structures are currently recognized from the Potomac Group (Albian–Cenomanian), two of which are assigned to the fossil genus *Platananthus* (Friis *et al.* 1988). Two species of *Platananthus* have also been recognized from the Santonian–Campanian floras of southern Sweden and North Carolina, USA (Friis *et al.* 1988). Pistillate inflorescences and flowers thought to be produced by the same plants as *Platananthus* are assigned to the extinct genus *Platanocarpus* (Friis *et al.* 1988).

Stamens of *Platanathus potomacensis* from the West Brothers locality, Maryland (late Albian), are broadly ellipsoidal with a short filament and basifixed, tetrasporangiate anther. The two pairs of pollen sacs are laterally arranged and pollen sacs are of equal length and diameter. Connective tissue is prominent between the pollen sacs, and is well developed below the thecae and at the apex, where it forms a rounded, cap-like protrusion (Fig. 10.4a). The anthers show well-developed valvate dehiscence. Each theca opens by two valves that split between the pollen sacs and along transverse slits at the apex and base of the theca (Figs 10.4a, 10.8d). *In situ* pollen grains are prolate, 8.5–12 μm in polar length, tricolpate, semitectate, and reticulate (Fig. 10.4b). The colpus has a verrucate sculpture. Exine thickness is about 0.8 μm. The tectum is moderately thick (0.2 μm); columellae are short and simple. The foot-layer is thick and uniform. The endexine is finely granular, relatively thick (0.2 μm), and thicker under the colpus (0.7 μm) (Fig. 10.4c).

Platananthus sp. from the Mauldin Mountain locality, Maryland (earliest Cenomanian), is similar to *P. potomacensis* in general structure, with tetrasporangiate anthers, laterally arranged pollen sacs, and valvate dehiscence, but stamens from Mauldin Mountain are larger and differ in their more elongated, narrower shape (Fig. 10.4d). *In situ* pollen grains are prolate, 10 μm in polar length, tricolpate, semitectate, and reticulate to foveolate (Fig. 10.4e). The colpus has a prominent verrucate sculpture. Exine thickness is about 1 μm. The tectum is thin; columellae are long and simple. The footlayer is thick and uniform. The endexine is granular, relatively thick, and thicker under the colpus (Fig. 10.4f).

Fig. 10.3 Stamens and pollen of *Spanomera* Drinnan *et al.* (Hamamelididae) from the mid-Cretaceous of the Potomac Group. (a)–(c), *S. mauldinensis* Drinnan *et al.* from the early Cenomanian Mauldin SEM, × 80. (b) *In situ* pollen, SEM, × 3200. (c) Pollen wall ultrastructure, TEM, × 4200. (d)–(f), *S. marylandensis* Drinnan *et al.* from the late Albian West Brothers locality, Maryland. (d) Stamen, SEM, × 120. (e), *In situ* pollen, SEM, × 3000. (f), Pollen wall ultrastructure, TEM, × 4200.

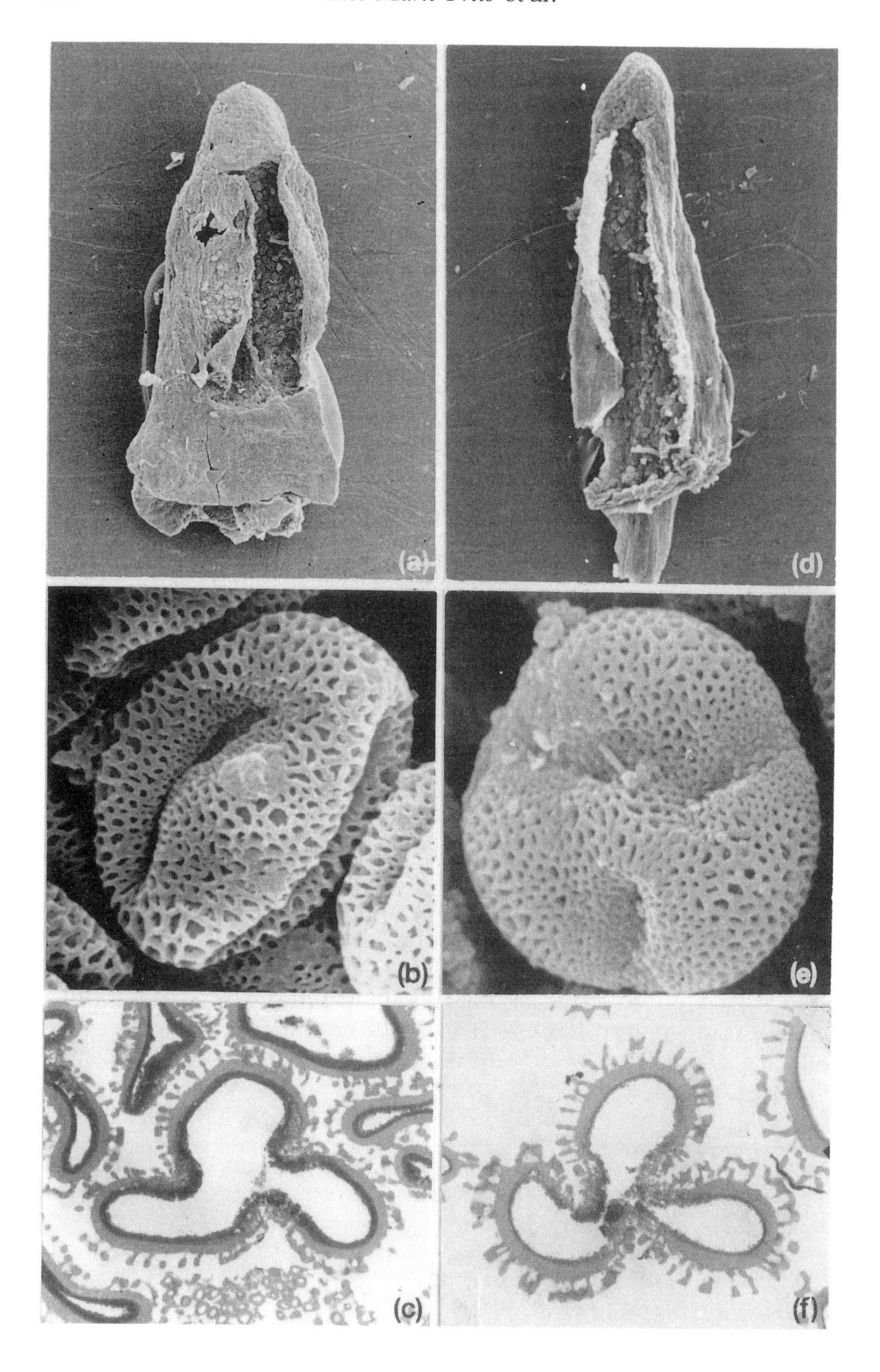
(a)
(d)
(b)
(e)
(c)
(f)

Ball-like inflorescences with unisexual flowers very similar to those of *Platananthus/Platanocarpus*, but differing in epidermal details of carpels and aperture characteristics of the pollen, have been recorded from the West Brothers locality (Friis *et al.* 1988), and are a characteristic component of the late Albian Bull Mountain flora, Maryland (Doyle and Hickey 1976, work in progress). At the Bull Mountain locality, dispersed stamens are particularly common. They have a short dorsifixed filament that grades into the connective tissue without a joint. Anthers are tetrasporangiate with dorsiventrally arranged pollen sacs. Connective tissue is restricted to the dorsal surface of the anther and is expanded into a peculiar and very characteristic apical protrusion with a hook-shaped tip that is bent downward towards the center of the flower (Figs 10.5e, 10.7a, 10.8e). The tip is typically short, but may extend almost to the base of the anther (Fig. 10.7a). Dehiscence is valvate with two valves over each theca (Fig. 10.8e). *In situ* pollen grains are spheroidal to prolate, 10 μm in polar length, tricolporate, semitectate and reticulate to foveolate (Fig. 10.5f). The colpus has a verrucate sculpture. Exine thickness is about 1 μm. The tectum is thin; columellae are long and simple. The footlayer is thick and uniform. The endexine is granular, relatively thick, and thicker under the colpus (Fig. 10.5g). The colporate apertures of the pollen prevent an assignment of these fossils to *Platananthus* and they may represent an early rosid line.

At the 'Bank near Brook' locality, Virginia (middle Albian; Fontain 1889; Doyle and Hickey 1976), ball-like inflorescences and dispersed stamens are associated with abundant leaves of *Sapindopsis*. The well-preserved inflorescences are all pistillate, but the dispersed stamens are believed to be derived from similar heads (work in progress). The stamens are distinguished from other platanoid stamens by having a filament about twice the length of the anther (Fig. 10.5a). The filament is broad and passes gradually into the connective tissue of the anther without a joint. The anther is basifixed and tetrasporangiate, with the two pair of pollen sacs in a lateral position. The pollen sacs are small

---

Fig. 10.4 Stamens and pollen of *Platananthus* (Hamamelididae) from the mid-Cretaceous of the Potomac Group. (a)–(c), *Platananthus marylandensis* Friis, Crane, and Pedersen from the late Albian West Brothers locality, Maryland. (a) Lateral view of stamen showing the two valves of a single theca and sterile tissue above and below the pollen sacs, SEM, × 100. (b) *In situ* pollen, SEM, × 5000. (c), Pollen wall ultrastructure, TEM, × 5600. d–f, *Platananthus* sp. from the early Cenomanian Mauldin Mountain locality, Maryland. (d) Lateral view of stamen showing two valves of a single theca, SEM, × 75. (e) *In situ* pollen, SEM, × 5000. (f) Pollen wall ultrastructure, TEM, × 5600.

and constitute only a minor part of the whole anther. In contrast, the central connective tissue between the pollen sacs is extensive and extends into a small apical protrusion. Dehiscence is distinctly valvate with two valves over each theca. The valves are laterally hinged and strongly curled back at dehiscence. *In situ* pollen grains are spheroidal to prolate, 10–15 μm in polar length, tricolpate, semitectate, and reticulate to foveolate (Fig. 10.5b). The colpus has a verrucate sculpture. Exine thickness is about 1 μm. The tectum is thin; columellae are long and simple. The footlayer is thick and uniform. The endexine is granular, relatively thick, and thicker under the colpus (Fig. 10.5d). Several of the *in situ* pollen grains appear immature with ektexine poorly developed (Fig. 10.5c).

Platanoid inflorescences and dispersed stamens are also common in later Cretaceous floras and include two species of *Platananthus* (e.g. Friis *et al.* 1988). Stamens of *Platananthus scanicus* from southern Sweden (Santonian–Campanian) have short filaments, basifixed tetrasporangiate anthers similar to the mid-Cretaceous forms but the arrangement of pollen sacs is more distinctly ventral, and sterile tissue between pollen sacs is less abundant and mostly restricted to the dorsal side of the anther. The apical protrusion of the connective tissue is extensive and may equal the length of the pollen sacs. Dehiscence is valvate and occurs along a longitudinal slit that bifurcates at the base and apex of the thecae. *In situ* pollen is prolate, 15–16.5 μm, tricolpate, semitectate, and reticulate. The colpus has a verrucate sculpture. Exine thickness is about 1.6 μm. The tectum is thin; columellae are short and simple. The footlayer is very thick (0.8 μm) and uniform. The endexine is granular, very thin, and thicker and layered under the colpi.

While stamens of the mid-Cretaceous *Platananthus* and of *P. scanicus* frequently occur dispersed in the fossil floras, those of *Platananthus hueberi* from Neuse River, North Carolina (Campanian) are always attached to the inflorescence and often densely packed. They are

---

Fig, 10.5 Stamens and pollen of platanoids (Hamamelididae) from the mid-Cretaceous of the Potomac Group. (a)–(d), Platanoid from the middle Albian 'Bank near Brooke' locality. (a) Stamen showing long filament, massive connective and valvate anther dehiscence, SEM, × 90. (b) *In situ* pollen, SEM, × 4000. (c) Pollen wall ultrastructure showing immature (?) grain consisting mainly of endexine material TEM, × 5600. (d) Pollen wall ultrastructure showing thick ektexine and endexine that is thickened under the aperture, TEM, × 5600. (e)–(g) Platanoid from the late Albian–earliest Cenomanian Bull Mountain locality. (e) Dorsal view of stamen showing dorsal and apical connective tissue, SEM, × 100. (f) *In situ* pollen, SEM, × 5000. (g) Pollen wall ultrastructure, TEM, × 5600.

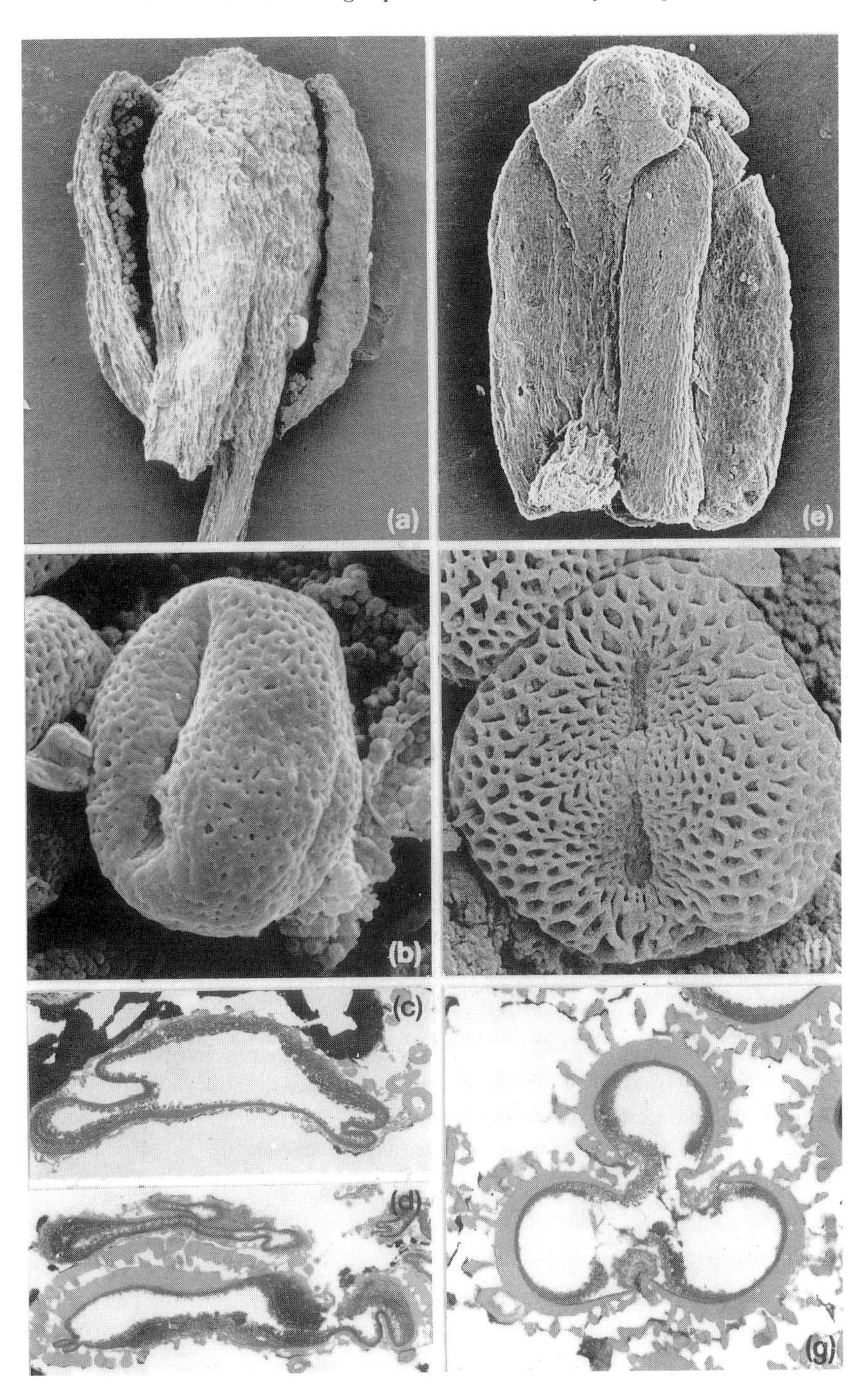
(a)
(e)
(b)
(f)
(c)
(d)
(g)

similar to the Scanian fossils in their dorsiventral arrangement. Filaments are slightly longer and the apical protrusion is a flat, cap-like structure. A longitudinal dehiscence slit is present, but it has not been established whether the slit bifurcates and therefore whether anther dehiscence is valvate as in other species of *Platananthus*. *In situ* pollen grains are prolate, 13–15 μm, tricolpate, semitectate, and reticulate. The colpus has a prominent verrucate sculpture. Exine thickness is about 1.2 μm. The tectum is thick, columellae are long and simple. The footlayer is moderately thick and uniform. The endexine is granular, very thick, and thicker and layered under the colpus.

Hamamelidaceae. Fossil flowers similar in some features to those of modern *Hamamelis* have been identified from the Santonian–Campanian of southern Sweden and assigned to the extinct genus *Archamamelis* (Friis 1985*b*; Endress and Friis 1991). Stamens are bisporangiante. Dehiscence is valvate with a single valve over each theca opening toward the center of the flower (Fig. 10.8f). The filament is relatively massive and grades into the connective tissue without a joint. Apically the connective tissue expands into a massive, rounded cap. Pollen grains observed *in situ* are spheroidal and tricolpate, about 15 μm in polar length, semitectate, and reticulate to foveolate. The colpus has a verrucate sculpture. Exine thickness is about 1.5 mm. The tectum is moderately thick; columellae are long and simple. The footlayer is moderately thick and uniform. The endexine is coarsely granular and thin, except under the apertures where it is thick and layered.

A single dispersed stamen very similar to that of the Late Cretaceous hamamelidaceous flower has been discovered from the late Albian (pollen zone IIB) Puddledock locality of the Potomac Group. In this specimen each theca opens by a single laterally hinged valve.

Normapolles plants. In the Late Cretaceous flowers of juglandalean/myricalean affinity containing pollen of the distinct Normapolles type are abundant. They are known from the Santonian–Campanian strata of southern Sweden, from several mid-European floras (Friis 1983; Knobloch and Mai 1986; Friis and Crane 1989), and have now also been discovered from Late Cretaceous (Campanian–Maastrichtian) floras of Portugal (work in progress). Most of the flowers studied so far are apparently bisexual, but unisexual forms also occur among the Portuguese fossils. All flowers in which stamens are known have a distinct filament, tetrasporangiate anthers with simple rather than valvate dehiscence, no apical extension of the connective, and little connective tissue between the pollen sacs. Pollen grains are typically oblate with triangular equatorial outline, triporate/tricolporate with strongly thickened apertural regions, and tectate.

Rosidae

Other stamen types distinct from the massive, valvate stamens of magnoliids and hamamelidids occur in the early Cenomanian flora of the Mauldin Mountain locality. They have distinct filaments, tetrasporangiate anthers, poorly developed connective tissue between the pollen sacs, and no apical extension of the connective. These stamens are believed to represent fossil plants of rosid affinity. One type has been recovered dispersed in the flora as well as attached to a small pentamerous flower. It has small, round, dorsified anthers. *In situ* pollen grains are subspheroidal, about 10 μm in polar length tricolpate, tectate, and weakly striate. The colpus has a verrucate sculpture. Exine thickness is about 0.8 μm. The tectum is very thick (0.5 μm) and homogeneous; columellae are very short (?granular). The footlayer is moderately thick and uniform. The endexine is granular and thin, except under the apertures. Orbiculae occur associated with the *in situ* pollen.

Two other stamen types from the Mauldin Mountain flora have more elongated anthers, but the *in situ* pollen grains are very similar, small, tectate and striate to rugulate. Similar stamens and pollen have also been recorded from a pentamerous flower of possibly rosid affinity from the Cenomanian Rose Creek locality, Nebraska (Basinger and Dilcher 1984; see also Upchurch and Dilcher 1990).

Among the Santonian–Campanian flowers from southern Sweden, stamens with distinct filament, small dorsifixed and tetraporangiate anthers, with poorly developed connective tissue, are predominant. Several of these are clearly of rosid affinity, such as *Scandianthus* (Friis and Skarby 1982) and *Silvianthemum* (Friis 1990), while others, such as *Actinocalyx*, may be of dilleniid affinity (Friis 1985*a*). Pollen in these later Cretaceous rosids/dilleniids are typically very small (6–12 μm), tectate, with a smooth or finely foveolate surface. They are often embedded in a pollenkitt-like substance.

Discussion

Stamen morphology and distribution of sterile tissue

A remarkable feature of Albian and Cenomanian angiosperm assemblages is the common occurrence of massive stamens with abundant sterile tissue and extensive apical protrusions of the connective. Based on floral, stamen, and pollen characters these stamens have been assigned to the Magnoliidae and 'lower' Hamamelididae (Fig. 10.6). They generally lack distinct differentiation between the anther and

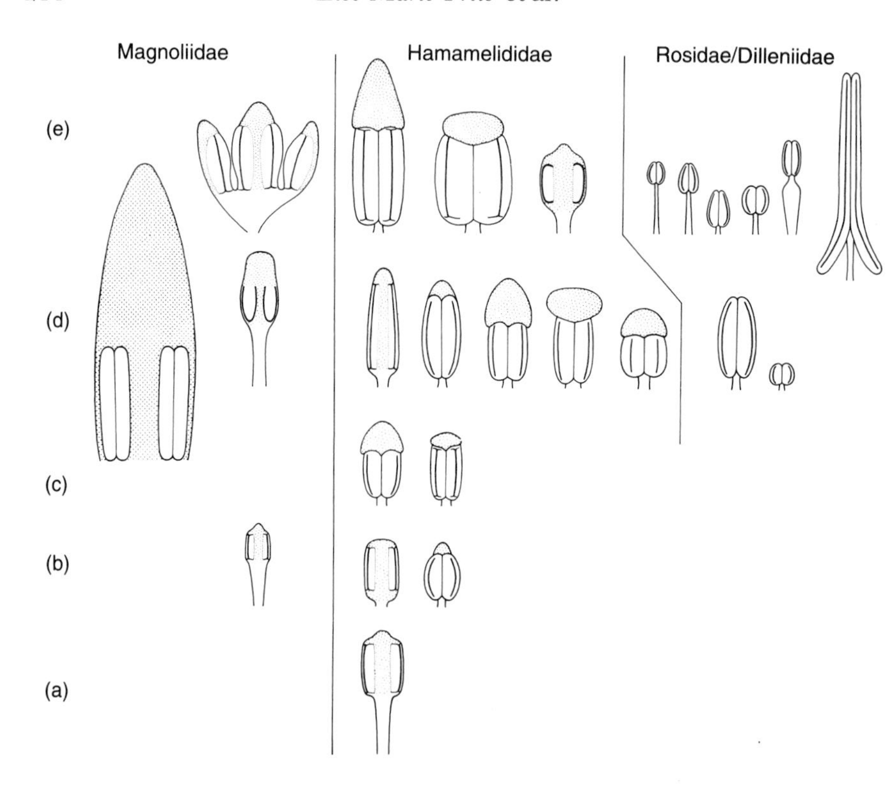

Fig. 10.6 Diagram summarizing the diversity of stamen structure among Cretaceous angiosperms in selected floras. (A) Bank near Brooke locality (middle Albian, lower part of pollen zone IIb). (B) West Brothers locality (late Albian, upper part of pollen zone IIb). (C) Bull Mountain (late Albian–early Cenomanian, middle part of pollen zone IIc). (D) Mauldin Mountain (early Cenomanian, lower part of pollen zone III). (E) Åsen, southern Sweden (Santonian–Campanian).

the basal sterile part (filament), and anthers are usually basifixed. In most species filaments are very short and thick. In specimens with longer filaments (platanoid from Bank near Brooke, lauraceous *Mauldinia* from Mauldin Mountain, chloranthoid from the West Brothers locality) the filament is relatively thick, probably originally, fleshy, and passes gradually into the massive connective of the anther. In the two species of *Spanomera*, anthers are dorsifixed but filaments also appear fleshy and grade into the connective tissue without a distinct joint. Anthers are typically tetrasporangiate, with four pollen sacs connected by abundant sterile tissue. Bisporangiate anthers are rare in Cretaceous angiosperms and have so far been recorded only in the early Cenomanian lauraceous flower *Mauldinia mirabilis*, in the

Santonian–Campanian hamamelidaceous flowers from Scania, and dispersed in the late Albian Puddledock flora. In several species (chloranthoid from the West Brothers locality, *Platananthus potomacensis, Platananthus* sp. from Mauldin Mountain, platanoid from Bank near Brooke) anthers are thick (radially as well as tangentially), have massive central connective tissue, and all four pollen sacs are of almost equal size. Other stamens (e.g. *Spanomera mauldinensis, S. potomacensis*, platanoid from Bull Mountain) have a more dorsiventral organization; connective tissue is mainly restricted to the dorsal side, and the dorsal and ventral pollen sacs differ in size. Only a single kind of stamen with a distinct laminar structure has so far been recognized. Pollen sacs in this stamen type are narrow and appear to be completely embedded in the extensive connective tissue.

A distinct apical protrusion of the connective is present in all Albian stamens studied so far. The shape of the protrusion varies considerably, and it also ranges in length from about one-fifth of the length of the thecae to considerably longer than the thecae. In several species the distal part of the connective forms a large, shield-like cap that is clearly distinct from the remaining anther and overhangs the apex of the thecae (*Platananthus potomacensis, Platananthus* sp. from Mauldin Mountain, platanoid stamens from Bull Mountain, and several unassigned anthers, Fig. 10.7). In the stamens from Bull Mountain the apical cap is extended into a pointed, hook-shaped tip that bends towards the ventral face. Usually this hook is rather short, but in some specimens may extend for almost the full length of the thecae (Fig. 10.7a). In other species there is no joint between the thecal region of the anther and the apical expansion. Smaller, pointed connective protrusions characterize the two species of *Spanomera*, the chloranthoid androecium from the West Brothers locality, and the stamens from Bank near Brooke. An extremely long connective protrusion, apparently longer than the thecae, occurs in the laminar stamen from Mauldin Mountain (Fig. 10.2a).

Massive stamens of magnoliidean and hamamelididean affinity, similar to those from the mid-Cretaceous, have also been recovered in Santonian–Maastrichtian floras (e.g. *Chloranthistemon endressii, Plantananthus hueberi, P. scanicus*, several undescribed platanoid fossils from southern Sweden and Portugal, the hamamelidaceous flower from Åsen). However, they constitute only a minor part of the total stamen diversity in these later Cretaceous floras. They show some variation in position of pollen sacs, shape and size of apical extension of connective, and amount of sterile tissue connecting the thecae, but generally sterile tissue is less prominent than in stamens from the mid-Cretaceous. An exception is the laminar stamens of *Chloranthistemon endressii* in which

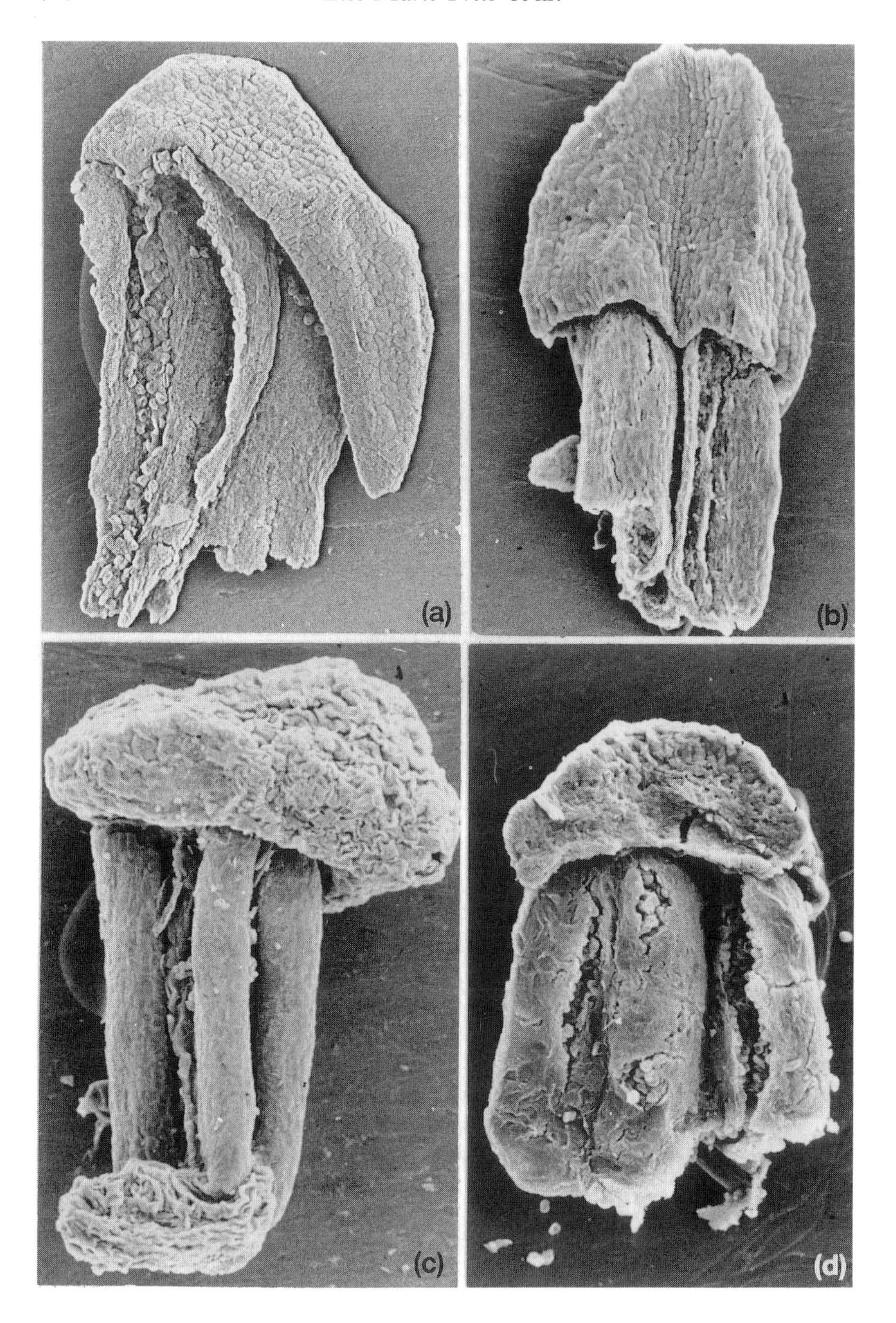
(a)
(b)
(c)
(d)

pollen sacs constitute only a very small proportion of the entire stamen and the sterile tissue forms a relatively large laminar lobed plate. These stamens all have very short filaments and anthers are basifixed.

In the Cenomanian floras, a few stamen types show a different morphological pattern (material from Mauldin Mountain, fossil flower from Rose Creek described by Basinger and Dilcher 1984). They have short filaments, but the transition between filament–anther is abrupt and marked by a distinct joint. The anthers are made up mostly by the thecae. Apical protrusions of the connective are lacking and sterile tissue between the thecae is poorly developed. Based on *in situ* pollen and floral morphology, these stamens are believed to be of rosid or possibly dilleniid affinity.

In Santonian–Maastrichtian floras, stamens with well-defined filament and anther dominate. Anthers are usually dorsifixed and the filament–anther transition is abrupt, as in the probable rosid stamens from the mid-Cretaceous. Filament length and form varies from short, fleshy types to long, thread-like or strip-like forms. Anthers are typically very small, more or less elliptic or rounded in outline (e.g. the rosid genera *Scandianthus* and *Silvianthemum* from Scania; Friis and Skarby 1982; Friis 1990), but narrow elongated anthers are also present (Normapolles flowers, thealean flower, several unassigned dispersed stamens). Connective tissue in these late Cretaceous higher hamamelidids, rosids, and dilleniids is generally poorly developed and does not form apical protrusions.

Secretory cavities in the connective have been observed in *Chloranthistemon endressii*, and stomata-like openings occur at the apical extension of the connective and at the connective proper in several platanoid taxa. These may have secreted scent or nectar and may have had an important attractive function.

Stamen dehiscence

Except for the stamens of *Spanomera*, which open by simple longitudinal slits, the massive magnoliids and hamamelidid stamens from the mid-Cretaceous show distinct valvate dehiscence (Fig. 10.8). In the lauraceous *Mauldinia*, an elongated elliptical flap is present over each

---

Fig. 10.7 Dispersed stamens from the mid-Cretaceous Potomac Group, showing variation in shape of apical expansions of the connective. (a) Platanoid stamen from the late Albian–earliest Cenomanian Bull Mountain locality with hook-shaped expansion of connective along the ventral side, SEM, × 150. (b)–(d) Three different unassigned stamens from the early Cenomanian Mauldin Mountain locality, Maryland showing prominent apical expansions of the connective, SEM, × 100.

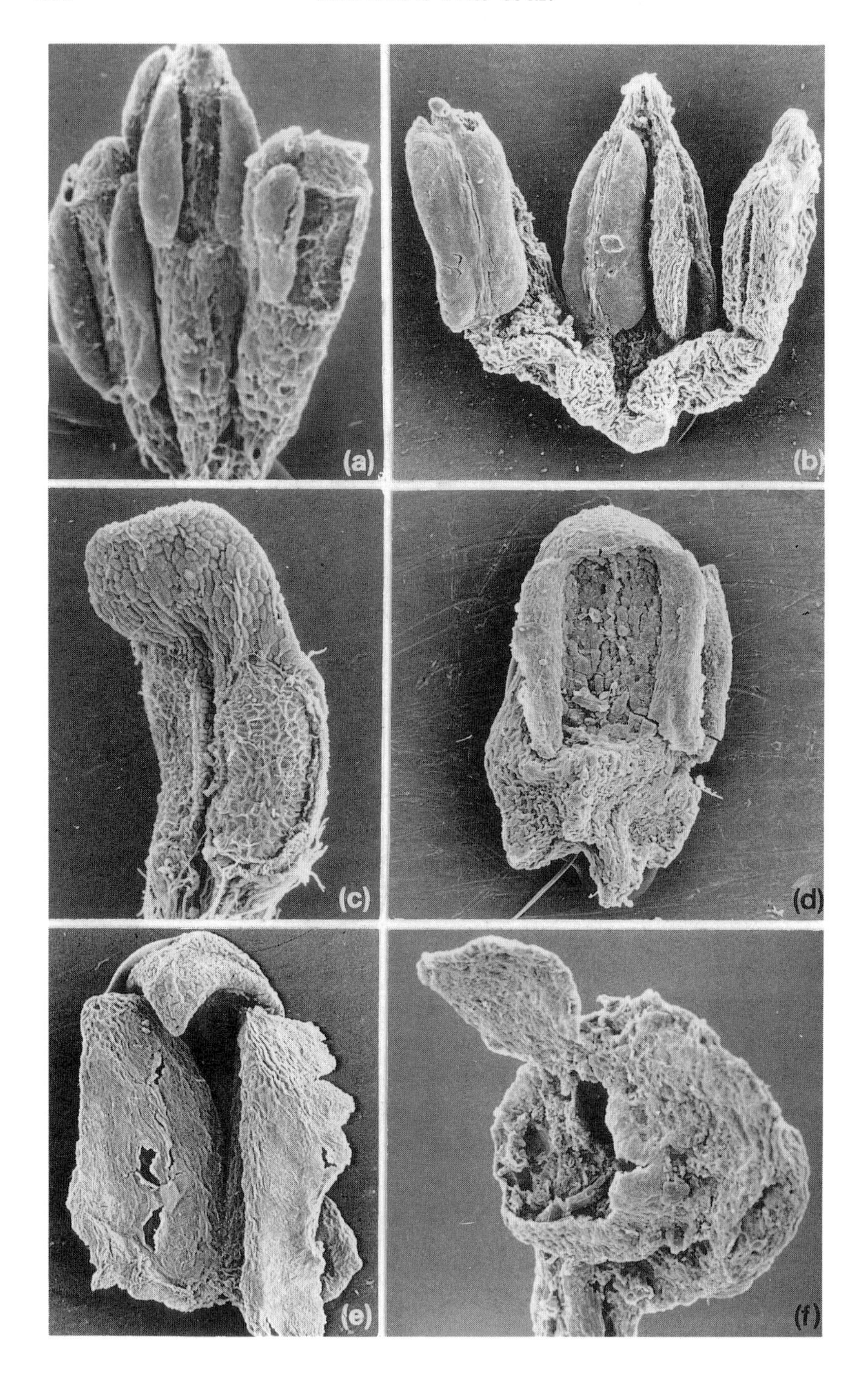
(a)
(b)
(c)
(d)
(e)
(f)

theca. The flap is hinged apically and rolls upwards at dehiscence. In the other mid-Cretaceous stamens with valvate dehiscence there are two valves over each theca and the valves are hinged longitudinally. In these stamens a longitudinal furrow extends between the pollen sacs of each theca and bifurcates to form transverse slits basally and apically. Valves are especially distinct in stamens with a prominent apical shield-like protrusion of the connective, as in the Bull Mountain specimens (Fig. 10.8e).

Valvate dehiscence is also present in the Late Cretaceous *Chloranthistemon endressii* and the various platanoid species from Scania. These taxa always have two valves per theca. In *Chloranthistemon* the basal bifurcations are distinct while apical bifurcations have not been observed. At dehiscence the valves curl back. Distinct short transverse furrows have been observed in *Chloranthistemon endressii* and in *Platananthus scanicus*. This is significant since this feature is much less clearly developed in related modern forms (*Chloranthus, Platanus*; Endress 1987; Endress and Hufford 1989; Hufford and Endress 1989). In *Archamamelis* from Åsen the anther is unusual in being bisporangiate with only a single valve over each theca. A similar situation occurs in *Mauldinia*, although the valve is hinged apically rather than laterally.

In the probable rosid stamens from the mid-Cretaceous, with poorly developed connective tissue, dehiscence is by simple longitudinal slits. This is also the case for the majority of later Cretaceous stamens (e.g. *Scandianthus, Silvianthemum, Actinocalyx*). Typically, there is a single slit per theca between the pollen sacs or over the ventral pollen sac. Dehiscence by apical pores have not been observed in any of the Cretaceous angiosperms.

Fig. 10.8 Dispersed stamens from the mid- and late Cretaceous showing valvate anther dehiscence. (a) Chloranthoid androecium from the late Albian West Brothers locality, Maryland, SEM, × 85. (b) Androecium of *Chloranthistemon endressii* Crane, Friis, and Pedersen (Chloranthaceae) from the Santonian-Campanian of Åsen, southern Sweden, with two thecae dehisced (left) and the other two closed, SEM, × 50. (c) Stamen of *Mauldinia mirabilis* Drinnan *et al.* (Lauraceae) from the early Cenomanian Mauldin Mountain locality, Maryland, showing one of the two apically hinged valves, SEM, × 100. (d) *Platananthus marylandensis* Friis, Crane, and Pedersen from the late Albian West brothers locality, Maryland; lateral view showing recurved valves, SEM, × 100. (e) Platanoid stamen from the late Albian Bull Mountain locality, Maryland, ventral view showing recurved valves, SEM, × 100. (f) stamen of *Archamamelis* from the Santonian-Campanian of Åsen, southern Sweden, showing one theca dehisced, SEM, × 100.

Pollen

Pollen grains recovered from the Albian and Cenomanian angiosperm floral structures and stamens are typically very small and reticulate, with ultrastructure characterized by a thick footlayer and often a very distinct endexine. Several monocolpate types have been recovered, but most of the stamens contain tricolpate or tricolporoidate/tricolporate pollen. In several taxa pollen grains have an opaque covering, believed to represent remains of pollenkitt. These pollen types also occur in Santonian–Maastrichtian magnoliids and 'lower' hamamelidids. Floral structures with small tricolpate, tectate–striate forms first appear in the Cenomanian, and the later Cretaceous floral structures pollen is typically very small, tectate–foveolate or smooth, often embedded in pollenkitt-like substance. Larger triporate/tricolporate forms of the Normapolles type that are common in some Late Cretaceous floras typically have complex modifications of the exine around the apertures, and a distinctly granular rather than columellate pollen wall architecture.

Conclusions

Stamens with abundant sterile tissue and valvate dehiscence occur in extant angiosperms only in members of the Magnoliidae and 'lower' Hamamelididae (Endress and Hufford 1989; Hufford and Endress 1989). The predominance and diversity of this type of stamen in mid-Cretaceous floras confirms the importance of these two subclasses in the early angiosperm radiation.

The massive stamens, typically coupled with an unspecialized and insignificant perianth, in most Albian and Cenomanian flowers suggest that stamens probably had other functions in early angiosperm floral biology, and that these early stamen types were most likely important in attracting pollinators to the flowers. The prominent apical expansions of the connectives, as well as the massive connective tissue, were possibly visually attractive, but may also have had a secretory function, as is indicated by stomata-like openings and bulging structures on the surface of some stamens. The presence of pollenkitt in several taxa also suggests that these early flowers were probably insect pollinated. The connective protrusion in early angiosperm flowers may also have had an important role in protecting the developing pollen sacs, as is probably the case in extant *Platanus* and Annonaceae (Endress 1985). Later stamens show much larger diversity. They commonly have well-developed filaments and often small anthers. Apical extensions of connectives are rare and the attractive and protective function probably shifted from anthers to perianth in the late Cretaceous flowers, which typically have well-differentiated calyx and corolla, and often prominent nectaries. This evolutionary direction in angiosperm flowers is also supported by new studies of archaic extant angiosperms (e.g. Friis and Endress 1990). Pollen so far extracted from the mid-Cretaceous stamens shows a less clear evolutionary pattern, but the mid-Cretaceous pollen typically has a very pronounced footlayer and a distinct endexine. Pollen from the later Cretaceous shows a wider variation in pollen wall ultrastructure but generally has a thinner footlayer and endexine than observed in the mid-Cretaceous forms.

The increasing diversity of stamen types through the mid- and late Cretaceous parallels the increasing diversity of pollen through the same interval. However, investigations of pollen *in situ* document the variety of taxa that produced broadly similar tricolpate/tricolporate pollen during the mid-Cretaceous, and underline the difficulties in determining the diversity of such probable mid-Cretaceous hamamelidids from studies of dispersed pollen alone.

Acknowledgements

We thank P.K. Endress and A.N. Drinnan for valuable discussions and their collaboration in some aspects of this study. We also thank B. Krunderup and B. Larsen for assistance in preparing TEM sections; Y. Arremo, J. Sommer, and I. Svensson for assistance in preparing illustrations. This work was supported in part by the Swedish Natural Science Research Council grants G-GU 9381-300 and G-GU 9381-303 (EMF), the NSF research grant 8708460 (P.R.C), support from the Field Museum Visiting Scientist Bass Fund, Department of Geology Visiting Scientist Program, and NATO Science Program (K.R.P.)

References

Basinger, J. F. and Dilcher, D. L. (1984). Ancient bisexual flowers. *Science* **224**, 511–13.

Brenner, G.J. (1963). The spores and pollen of the Potomac Group of Maryland. *State of Maryland Department of Geology, Mines and Water Resources Bulletin* **27**, 1–215.

Chlonova, A.F. and Surova, T.D. (1988). Pollen wall ultrastructure of *Clavatipollenites incisus* Chlonova and two modern species of *Ascarina* (Chloranthaceae). *Pollen et Spores* **30**, 29–44.

Crabtree, D.R. (1987). Angiosperms of the northern Rocky Mountains: Albian to Campanian (Cretaceous) megafossil floras. *Annals of the Missouri Botanical Garden* **74**, 707–47.

Crane, P.R. (1989). Paleobotanical evidence on the early radiation of non-magnoliid dicotyledons. *Plant Systematics and Evolution* **162**, 165–91.

Crane, P.R. and Dilcher, D.L. (1984). *Lesqueria*: An early angiosperm fruiting axis from the mid-Cretaceous. *Annals of the Missouri Botanical Garden* **71**, 384–402.

Crane, P.R., Friis, E.M., and Pedersen, K.R. (1989). Reproductive structure and function in Cretaceous Chloranthaceae. *Plant Systematics and Evolution* **165**, 211–26.

Couper, R.A. (1958). British Mesozoic microspores and pollen grains. *Palaeontographica* **103B**, 75–179.

Couper, R.A. (1960). New Zealand Mesozoic and Cainozoic plant microfossils. *Bulletin of the New Zealand Geological Survey, Palaeontology* **32**, 1–87.

Dilcher, D.L. and Crane, P.R. (1984). *Archaeanthus*: An early angiosperm from the Cenomanian of the Western Interior of North America. *Annals of the Missouri Botanical Garden* **71**, 351–83.

Doyle, J.A. (1969). Cretaceous angiosperm pollen of the Atlantic Coastal Plain and its evolutionary significance. *Journal of the Arnold Arboretum* **50**, 1–35.

Doyle, J.A. and Hickey, L.J. (1976). Pollen and leaves from the mid-Cretaceous Potomac Group and their bearing on early angiosperm evolution. In *Origin and early evolution of angiosperms*, (ed. C.B. Beck), 139–206. Columbia University Press, New York.

Drinnan, A.N., Crane, P.R., Friis, E.M., and Pedersen, K.R. (1990). Lauraceous flowers from the Potomac Group (mid-Cretaceous) of eastern North America. *Botanical Gazette* **151**, 370–84.

Drinnan, A.N., Crane, P.R., Friis, E.M., and Pedersen, K.R. (1991). Angiosperm flowers and tricolpate pollen of buxaceous affinity from the Potomac Group (mid-Cretaceous) of eastern North America. *American Journal of Botany* **78**, 153–76.

Endress, P.K. (1985). Stamenabzission und Pollenpräsentation bei Annonaceae. *Flora* **176**, 95–8.

Endress, P.K. (1987). The Chloranthaceae: reproductive structures and phylogenetic position. *Botanische Jahrbücher für Systematik, Pflanzengeschichte und Pflanzengeographie* **109**, 153–226.

Endress, P.K. and Friis, E.M. (1991). *Archamamelis*, hamamelidalen flowers from the Upper Cretaceous of Sweden. *Plant Systematics and Evolution* **175**, 101–14.

Endress, P.K. and Hufford, L.D. (1989). The diversity of stamen structures and dehiscence patterns among Magnoliidae. *Botanical Journal of the Linnean Society* **100**, 45–85.

Endress, P.K. and Stumpf, S. (1991). The diversity of stamen structures in Lower Rosidae (Rosales, Fabales, Proteales, Sapindales). *Botanical Journal of the Linnean Society* (in press).

Fontain, W.M. (1889). The Potomac or younger Mesozoic flora. *United States Geological Survey Monographs* **15**, 1–377.

Friis, E. M. (1983). Upper Cretaceous (Senonian) floral structures of juglandalean affinity containing Normapolles pollen. *Review of Palaeobotany and Palynology* **39**161–88.

Friis, E. M. (1984). Preliminary report of Upper Cretaceous angiosperm reproductive organs from Sweden and their level of organization. *Annals of the Missouri Botanical Garden* **71**, 403–18.

Friis, E. M. (1985*a*). *Actinocalyx* gen. nov., sympetalous angiosperm flowers from the Upper Cretaceous of southern Sweden. *Review of Palaeobotany and Palynology* **45**, 171–83.

Friis, E. M. (1985*b*). Structure and function in Late Cretaceous angiosperm flowers. *Biologiske Skrifter—Det Kongelige Danske Videnskabernes Selskab* **25**, 1–37.

Friis, E. M. (1988). *Spirematospermum chandlerae* sp. nov., an extinct species of Zingiberaceae from the North American Cretaceous. *Tertiary Research* **9**, 7–12.

Friis, E. M. (1990). *Silvianthemum suecicum* gen. et sp. nov., a new saxifragalean flower from the Late Cretaceous of Sweden. *Biologiske Skrifter—Det Kongelige Danske Videnskabernes Selskab* **36**, 1–35.

Friis, E. M. and Crane, P. R. (1989). Reproductive structures of Cretaceous Hamamelidae. In *Evolution, systematics, and fossil history of the Hamamelidae*, (ed. P. R. Crane and S. Blackmore) pp. 155–74. Oxford University Press, Oxford.

Friis, E. M., Crane, P. R., and Pedersen, K. R. (1988). Reproductive structures of Cretaceous Platanaceae. *Biologiske Skrifter—Det Kongelige Danske Videnskabernes Selskab* **31**, 1–56.

Friis, E. M. and Endress, P. K. (1990). Origin and evolution of angiosperm flowers. *Advances in Botanical Research* **17**, 99–162.

Friis, E. M. and Skarby, A. (1982) *Scandianthus* gen. nov., angiosperm flowers of saxifragalean affinity from the Upper Cretaceous of southern Sweden. *Annals of Botany* **50**, 569–83.

Hufford, L. D. and Endress, P. K. (1989). The diversity of stamen structures and dehiscence patterns among Hamamelididae. *Botanical Journal of the Linnean Society* **99**, 301–46.

Hughes, N. F. (1976). *Palaeobiology of angiosperm origins*. Cambridge University Press, Cambridge.

Kvaček, Z. (1990). Lauralean angiosperms in the Cretaceous. *3rd International Senckenberg Conference, Frankfurt am Main, Abstracts*.

Knobloch, E. and Mai, D. H. (1986). Monographie der Früchte und Samen in der Kreide von Mitteleuropa. *Rozpravy ústredního ústavu geologickénho, Praha* **47**, 1–219.

Muller, J. (1970). Palynological evidence on early differentiation of angiosperms. *Biological Review* **45**, 417–50.

Němejc, F. and Kvaček, Z. (1975). *Senonian plant macrofossils from the region of*

Zliv and Hluboká (near Ceské Budejovice) in South Bohemia. Univerzita Karlova, Praha.

Pedersen, K. R., Crane, P. R., Drinnan, A. N., and Friis, E. M. (1991). Fruits from the mid-Cretaceous of North America with pollen grains of the *Clavatipollenites* type. *Grana* (in press)

Upchurch, G. R. Jr (1984). Cuticle evolution in Early Cretaceous angiosperms from the Potomac Group of Virginia and Maryland. *Annals of the Missouri Botanical Garden* **71**, 522–50.

Upchurch, G. R. Jr and Dilcher, D. L. (1990). Cenomanian angiosperm leaf megafossils, Dakota Formation, Rose Creek locality, Jefferson County, southeastern Nebraska. *United States Geological Survey Bulletin* **1915**, 1–52.

Walker, J. W. and Walker, A. G. (1984). Ultrastructure of Lower Cretaceous angiosperm pollen and the origin and early evolution of flowering plants. *Annals of the Missouri Botanical Garden* **71**, 464–521.

Walker, J. W., Brenner, G. J., and Walker, A. G. (1983). Winteraceous pollen in the Lower Cretaceous of Israel: early evidence of a magnolialean angiosperm family. *Science* **220**, 1273–5.

11. Systematic implications of comparative morphology in selected Tertiary and extant pollen from the Palmae and the Sapotaceae

M.M. HARLEY
M.H. KURMANN
I.K. FERGUSON
Palynology Unit, The Herbarium, Royal Botanic Gardens, Kew, Richmond, Surrey, UK

Abstract

There are many records of pollen and spores from the Tertiary based on light microscope investigations. In some instances such studies provide relatively limited information which can restrict accuracy of identification. Pollen morphological studies of many extant families have made clear that the much finer detail of the exine revealed with electron microscopy can, with few exceptions, considerably improve taxonomic discrimination of dispersed pollen. Using Palmae and Sapotaceae as examples, it is shown how ultrastructural study of fossil pollen, assignable to these families, extracted from Tertiary (Eocene) sediments of England, can contribute to a better understanding of their affinities.

Introduction

Pollen morphological surveys have been incorporated into systematic treatments of plant families for a long time. With the advent of electron microscopy additional characters, not revealed by light microscopy (LM), have been recognized. Such surveys provide an important reference base for the palaeopalynologist, since it is common practice to

Pollen and Spores (ed. S. Blackmore and S.H. Barnes), Systematics Association Special Volume No. 44, pp. 225–238, Clarendon Press, Oxford, 1991.

identify fossil palynomorphs by comparison with extant pollen. These comparisons are mostly based on light microscopical investigations, and only recently have some comparative studies incorporated ultrastructural features (e.g. Walker and Walker 1984; Daghlian *et al.* 1985; Khlonova and Surova 1988). These later investigations have shown that the additional characters revealed by scanning and especially transmission electron microscopy allow for improved taxonomic discrimination of dispersed pollen.

Pollen morphological surveys using light, scanning electron and transmission electron microscopy of the largely tropical and subtropical families, Palmae (Ferguson *et al.* 1983, 1987, 1988; Frederiksen *et al.* 1985; Ferguson 1986; Mendis *et al.* 1987; Harley 1989, 1990*b*; Harley and Hall 1991; Dransfield *et al.* 1990) and Sapotaceae (Harley 1986, 1990*a*, 1991*a,b*), have been carried out at the Royal Botanic Gardens, Kew. These were executed in connection with the taxonomic revisions of the Palmae (Dransfield and Uhl 1986; Uhl and Dransfield 1987) and of the Sapotaceae (Pennington 1991). Pollen produced by members of these families has also been recorded, based on light microscopical investigations, by Pallot (1961), Sein (1961), and Gruas-Cavagnetto (1976) from Tertiary sediments of southern England.

This chapter forms a logical extension of the pollen morphological studies carried out on the extant pollen at Kew. It reports on the ultrastructural investigations of some of the fossil pollen types and discusses their importance for systematics.

Materials and methods

Extant pollen of the Palmae and the Sapotaceae was collected and prepared as described in detail by Ferguson (1986) and Harley (1986), and is summarized here. The material was acetolysed using the method of Erdtman (1960). Pollen for scanning electron microscopy (SEM) was sputter-coated with platinum. For transmission electron microscopy (TEM) the pollen samples were fixed in OsO_4, pre-stained in uranyl acetate (UA) and embedded in epon araldite. The thin sections were subsequently stained with UA and lead citrate in an LKB Ultrostainer.

Samples containing fossil pollen were collected from Alum Bay, Isle of Wight, UK. They are from the Younger Leaf Beds (Crane 1977) of the Bracklesham Beds and are Middle Eocene in age.

The clayey sediments were processed by standard procedures (e.g. Traverse 1988), using 15 per cent hydrochloric acid overnight and 40 per cent hydrofluoric acid for 2–3 months. Samples were sieved through a 90 μm mesh copper sieve and the particles smaller than 90 μm were further treated. They were density separated with zinc chloride (sp. gr. 2.0) and oxidized for 2 min in Schulze's solution. A few drops of 5 per cent ammonia were added to release humic acids. The density separation was repeated to further concentrate the palynomorphs. A part of the sample was stored in a 50 per cent glycerol–water mixture.

A drop of the sample in the glycerol–water mixture was spread on a slide and scanned under low magnification with a light microscope. Pollen grains of the desired type were picked up with a small piece of glycerol jelly on the end of a fine entomological pin, mounted on an applicator stick. Each grain was placed on a 10 mm coverslip on a warming plate, covered by a second cover slip and photographed from both sides.

For TEM, the pollen sandwich was placed on the warming plate and separated. The grain was located and a drop of water placed on it. Using a fine pin, the pollen grain was teased off the jelly to float in the water. The coverslip was then inverted on to a millipore filter on an aspirator and the grain subsequently sucked on to the filter. The filter and adhering pollen grain were sandwiched between 1 per cent agar and processed in the same way as the extant pollen.

For SEM a drop of the sample in water was spread over a 10 mm diameter coverslip fixed on to an SEM stub using silver conductive paint. Alternatively, a pollen grain on a coverslip, picked out of the glycerol jelly mixture, was washed with distilled water and placed on an SEM stub with silver conducting paint.

Observations

Palmae

Spinizonocolpites Muller (1968). [Fig. 11.1] Due to the zonosulcate morphology, these pollen grains easily separate into two halves and therefore complete fossilized grains are uncommon. Shape, more or less spheroidal, zonosulcate. Size, long axis: (42-) 46.0 (-50) μm. Surface finely reticulate with supratectal spines (Fig. 11.1a,b,f). Spine length ranging from 2 to 12.5 μm, bases usually not swollen, apices mostly blunt (Fig. 11.1b,e,f) but sometimes pointed (Fig. 11.1a,b). Spines can be divided into three groups based on their length: 2–3.5 μm; 5–7 μm; and 8–12.5 μm. Exine thickness (excluding spines), 1.5–2 μm. Supratectal spines solid (Fig. 11.1c,e). Tectum, 0.4–0.6 μm. Infratectum regularly columellate, up to 0.9 μm (Fig. 11.1c,e). Footlayer, 0.4–0.6 μm, subtended by a narrow, lamellated layer (Fig. 11.1e, arrow). Lamellated layer prominent in aperture region (Fig. 11.1d).

The appearance of these grains is remarkably similar to *Nypa fruticans* Wurmb. (Fig. 11.2). *Nypa* differs in the following characters: Size, long

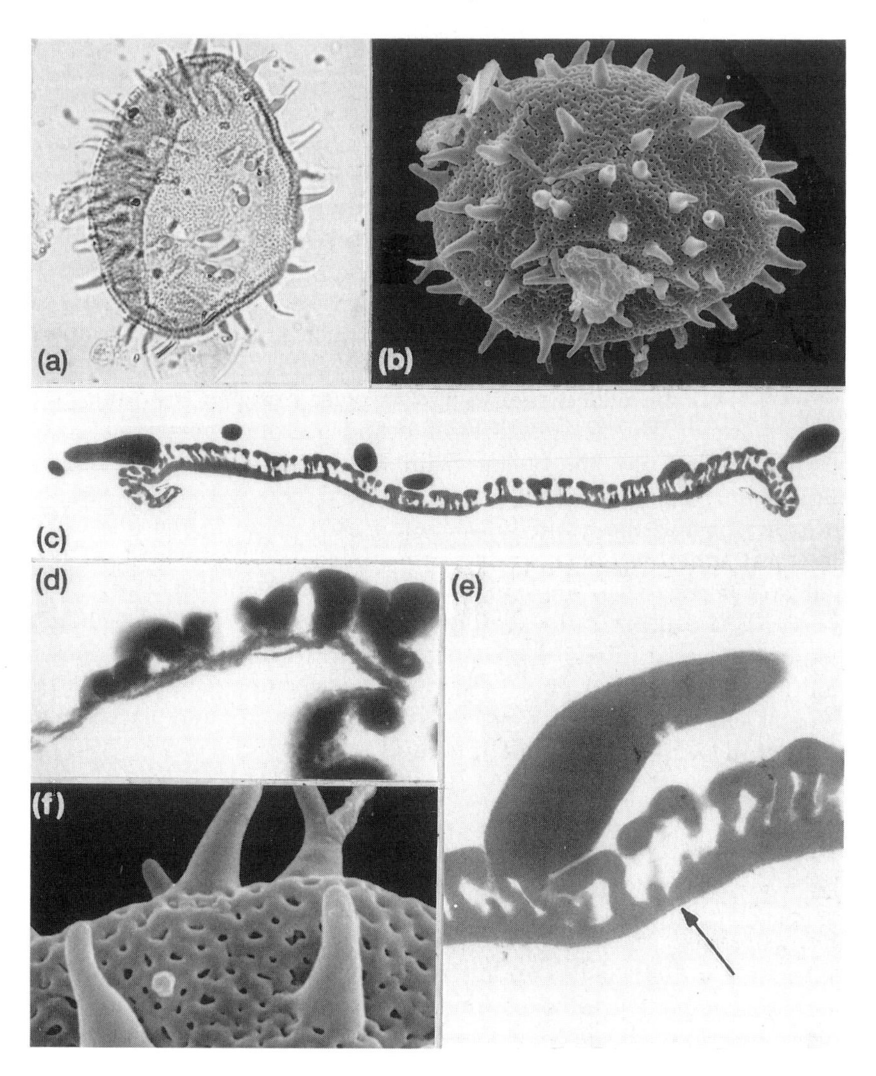

Fig. 11.1 *Spinizonocolpites*. (a) LM of pollen grain in semi-equatorial view, × 800. (b) SEM of pollen grain in polar view, × 1200. (c) TEM of half a pollen grain showing exine stratification, × 2600. (d) TEM of apertural region showing decreasing wall thickness and prominent lamellae, × 20 000. (e) TEM showing solid, supratectal spine; thin lamella subtending footlayer (arrow), × 6000. (f) SEM showing finely reticulate surface, × 5000.

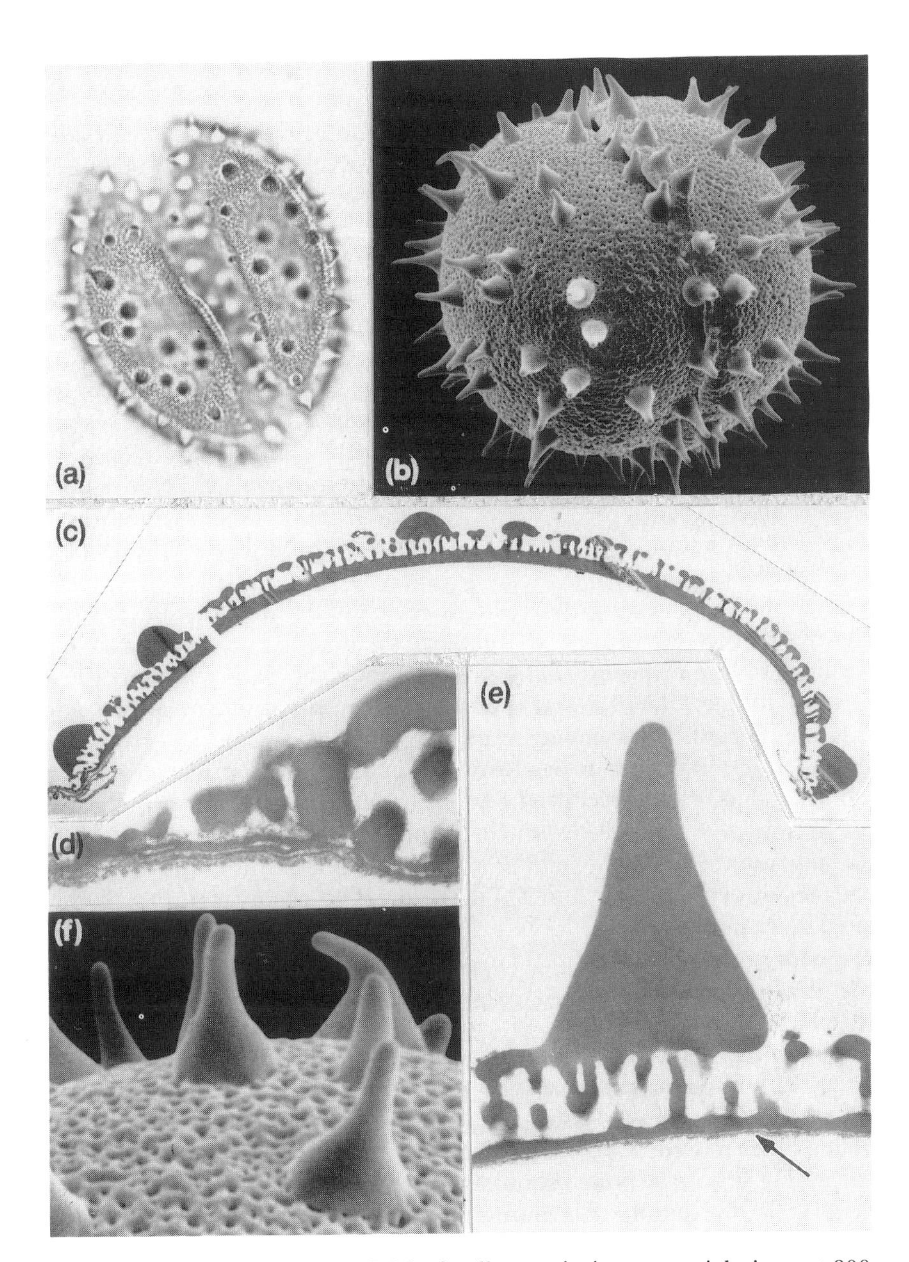

Fig. 11.2 *Nypa fruticans*. (a) LM of pollen grain in equatorial view, × 800. (b) SEM of pollen grain in semi-equatorial view, × 1200. (c) oblique TEM section of half a pollen grain showing exine stratification, × 2500. TEM of apertural region showing decreasing wall thickness and very distinct lamellae, × 20 000. (e) TEM showing solid, broad-based, supratectal spine; lamellae subtending footlayer (arrow), × 6000. (f) SEM of finely reticulate surface, × 5000. (a, from Bryan Jr 1277, Guam; b, e, from Whitmore 596, Thailand; c, d, f, from Tralau 3333, Sri Lanka.)

axis: (55-) 59.6 (-70) μm. Spines 4–7 μm long, bases usually broader and frequently swollen (Fig. 11.2a,b,e,f). Exine thickness (excluding spines), 1.5–3 μm. Tectum, 0.4–0.7 μm, infratectum up to 1.2 μm. Footlayer, 0.5–1 μm, subtended by a narrow lamellated layer (Fig. 11.2e, arrow). Lamellated layer very distinct in aperture region (Fig. 11.2d).

The size of the fossil pollen grains is approximately 20 per cent less than that of the extant grains. This is consistent with the findings of other workers, such as Buell (1946). The spines of the first group, 2–3.5 μm, are shorter than those of *Nypa fruticans*. Those of the second group, 5–7 μm, are comparable with extant *N. fruticans* and also with Muller's (1968) fossil species, *Spinizonocolpites echinatus*; while the spines of the third group, 8–12.5 μm, are noticeably longer than those recorded for extant *Nypa* and comparable with Muller's (1968) fossil species, *S. baculatus*.

Nypa fruticans, a mangrove palm of Asia and the west Pacific, is sufficiently distinct for almost all modern systematists to have placed it within the monospecific subfamily, Nypoideae (e.g. Tralau 1964; Dransfield and Uhl 1986). It is a morphologically uniform species and this is reflected in its pollen morphology. The considerable range in spine length apparent in the fossil grains is of interest in comparison with the small range recorded for recent *Nypa*. Variation in the shape of the spines is also apparent. Some of the spines appear hooked. As already pointed out by Babajide-Salami (1985), this is probably an artefact of compression and fossilization. The occurrence of a lamellated layer underlying the footlayer, observed in these grains, is unusual in pollen grains within the Palmae, it has only been recorded within the subfamily Calamoideae Griff., in the tribe Calameae Drude (Frederiksen *et al.* 1985) and in the tribe Lepidocaryeae Mart. (Sowunmi 1972; Ferguson 1986).

Nypa-like pollen is traceable to the Upper Cretaceous (Jardiné and Magloire 1965; Muller 1968), making it one of the oldest fossil palm pollen types recorded. In view of its easily recognized pollen, it is hardly surprising that *Nypa* pollen should occupy this niche in the evolutionary history for the family. Undoubtedly there are pollen grains of other palm species in the Cretaceous, with less distinctive pollen, not so far identified. It is interesting to note that it appears to have been more diverse and enjoyed a far more cosmopolitan distribution than the extant species. *Nypa* is considered to have evolved separately from a protopalm stock along its own evolutionary line (Moore 1973). It is possible that *N. fruticans* is a relict species of a previously more diverse group. That we should find notable diversity, including Muller's two form species, in a small Tertiary sample, as opposed to the almost unvarying morphology of the pollen of the extant species, provides good evidence for such a hypothesis.

Monocolpopollenites. Pflug and Thomson in Thomson and Pflug (1953). [Fig. 11.3a–e] A large number of grains in the sample were monsulcate with possible palm-like affinities. Shape, ellipsoid, monosulcate, sulcus *c.* the same length as long axis, acuminate. Size, long axis: 31–43 μm; short axis: 20–28 μm. Surface appears perforate, scabrate or finely rugulate.

One of the grains, embedded and thin sectioned, is illustrated (Fig. 11.3a). Shape, ellipsoid, monosulcate, sulcus, *c.* the same length as long axis, acuminate. Size, long axis: 37 μm; short axis; 26 μm. Surface finely rugulate (Fig. 11.3a). Exine thickness: 1–1.25 μm. Tectum, 0.4–0.6 μm, uneven. Infratectum, 0.2–0.4 μm, comprised of irregularly shaped and spaced columellae (Fig. 11.3c). Footlayer, 0.2–0.4 μm, irregular surface at base of columellae (Fig. 11.3e). A fossil grain isolated from a strew stub, studied with SEM, which has similar characteristics to those of the thin-sectioned grain, is illustrated in Fig. 11.3b and d.

Very similar wall stratification has been recorded within the recent Palmae. The most striking resemblances are found within the subfamily Arecoideae: tribe Areceae: within the subtribe Iguanurinae. This group, in which variation in pollen morphology is more or less limited to pollen wall ultrastructure, is othewise noted for its diversity of habit. The pollen wall stratification of the monospecific genus, *Satakentia liukiuensis* (Hatusima) H.E. Moore (Fig. 11.3f–h), from the Ryuku Islands in the western Pacific Ocean, has notable similarities. It only differs in the following characters: Size, long axis: (43–) 44.2 (–45) μm; short axis: (26–) 26.8 (–30) μm. Surface more or less smooth, perforate. Exine thickness: *c.* 1 μm. Tectum, 0.3–0.4 μm. Infratectum, 0.2–0.3 μm. Footlayer, 0.3–0.4 μm. Such close similarity has not been observed in the pollen wall morphology of the numerous examples of extant coryphoid genera (Dransfield *et al.* 1990).

The majority of genera cited for the north European fossil palm record are coryphoid, generally considered to be the least specialized group of palms. This undoubtedly relates to the macrofossil records which contain many examples of costapalmate leaves, particularly associated with the tribe Corypheae. The tribe also contains a majority of genera with simple, tectate, monosulcate pollen grains. However, Read and Hickey (1972) question the wisdom of making identifications based on leaves alone, since this is a very difficult character with which to identify even extant palm genera.

From an analysis of the diversity among the palms, Moore and Uhl

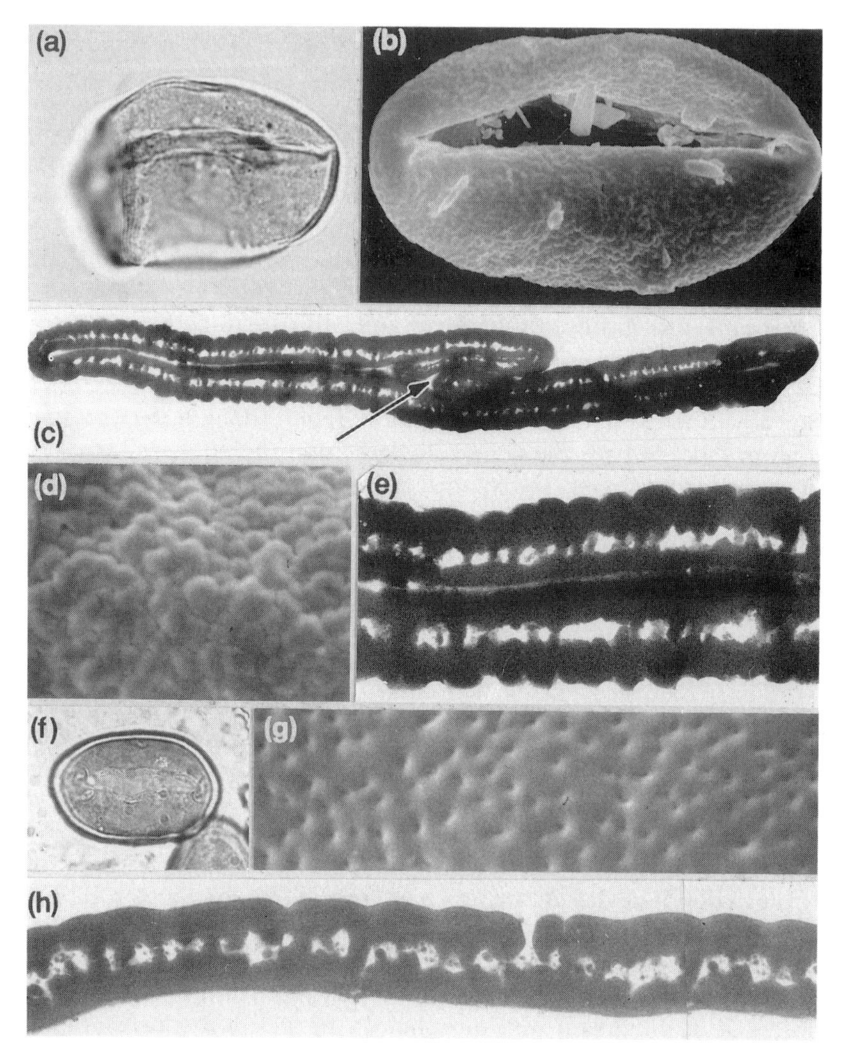

Fig. 11.3 a–e, *Monocolpopollenites*. (f–h) *Satakentia liukiuensis* (from Moore 9382, Ryuku Islands). (a) LM of pollen grain in polar view, × 1000. (b) SEM of pollen grain in polar view, × 2400. (c) TEM, transverse section of whole grain, apertural area enfolded (arrow), × 6000. (d) SEM of the finely rugulate surface, × 8000. (e) TEM showing exine stratification, × 15 000. (f) LM of pollen grain in polar view, × 500. (g) SEM of finely perforate surface, × 8000. (h) TEM showing exine stratification, × 9000.

(1982) considered that something like 5–7 major types derived very early from a protopalm stock and have evolved separately. It is probable that quite highly evolved palm genera were already in existence by the mid-Tertiary. The result of this study suggests that the simple monosulcate fossil pollen type also has affinities with genera in some of the more specialized subfamilies where this pollen type is also frequent (Harley 1990*b*). This should be taken into consideration when identifying this type of fossil pollen.

Sapotaceae

Tetracolporopollenites. Pflug and Thomson in Thomson and Pflug (1953). [Fig. 11.4] Shape, prolate-spheroidal, subprolate or prolate. Four-aperturate, colpi 2/3–5/6 polar length. Broadly or narrowly lalongate endoapertures. Size, polar axis: 25–37 μm; equatorial axis: 15–30 μm. Surface more or less smooth. Exine thickness: 1–2 μm.

One of the grains, subsequently embedded and thin-sectioned, is illustrated in Fig. 11.4a,b. Shape, subprolate, four-colporate, colpi 2/3 length of polar axis, endoapertures broadly lalongate. Size, polar axis: 30 μm; equatorial axis: 26 μm. Surface more or less smooth, finely perforate, perforations coarser in apocolpia. Exine thickness, 1.75–2 μm. Tectum thickness, 0.5–0.7 μm, interrupted in mesocolpia by fine channels (Fig. 11.4d,e) and in apocolpia by somewhat wider channels (Fig. 11.4d,g). Infratectum ranges from 0.2 to 0.3 μm in mesocolpia and 0.3–0.4 μm in apocolpia. The areas between the very short, irregularly shaped and spaced columellae contain loosely 'packed', irregularly sized granules of ectexinous material (Fig. 11.4e,g). Footlayer, 0.5–0.7 μm, thick, solid, and more or less smooth at the ectexine–endexine interface, much-reduced and granular in apertural area. Endexine very thick with slightly spongy appearance in endoapertural region (Fig. 11.4d,e), much thinner in mesocolpia, thinned towards and absent in apocolpia (Fig. 11.4d, arrows). A fossil grain studied with SEM, isolated from a strew stub, which has similar characteristics to those of the thin-sectioned grain is illustrated in Fig. 11.4c,f.

The fossil pollen grains show some notable similarities with those of the recent west African rainforest species *Tieghemella heckelii* Pierre ex Chevalier. [Fig. 11.5]. *Tieghemella* pollen differs in the following characters: four- or five-colporate, colpi 2/3–3/4 length of polar axis. Size, polar axis: (35–) 36.7 (–38) μm; equatorial axis: (28–) 24.4 (–34) μm. Surface finely granular (Fig. 11.5f). Exine thickness, 1.5–2 μm (Fig. 11.5g). Tectum, 0.5–0.7 μm. Infratectum ranges from 0.3 to 0.4 μm in mesocolpia and 0.4–5 μm in apocolpia. Footlayer, 0.6–0.8 μm.

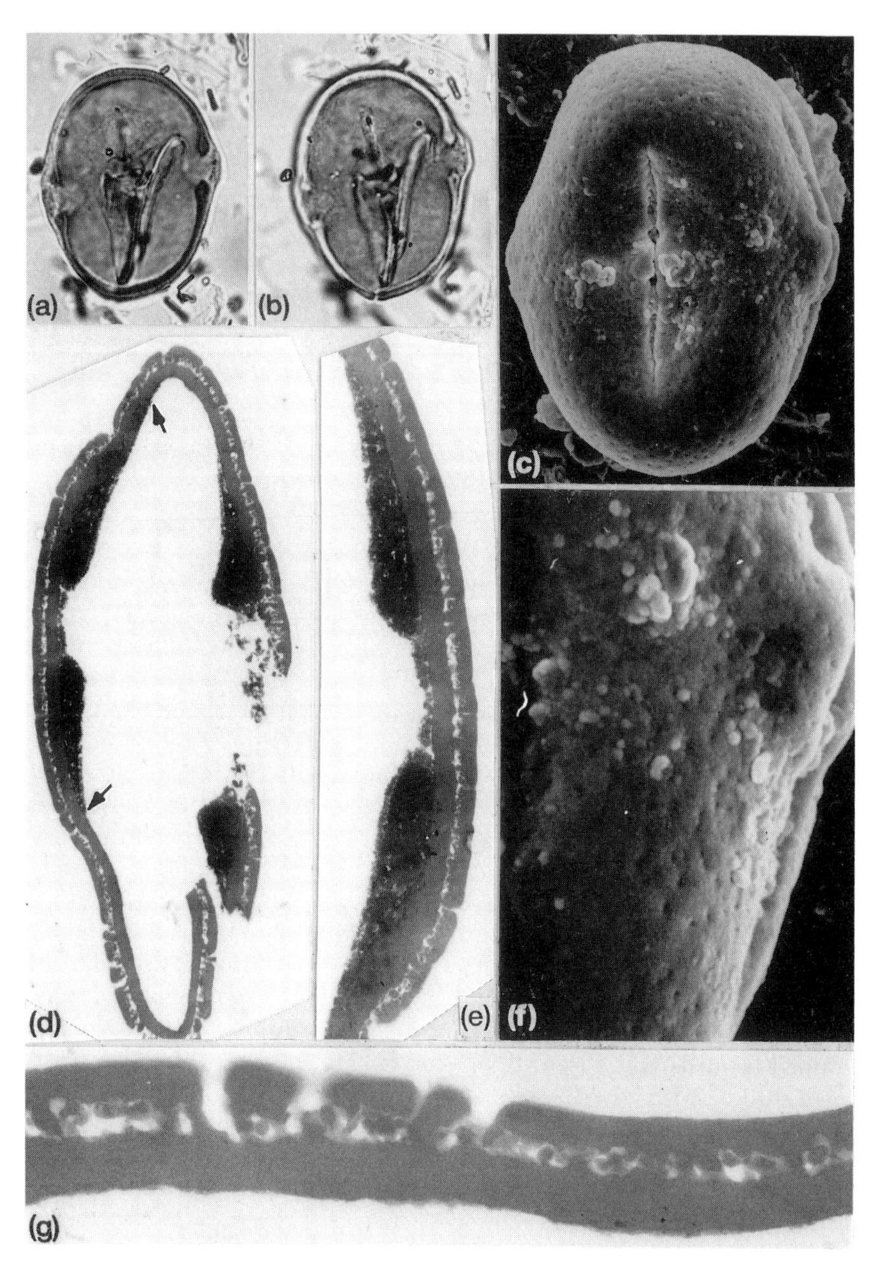

Fig. 11.4 *Tetracolporopollenites*. (a) High focus of pollen grain in equatorial view, × 800; (b) Mid-focus, × 800. (c) SEM of pollen grain in equatorial view, × 2400. (d) TEM of whole pollen grain, endexine between arrows, × 4000. (e) TEM, longitudinal section cut through endoaperture, × 7000. (f) SEM of more or less smooth, perforate surface, × 6400. (g) TEM detail of exine in apocolpial region, × 19 000.

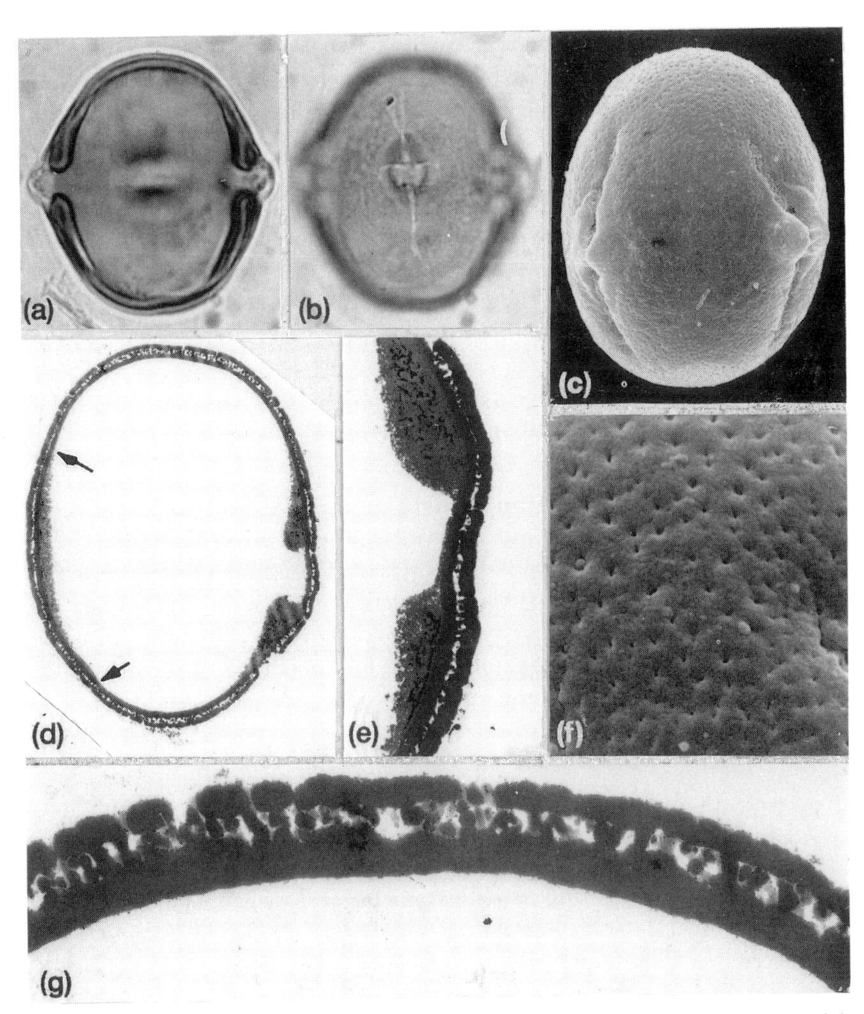

Fig. 11.5 *Tieghemella heckelii* (from Leeuwenberg 2752, Ivory Coast). (a), LM optical section of pollen grain in equatorial view, × 1000; (b) LM low focus, × 1000. (c) SEM of pollen grain in equatorial view, × 2200. (d) TEM of pollen grain in polar plane, endexine between arrows, × 2000. (e) TEM, longitudinal section through endoaperture, × 5400. (f) SEM of more or less smooth, perforate surface, × 10 000. (g) TEM detail of exine in apocolpial region, × 12 000.

With few exceptions, Sapotaceae pollen is easily recognized based on its morphology. Exine stratification together with shape and apertures characterize the pollen of this family. Harley (1991*a*) described 12 pollen types for the family. Although four of the pollen types are restricted to a few of the smaller genera or sections of larger genera, the pollen may not always be taxonomically useful at the generic level. *Tieghemella heckelii* is placed in Pollen Type 1, which occurs predominantly in tribes Mimusopeae and Isonandreae (Pennington 1991).

Fossil pollen records for the Sapotaceae are widespread, suggesting a cosmopolitan distribution during the early Tertiary. Fossil pollen grains for all the major types described for the recent Sapotaceae have been published (see Harley 1991*b* for references).

Conclusion

Despite the few taxa illustrated in this chapter, the study demonstrates how careful examination and comparison of dispersed pollen with that of extant species, especially using transmission electron microscopy, allows for more accurate identification. Such studies are of fundamental importance because they provide an insight into the affinities of dispersed fossil pollen grains. Furthermore, this contributes to a better understanding of the evolution of angiosperms and facilitates palaeoecological interpretation.

Acknowledgements

We would like to express our sincere gratitude to Dr C. R. Hill, Department of Palaeontology, The Natural History Museum, for his invaluable guidance to Alum Bay. We are grateful to Peter Cade for preparing the pollen of *Nypa*, to Ian Flawn for preparing pollen of *Satakentia*, and to Sue Ellison for skilful printing of the photographs.

References

Babajide-Salami, M. (1985). Upper Senonian and lower Tertiary pollen grains from the southern Nigeria sedimentary basin. *Revista Española Micropaleontologia* **17**, 5–26.

Buell, M. F. (1946). Size frequency study of fossil pine pollen compared with herbarium preserved pollen. *American Journal of Botany* **33**, 510–16.

Crane, P. R. (1977). The Alum Bay plant beds. *Tertiary Research* **1**, 95–9.

Daghlian, C. P., Skvarla, J. J., Pocknall, D., and Raven, P. H. (1985). *Fuchsia* pollen from the early Miocene of New Zealand. *American Journal of Botany* **72**, 1039–47.

Dransfield, J. and Uhl, N. W. (1986). An outline of a classification of palms. *Principes* **30**, 3–11.

Dransfield, J., Ferguson, I. K., and Uhl, N. W. (1990). The coryphoid palms: patterns of variation and evolution. *Annals of the Missouri Botanical Garden* **77**, 802–15.

Erdtman, G. (1960). The acetolysis method. A revised description. *Svensk Botanisk Tidskrift* **54**, 561–4.

Ferguson, I. K. (1986). Observations on the variation in pollen morphology of Palmae and its significance. *Canadian Journal of Botany* **64**, 3079–90.

Ferguson, I. K., Dransfield, J., Page, F. C., and Thanikaimoni, G. (1983). Notes on the pollen morphology of *Pinanga* with special reference to *P. aristata* and *P. pilosa* (Palmae: Arecoideae). *Grana* **22**, 65–72.

Ferguson, I. K., Havard, A. J., and Dransfield, J. (1987). The pollen morphology of tribe Borasseae (Palmae: Coryphoideae). *Kew Bulletin* **42**, 405–22.

Ferguson, I. K., Dransfield, J. and Flawn, I. (1988). A review of the pollen morphology and systematics of the genera *Ravenea* and *Louvelia* (Ceroxyleae: Ceroxyloideae: Palmae). *Journal of Palynology* **23–24**, 65–72.

Frederiksen, N. O., Wiggins, V. D., Ferguson, I. K., Dransfield, J., and Ager, C. M. (1985). Distribution, paleoecology, paleoclimatology, and botanical affinity of the Eocene pollen genus *Diporoconia* n.gen. *Palynology (Dallas)* **9**, 37–60.

Gruas-Cavagnetto, C. (1976). Étude palynologique du Paléogène du sud de l'Angleterre. *Cahiers de Micropaléontologie* **1**, 1–49.

Harley, M. M. (1986). Distinguishing pollen characters for the Sapotaceae. *Canadian Journal of Botany* **64**, 3091–100.

Harley, M. M. (1989). Pollen morphology of *Voanioala gerardii* J. Dransf. (Palmae: Arecoideae: Cocoeae: Butiinae) from Madagascar. *Kew Bulletin* **44**, 199–205.

Harley, M. M. (1990*a*). Pollen. In Pennington, T. D. Sapotaceae. Flora Neotropica Monograph 52, pp. 11–29, 710–41. New York Botanical Garden.

Harley, M. M. (1990*b*). Occurrence of simple tectate, monosulcate or trichotomosulcate pollen grains within the Palmae. *Review of Palaeobotany and Palynology* **64**, 137–47.

Harley, M. M. (1991*a*). Pollen morphology of the Sapotaceae. In Pennington, T. D. The Genera of Sapotaceae pp. 23–50. Royal Botanic Gardens, Kew.

Harley, M. M. (1991*b*). The pollen morphology of the Sapotaceae. *Kew Bulletin* 46, in press.

Harley, M. M. and Hall, D. H. (1991). Pollen morphology of the African palms. *Palaeoecology of Africa and the Surrounding Islands*, in press.

Jardiné, S. and Magloire, L. (1965). Palynologie et Stratigraphie du Crétacé des bassins du Sénégal et de Côte d'Ivoire. *Mémoires du Bureau de Recherches Géologiques et Minières* **32**, 187–245.

Khlonova, A. F. and Surova, T. D. (1988). Comparative analysis of sporoderm ultrastructure of *Clavatipollenites incisus* and two species of *Ascarina*

(Chloranthaceae). *Botanicheskii Zhurnal* **73**, 305–14.

Mendis, N.M., Ferguson, I.K., and Dransfield, J. (1987). The pollen morphology of the subtribe Oncospermatinae (Palmae: Arecoideae: Areceae). *Kew Bulletin* **42**, 47–63.

Moore, H.E. Jr (1973). Palms in the tropical forest ecosystems of Africa and South America. In (ed. B.J. Meggers, E.S. Ayensu, and W.D. Duckworth), *Tropical forest ecosystems in Africa and South America: A comparative review*, pp. 63–88. Smithsonian Institution Press, Washington, DC.

Moore, H.E. Jr and Uhl, N.W. (1982). Major trends of evolution in palms. *Botanical Review* **48**, 1–69.

Muller, J. (1968). Palynology of the upper Pedawan and Plateau sandstone formations (Cretaceous–Eocene) in Sarawak, Malaysia. *Micropalaeontology* **14**, 1–37.

Pallot, J.M. (1961). Plant microfossils from the Isle of Wight. Unpublished Ph.D. thesis. University of London.

Pennington, T.D. (1990). Sapotaceae. *Flora Neotropica Monograph* 52, pp. 1–10, 30–709, 742–70. New York Botanical Garden.

Pennington, T.D. (1991). The genera of Sapotaceae Royal Botanic Gardens, Kew.

Read, R.W. and Hickey, L.J. (1972). A revised classification of fossil palm and palm-like leaves. *Taxon* **21**, 129–37.

Sein, M.K. (1961). Fossil spores of the London Clay. Unpublished Ph.D. thesis. University of London.

Sowunmi, M.A. (1972). Pollen morphology of the Palmae and its bearing on taxonomy. *Review of Palaeobotany and Palynology* **13**, 1–80.

Thomson, P.W. and Pflug, H. (1953). Pollen und Sporen des mitteleuropäischen Tertiärs. *Palaeontographica* **94B**, 1–138.

Tralau, H. (1964). The genus *Nypa* van Wurmb. *Kungliga Svenska Vetenskapsakademiens Handlingar, Fjärde Serien* **10**, 5–29.

Traverse, A. (1988). *Paleopalynology*. Unwin Hyman, Boston.

Uhl, N.W. and Dransfield, J. (1987). *Genera Palmarum. A classification of palms based on the work of H. E. Moore, Jr*. International Palm Society and The L. H. Bailey Hortorium, Kansas, USA.

Walker, J.W. and Walker, A.G. (1984). Ultrastructure of lower Cretaceous angiosperm pollen and the origin and early evolution of flowering plants. *Annals of the Missouri Botanical Garden* **71**, 464–521.

12. Determining character polarities in pollen

M.S. ZAVADA
Department of Biology, University of Southwestern Louisiana, Lafayette, USA

Abstract

Four widely used methods for establishing the polarity of character transformations, namely the outgroup method, the parsimony method, the palaeontological method, and the ontogenetic method, are considered in relation to characters relevant to discussions of seed plant phylogeny. Special emphasis is given to discussions of ontogenetic data. Two basic patterns of development of the primexine into sexine are described, and it is suggested that in the ancestral condition for seed plants primexine deposition is immediately followed by deposition of protosporopollenin on sites corresponding to the future tectum, and then the infrastructural layer. Problems arising from the occurrence of pre-angiospermous dispersed fossil pollen that possesses features considered characteristic of angiosperms are discussed in detail.

Introduction

Evolutionary botanists have investigated the morphology, development, and fossil history of various plants and plant organs, to gain insight into their phylogenetic relationships. Over the past 30 years pollen has been added to the list of plant organs that may provide insight into evolutionary relationships of various plant groups. Pollen has been shown to be useful in determining taxonomic position at various taxonomic levels depending on the group under consideration (Wodehouse 1935; Traverse 1988). In addition, the fossil pollen record appears to be more complete than it is for other dispersed plant parts (e.g. leaves, wood, and reproductive structures). The fossil pollen record may lead

Pollen and Spores (ed. S. Blackmore and S.H. Barnes), Systematics Association Special Volume No. 44, pp. 239–256, Clarendon Press, Oxford, 1991.

to an understanding of the timing of the origin of various plant groups (Doyle 1969), their palaeogeographical distribution (Herngreen and Chlonova 1981), and the temporal derivation of morphological characters (Chaloner 1967). Pollen may also be useful for determining evolutionary relationships among major groups of plants. One botanical problem that pollen studies may contribute to is our understanding of the origin of angiosperms (Doyle 1969). However, determining the polarities among pollen characters is a necessary prerequisite for evaluating the significance of pollen for understanding evolutionary relationships. This must be done independently of the establishment of polarities for other plant organs that are associated with the pollen. The fact that a taxon possesses a suite of characters with established polarities in one organ is not support for the assumption that the polarities of the pollen characters will have the same directionality (e.g. *Magnolia* pollen can not be assumed to be primitive based on the occurrence of primitive characters in other organs).

There are four widely used methods for establishing character polarities:

(1) the outgroup method (Watrous and Wheeler 1981; Farris 1982);
(2) the parsimony method (Maddison *et al.* 1984);
(3) the palaeontological method (Donoghue *et al.* 1989); and
(4) the ontogenetic method (Nelson 1978; Nelson and Platnik 1981; Patterson 1982, 1983).

The outgroup method considers a character to be ancestral if it occurs in the outgroup and the ingroup (i.e. common equals primitive). Estimating ancestral states with the parsimony method assumes that character *x* found in the outgroup transformed to *y* in the ingroup is derived. However, character *x* found in the outgroup remaining *x* in the ingroup is primitive. In the palaeontological method, if character *x* precedes character *y* in time (character *y* believed to be derived from character *x*), character *x* is ancestral. The ontogenetic method states, given that ancestral characters are retained in descendant ontogenies, ancestral characters are more general than derived characters. All of these methods have their strengths and weaknesses (for discussion see Farris 1982; Maddison *et al.* 1984; De Queiroz 1985). The present contribution applies these methods to determining pollen character polarities among the angiosperms and gymnosperms (*sensu lato*) that may have played a role in the origin of angiosperms. Special consideration is given to the ontogenetic data, and its usefulness for determining pollen character polarities.

General features of pollen development

Pollen ontogeny exhibits a number of general features, i.e. features common to the developmental patterns of gymnosperm and angiosperm pollen (Dickinson 1976). The initiation of the pollen mother cell (PMC), followed by meiosis which results in four haploid microspores, each surrounded by a cell wall (intine) is common to all seed plants. Concomitantly with the process of reduction division is the development of a sporopolleninous wall (exine) prior to the development of the intine (cell wall). The transition from aquatic to terrestrial habitat was probably accompanied by the evolutionary origin of the exine. The occurrence of a sporopolleninous wall surrounding the microspores is common (general) to almost all seed plants. In seed plants, the sequence of developmental events associated with exine development are similar. Initially there is formation of the callose special wall (CSW) around the PMC and subsequently around each of the microspores in the tetrad following meiosis. Concomitantly or immediately following CSW formation around each of the microspores is the development of the primexine on which the sexine (tectum and infrastructural layer) pattern is established, followed by sporopollenin polymerization. Just prior to, or immediately following, the completion of the sexine, the basal layer begins development (footlayer and endexine) on short or long unit-like membranes. After completion of the basal layer, the intine (cell wall) develops and is completed by the time of anthesis. This general pattern of wall development varies little among gymnosperms and angiosperms. A general developmental paradigm common to a large group of organisms (e.g. seed plants) is expected. Drastic changes in the overall developmental pattern, or a completely new developmental paradigm, would be unexpected. Most evolutionary changes occur within the constraints of an already existing developmental paradigm (Wolpert 1983). During the course of evolution in pollen it is reasonable to assume that changes will occur in the timing of developmental events, the position of individual components of the developing structure, and prolongation or truncation of the developmental sequence. Sexine structure varies considerably; however, this variation can be attributed to changes in the position of recognition sites for sporopollenin polymerization on the primexine. Likewise, the relationship between the timing of primexine development and sexine development might also vary, e.g. the formation of the primexine is sequentially followed by sexine development, or the primexine is fully developed then followed by sexine development. This type of variation is due to a change in the relative timing of these events, not a major change in the overall developmental sequence.

The mode of development in the footlayer and endexine is similar (Blackmore and Crane 1987). Footlayer is distinguished from endexine based on differences in its staining properties with basic fuchsin, which may reflect differences in the chemistry of these two wall layers (Faegri and Iversen 1950; Traverse 1988). The presence or absence of footlayer and endexine can be attributed to changes in the timing of chemical changes in sporopollenin or sporopollenin-associated compounds deposited on short or long unit-like membranes during development of the basal layer. Alternatively, it can be attributed to the prolonged development of the basal layer accompanied by a chemical change in the sporopollenin or sporopollenin-associated compounds.

Variation in tetrad symmetry, aperture morphology, and distribution are due to positional changes in the control mechanisms of these particular features and do not reflect major changes in the developmental paradigm of pollen. The pattern of pollen development can be considered conservative; however, aspects of timing of developmental events and the positional arrangement of the structural elements that comprise the pollen wall, apertures, and tetrad symmetry appear to be highly variable based on morphological studies of pollen.

Summary of developmental patterns in selected gymnosperms and angiosperms

The wall development of pollen of various gymnosperm and angiosperm families is considered. These families are: Ginkgoaceae (Rohr 1977), Cycadaceae (Audran 1981; Zavada 1983*a*), Taxodiaceae (Kurmann 1986), Taxaceae (Pennel and Bell 1986), Pinaceae (Dickinson 1971), Gnetaceae (Gabarayeva and Zavada, in preparation), Magnoliaceae (Gabarayeva 1986*a*, *b*, 1987*a*, *b*), Annonaceae (Waha 1987), Austrobaileyaceae (Zavada 1984*a*), Ulmaceae (Rowley and Rowley 1986), and Liliaceae (Heslop-Harrison 1968). The studies on the pollen wall development of gymnosperms represent a diversity of pollen types that include pollen with granular and alveolar wall structure, and saccate and non-saccate types. The angiosperm families have pollen with presumably ancestral and derived characteristics.

Common to the development of seed plant pollen is the formation of a primexine (pre-exine or glycocalyx) once the microspores have been sequestered by the CSW. The primexine is presumably deposited by the microspore early in the tetrad phase. It has been suggested that the primexine has recognition sites for sporopollenin polymerization that corresponds to the pattern of the sexine (Rowley and Rowley 1986). There may be two basic patterns of primexine–sexine development:

1. Primexine deposition begins and is immediately followed by

deposition of proto-sporopollenin on the recognition sites of the future tectum, and subsequently on the infrastructural layer. This mode of development gives the impression that the sexine is developed successively from the outside, inward.

2. In the second type, primexine deposition is completed prior to the deposition of proto-sporopollenin on any of the sexinal recognition sites; however, the first recognizable electron-dense sexine elements are in the future infrastructural layer. This is immediately followed by the lateral accretion of proto-sporopollenin to form the tectum.

In both modes of development the sexine is completed by time the microspores are released from the callose special wall.

The distribution of these two modes of sexine development, among the taxa considered, is difficult to determine. This is due to the inconsistency with which different authors describe similar developmental events. Thus, the presence or absence of one of these developmental modes is subject to some interpretation. Patterns of sexine development similar to the first type have been reported in the Cycadaceae (*Ceratozamia mexicana*, Audran 1981; *Zamia floridana*, Zavada 1983*a*), Ginkgoaceae (*Ginkgo biloba*, Rohr 1977), Pinaceae (*Abies concolor*, Kurmann 1986; *Pinus banksiana*, Dickinson 1971; *Tsuga canadensis*, Kurmann 1986), Gnetaceae (*Welwitschia mirabilis*, Gabarayeva and Zavada, in preparation), Magnoliaceae (*Michelia fuscata*, Gabarayeva 1986*a*, *b*), and the Annonaceae (*Asimina triloba*, Waha 1987). The successive development of the sexine from the outside in, on a partially or near complete primexine, is a feature of all these taxa during the early to mid-tetrad phase.

The second type of sexine development, i.e. formation of the primexine followed by polymerization of proto-sporopollenin on the infrastructural elements, may be restricted to angiosperms. This type is found in the Magnoliaceae (*Mangletia tenuipes*, Gabarayeva 1987*a*, *b*, *c*), Austrobaileyaceae (*Austrobaileya maculata*, Zavada 1984*a*), Ulmaceae (*Ulmus* sp., Rowley and Rowley 1986), and Liliaceae (*Lilium longifolium*, Heslop-Harrison 1968).

A third type of sexine development, which involves the deposition of the sexine by tapetal tissues, occurs in *Taxodium distichum* (Kurmann 1986) and *Taxus baccata* (Pennel and Bell 1986). In these taxa the nexine begins development in the tetrad phase and, upon release of the microspores from the CSW, spherical to irregular shaped granules from the tapetum are apposed to the outer layer of nexine, forming the sexine. This type of development is associated with sexines comprised of compacted granules. Despite the tapetal origin of the sporopollenious granules, the position of the granules on the outer surface of the nexine may be due to the deposition of the recognition sites by the microspores prior to nexine development in the tetrad phase.

In the first two developmental modes described above, neither is associated with one type of wall structure. The occurrence of the successive type of development (1) in both gymnosperms and angiosperms and the restriction of the second type (2) to angiosperms suggests that (1) is the ancestral type.

Following the formation of the sexine is the development of the basal layer on short or long unit-like membranes sequentially appressed to the base of the sexinal elements. The basal layer can be composed of two distinctive layers, based on their staining properties with basic fuchsin, the footlayer and endexine (nexine 1 and nexine 2, respectively). The footlayer is usually contiguous in non-apertural regions. The endexine may be restricted to the region adjacent to the apertures, contiguous, or completely absent. The recognition of these layers with electron microscopic stains (e.g. uranyl acetate and lead citrate or permanganate) is variable, and no reliable standard has been established. The footlayer generally stains similarly to the sexine. The endexine, when present, may appear more electron dense or electron translucent than the footlayer. Investigators have recognized the presence of endexine based on subtle changes in staining of the basal layer, differences in staining that other workers may consider insignificant. Thus, there is no agreement on what actually constitutes an endexine in pollen investigated with the electron microscope (Zavada 1984*b*).

The development of the footlayer is usually completed by the time the four microspores are released from the callose special wall. In the Cycadaceae (Audran 1981; Zavada 1983), Ginkgoaceae (Rohr 1977), Pinaceae (Dickinson 1971), Taxodiaceae (Kurmann 1986), Taxaceae (Pennel and Bell 1986), Annonaceae (Waha 1987), and Liliaceae (Heslop-Harrison 1968) the sporopolleninous wall layers (sexine and nexine, *sensu lato*) complete their development by the time the microspores are released from the CSW. None of these taxa exhibit staining differences in the basal layer that are comparable to the staining differences with basic fuchsin in pollen taxa with an endexine observed with light microscopy. The development of the inner portion of the basal layer continues in the free spore phase in the Gnetaceae (Gabarayeva and Zavada, in preparation), Magnoliaceae (*Michelia fuscata*, Gabarayeva 1986*a*, *b*; *Manglietia tenuipes*, Gabarayeva 1987*a*, *b*, *c*), Austrobaileyaceae (Zavada 1984*a*), and Ulmaceae (*Ulmus* sp., Rowley and Rowley 1986). Endexine has been recognized in the magnoliaceous taxa, based on the time and mode of formation and resistance to solubility in 2-aminoethanol (Gabarayeva 1987*b*). However, no significant difference in staining is evident in these taxa.

Austrobaileya maculata is the only taxon that exhibits a significant difference in staining between the outer portion of the basal layer (foot-layer) and the inner portion of the basal layer (endexine). Although chemical differences may exist between the innermost portion of the basal layer that develops during the free spore phase and the outer portion of the basal layer that develops during the tetrad phase, based on resistance to degradation by 2-aminoethanol, the primary criterion for determining the presence of endexine is its differential staining, in comparison to the ektexine, with basic fuchsin (Faegri and Iversen 1950). The use of basic fuchsin and resistance to 2-aminoethanol degradation as indicative of endexine assumes that the cause of the staining difference between the endexine and ektexine, and resistance to 2-aminoethanol degradation, are chemically related phenomena. The occurrence of a differentially staining endexine appears to be restricted to angiosperms. The presence of a homogeneously staining basal layer, a characteristic that occurs in gymnosperms and angiosperms, is considered to be the ancestral condition.

Following the development of the sporopolleninous basal layer is the fibrillar intine, during the mid–late free-spore phase. The intine has been ignored by many pollen morphologists (see Kress and Stone 1983*a*, *b*). It is chemically removed by acetolysis and is rarely preserved in fossil pollen. However, morphological variation in the intine may have phylogenetic significance. The intine usually appears as a single fibrillar layer. In a number of angiosperm taxa the intine may be structurally elaborated (e.g. pollen of Cannaceae, Kress and Stone 1982) or multilayered (e.g. pollen of Magnoliaceae, Gabarayeva 1986*b*, 1987*c*, Orchidaceae, Zavada 1990*a*). Save the magnoliaceous taxa considered in the present contribution, all taxa have a single layered fibrillar intine. The occurrence of this type of intine in gymnosperms and angiosperms suggests that the single layered fibrillar intine is the ancestral state.

Character polarities of pollen in gymnosperms and angiosperms: a discussion

The purpose of this paper is not to discuss the relative merits of the various methods for determining character polarities, but to apply the most frequently used methods to determine the polarities of pollen characters in angiosperms and the various gymnosperm groups (*sensu lato*) that may have played a role in the origin of the angiosperms. The outgroup and parsimony methods may be considered mutually exclusive methods. The determination of polarities based on one method does not necessarily corroborate or refute the polarities determined by the other method. The palaeontologic and ontogenetic methods may be considered alternative methods that may lend support to polartities determined by the outgroup or parsimony methods. Although the four methods were initially introduced as equally significant alternatives to deriving workable hypotheses about polarities, I consider the palaeontologic and ontogenetic methods subordinate to the outgroup and parsimony methods. The outgroup and parsimony methods can be used to determine polarities in large groups of taxa, regardless of the type of character data. In contrast, a data set compiled for the exclusive use of the palaeontologic or ontogenetic method might have significant gaps. There are few fossil plant group that are completely known. Likewise, the development of various organs in extant plants have not been studied for a wide range of taxa. Despite the difficulties in using only the palaeontologic or ontogenetic data for determining character polarities, this data may provide significant supplementary information that may aid in determining character polarities with the outgroup or parsimony method.

In the discussion that follows, the suggested polarities are primarily determined by the outgroup or parsimony methods, and the palaeontologic and ontogenetic methods will be used to support or refute these polarity determinations (Donoghue *et al.* 1989).

In this discussion a character is considered to be a group of states that are modifications or alternative forms of the same thing (i.e. are homologous). The pollen characters considered here include; pollen unit, pollen shape, aperture type, sculpturing, and pollen wall structure. The functional taxonomic groups that will be considered include the Magnoliidae, Liliopsida (after Cronquist 1981), the Gnetales, Corystospermales, Peltaspermales, Glossopteridales, Bennettitales, and Caytoniales. All of the fossil groups are associated with other plant organs that may provide additional character suites useful for determining their phylogenetic relationships with the angiosperms. The angiosperm taxa are considered the ingroup, and the other taxa considered as an unresolved outgroup (the Gnetaceae and Mesozoic pteridosperms). Each of the individual taxa in the unresolved outgroup can be considered a functional outgroup (e.g. Gnetales, Corystospermaceae). Recently, the Gnetales have been considered to be the angiosperm sister group (Doyle and Donoghue 1986). The occurrence of vessels like those found in angiosperms (MacDuffie 1921) and double fertilization (Friedman 1990) has lent support to this assertion. In addition to the fossil taxa noted above, the bearing that the fossil dispersed pollen may have on determining pollen character polarities is also considered.

The primitive states of two characters, pollen unit and aperture type, have been widely recognized. Pollen shed as a monad and the monosulacte aperture type have been considered primitive character states in angiosperms prior to modern phylogenetic methods (Wodehouse 1935). The polarities of these characters are the same using the outgroup, parsimony, and palaeontologic methods. Despite the fact that the Mesozoic pteridosperms and the Gnetales are considered an unresolved outgroup, the monosulcate aperture and pollen shed as a monad occur in all of these taxa, and in many Magnoliidae and the Liliopsida. Based on the occurrence of the monosulcate aperture type, the primitive dicots appear to reside in the Magnoliidae. However, these characters provide little insight into the relationship of the monocots to the Magnoliidae, the Mesozoic pteridosperms, and the Gnetales, due to the frequent occurrence of these character states in pollen of all the subclasses of the Liliopsida (Zavada 1983*b*). The ontogenetic method offers little insight into the character polarities of aperture types. The mechanism by which aperture position is determined during development is little known.

Intuitively, tetrads might be considered the primitive pollen unit based on ontogeny. The tetrad as a pollen unit might be interpreted as cohesion of the microspores in the meiotic tetrad. Recent studies have suggested that cohesion of the pollen tetrad involves modifications of the sporopolleninous exine, or envelopment of the meiotic tetrad in a common sporopolleninous wall (Knox and McConchie 1986). Dispersal of exineless meiospores in lower plants are invariably monads. Thus, the development of tetrads or polyads in which the sporopolleninous exine is involved in the cohesion of the pollen unit occurred after the derivation of the sporopolleninous exine. The primitive nature of the monad and the monosulcate aperture type can be considered decisive polarities.

The shape of monosulcate pollen can not be described in a similar manner to the isopolar pollen found in angiosperms. Walker and Doyle (1975) have suggested that 'boat-shaped' pollen is primitive in angiosperms. The boat shape is determined by viewing the pollen in distal view, where the axis parallel to the long axis of the sulcus (length in distal view) is greater than the axis perpendicular to the long axis of the sulcus (width in distal view), i.e. the ratio of length to width is greater than one. Pollen of the Gnetales, Peltaspermales, Bennettitales, and various Magnoliidae and Liliopsida exhibit this shape. However, the saccate pollen of the Glossopteridales, Corystospermales, and Caytoniales have a length/width ratio of less than one if the sacci are considered in the measurement, and a length/width ratio of near one if only the corpus is considered. If any of the taxa with saccate pollen are considered the functional outgroup, pollen shape following the loss of the sacci would be circular to slightly oval (sacci, like those found in gymnosperms, are not present in angiosperms; however, see Zavada and

Taylor 1986; Zavada and Benson 1987). The circular to slightly oval pollen shape occurs in monosulcate pollen of angiosperms. This character state can be considered primitive using the outgroup method, i.e. this shape (minus the sacci) is common to the outgroup and the angiosperms. However, it would be more parsimonious to assume that monosulacte, non-saccate, 'boat-shaped' pollen is primitive, all of these characters are present in taxa of the outgroup and require no additional changes (assuming that the circular to oval shape occurs invariably with the saccate condition). Thus, two character states ('boat-shaped' and circular to slightly oval) might be considered primitive, depending on the method used; however, the parsimony method decisively favours the monosulcate, non-saccate, 'boat-shaped' pollen of the Gnetales, Peltaspermales, or Bennettitales.

Pollen sculpture types in the angiosperms are diverse, in contrast to the taxa in the unresolved outgroup. In the Mesozoic pteridosperms and Gnetales the tectum may be psilate, scabrate, or rugulate. In the Gnetales, pollen is polyplicate (*Ephedra*), echinate (*Gnetum*), or striate (*Welwitschia*), the striae being parallel to the long axis of the sulcus. In all of these taxa the tectum is predominantly imperforate. All of these various sculpturing types occur in the Magnoliidae and Liliopsida, depending on the taxon under consideration. Using the outgroup, parsimony, and/or palaeontologic methods, support for any of these sculpturing types as primitive is possible. The ontogenetic method provides little insight into the polarities of sculpturing types. Much like aperture type and position, the mechanism for the determination of sculpturing type and distribution during development is unknown. The position and nature of the apertures and sculpturing elements are probably related to the positioning of recognition sites for the accumulation of sporopollenin on the primexine during the tetrad phase. In general, imperforate, psilate, scabrate, or rugulate exine sculpturing can be considered primitive.

With the advent of electron microscopy in palynology and the investigation of pollen wall ultrastructure, a new suite of characters in addition to pollen unit, aperture type, shape, and sculpturing have emerged that may be useful for determining phylogenetic relationships. Pollen wall structure can be classified into four types, based primarily on their infrastructural layer (Doyle *et al.* 1975); these are the columellate type, alveolar type, granular type, and homogeneous type (atectate or intectate type). The columellate wall structure type is comprised of an outer perforate or imperforate tectum and an infrastructural layer of cylindrical-shaped columns of sporopollenin, which may or may not be fused to the overlying tectum. The bases of the columns are generally fused to the basal layer. This wall structure type is restricted to angiosperms in the groups under consideration and can not be considered as a primitive characteristic, given the occurrence of the homogeneous and granular wall structures in the unresolved outgroup and in angiosperms. Likewise, the alveolar wall structure is restricted to the Glossopteridaceae (Zavada, in preparation) and Caytoniales (Zavada and Crepet 1986). The alveolar or columellate wall structure types can not be considered primitive using the parsimony or outgroup methods. The alveolar and columellate wall structure types, being restricted to the outgroup or ingroup, require additional morphological changes, a situation less parsimonious than if the granular or homogeneous wall structure are considered primitive (which require no morphological changes). In addition, each of these wall structure types are restricted to the outgroup or ingroup, and can not be considered common (outgroup method). The homogeneous wall structure is present in the Bennettitales (Taylor 1973) and possibly the peltasperms (Zavada, work in progress), and infrequently occurs in the angiosperm taxa considered in this analysis (e.g. *Degenaria*, Dahl and Rowley 1965). The granular wall structure also occurs in two taxa in the unresolved outgroup, the Gnetales (Zavada, unpublished data) and possibly the Corystosperms (Taylor *et al.* 1984; Zavada and Crepet 1985). Support for the granular or homogeneous wall structure as primitive, using the parsimony or outgroup method, is possible. Although a distinction between the granular and homogeneous (atectate or intectate) wall structure types is often made, developmental data suggests that the two wall structures may represent variation of the same thing. The homogeneous wall of *Degenaria* may result from the compaction of sporopolleninous granules during development (Dahl and Rowley 1965). The thick homogeneous tectum underlying the ridges of *Welwitschia* is a result of the compaction of sporopolleninous granules during development Zavada and Gabarayeva in 1991), after the formation of a thin homogeneous tectum. Although the homogeneous and granular wall structures may appear different at maturity, they are the result of very similar developmental processes, and the differences can be attributed to the degree of compaction of the sporopolleninous granules. Using the outgroup or parsimony method, the granule wall structure (including the homogeneous type) is primitive and is present in the unresolved outgroup and in some angiosperms. The fossil record has little to offer, angiosperm granular walled pollen is indistinguishable from gymnosperm granular walled pollen (Zavada 1984*b*). The first recognizable angiosperm pollen in the fossil record has the columellate wall structure type (Doyle *et al.* 1975). The pattern of pollen wall development in *Welwitschia* (granular wall) (Gabarayeva and Zavada, in preparation) and in some alveolar walled gymnosperms (e.g. *Pinus*, Dickinson 1971;

Ceratozamia, Audran 1981; *Zamia*, Zavada 1983*a* is similar to that observed in some magnoliaceous (Gabarayeva 1986*a*, *b*) and annonaceous (Waha 1987; Gabarayeva 1988) taxa, and also supports the primitive nature of the granular wall.

The basal layer is generally divided into two layers, the outer footlayer and the inner endexine. These layers are distinguishable from one another, based on their staining properties with basic fuchsin (Faegri and Iversen 1950). However, distinguishing these layers with electron microscopic stains (e.g. uranyl acetate and lead citrate) is often difficult and does not always correlate with differences observed with basic fuchsin staining and light microscopy (Zavada 1984*b*; Zavada, unpublished data). Other characteristics have been used to distinguish endexine from footlayer, e.g. the differential solubility of these wall layers in 2-aminoethanol and the timing of their development (Zavada 1984*b*). Although the mode of development is similar in these wall layers (Blackmore and Crane 1987), footlayer generally develops during the tetrad phase and endexine early in the free-spore phase. The difference in the timing of development of the endexine and its greater resistance to solubility in 2-aminoethanol is not always accompanied by a difference in staining with EM stains (Gabarayeva 1987*b*). Despite these difficulties in defining endexine, a more pragmatic approach is taken here in determining the presence or absence of endexine, i.e. endexine is not considered present if the staining with EM stains is not of the same magnitude observed with basic fuchsin and light microscopy. Thus, no differential staining layer is present in any member of the unresolved outgroup. Recognizable endexine is present in the Magnoliidae and Liliopsida; however, it is most commonly unrecognizable or absent. Absence of a recognizable endexine is the primitive condition. This is supported by the outgroup, parsimony, and palaeontologic methods. In addition, many gymnosperms complete formation of the basal layer in the tetrad phase. If the time of development of the endexine is considered a valid criterion, this suggests that endexine is absent in most gymnosperms. In *Welwitschia*, the inner portion of the basal layer completes its development during the early free-spore phase (Gabarayeva and Zavada, in preparation). Although this wall layer exhibits no difference in staining to the outer portion of the basal layer, it may be considered endexine if the timing of development of the different portions of the basal layer is considered a valid criterion to characterize footlayer and endexine.

Using the outgroup or parsimony method, and considering the Bennettitales, Caytoniales, Corystospermales, Glossopteridales, Gnetales, and Peltaspermales as an unresolved outgroup to the angiosperms (Magnoliidae and Liliopsida), it is reasonable to consider the monosulcate aperture type, pollen shed in monads, elliptical ('boat-shaped') or circular pollen, imperforate psilate, scabrate, or rugulate exine sculpturing, and the granular wall structure with an undifferentially staining basal layer as the primitive character states. Among the taxa in the unresolved outgroup there has been renewed interest in the Gnetales as the angiosperm sister group. It is interesting that the pollen of *Welwitschia* has all of the above characters. The only unusual feature is the striate exine—an exine sculpturing type that is similar to some lauraceous and araceous taxa but that differs in the details of their pollen wall structure (Zavada 1990*b*). In addition, the recent report of a fossil gnetalean plant similar to *Welwitschia* places the gnetales temporally among some of the earliest angiosperms (Crane and Upchurch 1987).

The acceptance of the above characters as primitive is contingent upon the members of the outgroup being an adequate representation of the diversity of the taxa and the pollen types that can be included in this group (extant and fossil). However, if this outgroup represents only a small sample of the actual diversity of the taxa and the pollen that can be included in this group (due to gaps in the fossil record), the use of these polarity determinations for elucidating the phylogenetic relationships between the angiosperms and the outgroup should be used with caution. Recent studies of pre-Cretaceous dispersed pollen suggests that a diverse assemblage of pollen types, many with angiospermous characters that can be considered derived, predated the earliest recognizable angiosperms (Zavada 1984*b*; Pocock and Vasanthy 1988; Cornet 1989). The taxonomic position of fossil dispersed pollen is difficult to determine based on pollen characters alone (Zavada 1984*b*). Thus, the taxonomic relationships of the fossil dispersed angiosperm-like pollen to the taxa of the unresolved outgroup is unknown. Our inability to determine their taxonomic position does not preclude the possibility that they are, in fact, close relatives of the members of the unresolved outgroup. Any phylogenetic analysis that uses only the taxa in the unresolved outgroup must explicitly present the caveat that the validity of the analysis rests on the fact that these groups adequately represent the taxonomic and palynological diversity of this group. If these fossil dispersed pollen grains do prove to be members of the Mesozoic pteridosperms or Gnetales, many of the polarities based exclusively on the unresolved outgroup would have to be re-evaluated. Operating under the assumption that the pre-Cretaceous angiosperm-like pollen is related to members of the unresolved outgroup, the polarities previously determined for shape, sculpturing type, and wall structure may be considered equivocal. This would undermine the value of pollen for determining phylogenetic relationships among these taxa.

The application of modern phylogenetic methods for elucidating the origin of the angiosperms (e.g. Crane 1985; Doyle and Donoghue 1986; Zimmer *et al.* 1989), coupled with the recent developmental and morphological studies of gnetalean taxa (Friedman 1990; Zavada and Gabarayeva 1991) and the mesozoic pteridosperms, suggests that the Gnetales are the premier angiosperm sister group. The importance of these studies can not be underestimated; however, it is the diversity and continuity of the dispersed fossil pollen record, in comparison to the megafossil record, that brought our attention to the possibilities of the fossil pollen record providing new insight into the origin of the angiosperms. It is ironic that the inclusion of the dispersed fossil pollen data in a phylogenetic analysis obscures the relationships of the taxa it was meant to clarify. Although our current favour rests on the Gnetales as the angiosperm sister group, we should not entirely dismiss the unsettling question of the significance of the dispersed angiosperm-like pollen.

Acknowledgement

The author graciously thanks Nina Gabarayeva for her advice and many helpful suggestions.

References

Audran, J.-C. (1981). Pollen and tapetum development in *Ceratozamia mexicana* (Cycadaceae): sporal origin of the exinic sporopollenin in cycads. *Review of Paleobotany and Palynology* **33**, 315–46.

Blackmore, S. and Crane, P. R. (1987). The systematic implications of pollen and spore ontegeny. In *Ontogeny and systematics*, (ed. C.J. Humphries), pp. 83–115. Columbia University Press, New York.

Chaloner, W. G. (1967). Spores and land plant evolution. *Review of Palaeobotany and Palynology* 1, 83–93.

Cornet, B. (1989). Late Triassic angiosperm-like pollen from the Richmond Rift Basin of Virginia, U.S.A. *Palaeontographica B* **213**, 37–87.

Crane, P.R. (1985). Phylogenetic analysis of seed plants and the origin of angiosperms. *Annals of the Missouri Botanical Garden* **72**, 716–93.

Crane, P. R. and Upchurch, G. R. (1987). *Drewia potomacensis* gen. et sp. nov., an early Cretaceous member of the Gnetales from the Potomac Group of Virginia. *American Journal of Botany* **74**, 1722–36.

Cronquist, A. (1981). *An integrated system of classification of flowering plants.* Columbia University Press, New York.

Dahl, A. O. and Rowley, J. R. (1965). Pollen of *Degenaria vitiensis. Journal of the Arnold Arboretum* **46**, 308–23.

De Queiroz, K. (1985). The ontogenetic method for determining character polarity and its relevance to phylogenetic systematics. *Systematic Zoology* **34**, 280–99.

Dickinson, H. G. (1971). The role played by sporopollenin in the development of pollen in *Pinus banksiana*. In *Sporopollenin*, (ed. J. Brooks, P. K. Grant, M. Muir, P. Van Gijzel, and G. Shaw), pp. 31-67. Academic Press, London.

Dickinson, H. G. (1976). Common factors in exine deposition. In *The evolutionary significance of the exine* (ed. I. K. Ferguson and J. Muller), pp. 27–38. Academic Press, London.

Donoghue, M.J., Doyle, J.A., Gauthier, J., Kluge, A.G., and Rowe, T. (1989). The importance of fossils in phylogeny reconstruction. *Annual Review of Ecology and Systematics* **20**, 431–60.

Doyle, J. A. (1969). Cretaceous angiosperm pollen from the Atlantic Coastal Plain and its evolutionary significance. *Journal of the Arnold Arboretum* **50**, 1–35.

Doyle, J. A. and Donoghue, M. J. (1986). Seed plant phylogeny and the origin of angiosperms: An experimental cladistic approach. *Botanical Review* **52**, 321–431.

Doyle, J. A., Van Campo, M., and Lugardon, B. (1975). Observations on exine structure of *Eucommiidites* and Lower Cretaceous angiosperm pollen. *Pollen et Spores* **17**, 429–86.

Faegri, K. and Iversen, J. (1950). *Textbook of modern pollen analysis*. Munksgaard, Copenhagen.

Farris, J. S. (1982). Outgroups and parsimony. *Systematic Zoology* **31**, 328–34.

Friedman, W. E. (1990). Double fertilization in *Ephedra*, a nonflowering seed plant: Its bearing on the origin of angiosperms. *Science* **247**, 951–4.

Gabarayeva, N. I. (1986*a*). The development of the exine in *Michelia fuscata* (Magnoliaceae) in connection with the changes in cytoplasmic organelles of microspores and tapetum. *Botanischeski Zhurnal* **71**, 311–22.

Gabarayeva, N. I. (1986*b*). Ultrastructure analysis of the intine development of *Michelia fuscata* (Magnoliaceae) in connection with the changes of cytoplasmic organelles of microspores and tapetum. *Botanischeski Zhurnal* **71**, 416–28.

Gabarayeva, N. I. (1987*a*). Ultrastructure and development of sporoderm in *Manglietia tenuipes* (Magnoliaceae) during tetrad period: The primexine formation in connection with cytoplasmic organelle activity. *Botanischeski Zhurnal* **72**, 281–90.

Gabarayeva, N. I. (1987*b*). Ultrastructure and development of lamellae of endexine in *Manglietia tenuipes* (Magnoliaceae) in connection with the question of endexine existence in primitive angiosperms. *Botanischeski Zhurnal* **72**, 1310–17.

Gabarayeva, N. I. (1987*c*). Ultrastructure and development of pollen grain wall in *Manglietia tenuipes* (Magnoliaceae): The formation of intine in connection with the activity of cytoplasmic organelles. *Botanischeski Zhurnal* **72**, 1470–8.

Gabarayeva, N.I. (1988) Investigation of the ontogeny of various species of Magnoliaceae and Annonaceae for elucidating the structure and phylogeny of the mature sporoderm. In *Palynology in the USSR, Articles of Soviet Palynologists at the 8th International Palynological Congress, Brisbane, Australia*, (ed. A.F. Chlonova), pp. 48–52. Academy of Science, Novosibirsk, USSR.

Herngreen, G.F.W. and Chlonova, A.F. (1981). Cretaceous microfloral provinces. *Pollen et Spores* **23**, 441–555.

Heslop-Harrison, J. (1968). Wall development within the microspore tetrad of *Lilium longiflorum*. *Canadian Journal of Botany* **46**, 1185–92.

Knox, R.B. and McConchie, C.A. (1986). Structure and function of compound pollen. In *Pollen and spores: form and function*, (ed. S. Blackmore and I.K. Ferguson), pp. 265–82. Academic Press, London.

Kress, W.J. and Stone, D.J. (1982). Nature of the sporoderm in monocotyledons, with special reference to the pollen grains of *Canna* and *Heliconia*. *Grana* **21**, 129–48.

Kress, W.J. and Stone, D.J. (1983*a*). Morphology and phylogenetic significance of exine-less pollen of *Heliconia* (Heliconiaceae). *Systematic Botany* **8**, 149–67.

Kress, W.J. and Stone, D.J. (1983*b*). Pollen intine structure, cytochemistry and function in monocots. In *Pollen: biology and implications for plant breeding*, (ed. D.L. Mulcahy and E. Ottaviano), pp. 159–63. Elsevier, New York.

Kurmann, M.H. (1986). Pollen wall ultrastructure and development in selected gymnosperms. Unpublished Ph.D. thesis, Ohio State University.

MacDuffie, R.C. (1921). Vessels of the Gnetalean type in angiosperms. *Botanical Gazette* **71**, 438–45.

Maddison, W.P., Donoghue, M.J., and Maddison, D.R. (1984). Outgroup analysis and parsimony. *Systematic Zoology* **33**, 83–103.

Nelson, G. (1978). Ontogeny, phylogeny, paleontology and the biogenetic law. *Systematic Zoology* **27**, 324–45.

Nelson, G. and Platnik, N. (1981). *Systematics and biogeography*. Columbia University Press, New York.

Patterson, C. (1982). Morphological characters and homology. In *Problems of phylogenetic reconstruction*, Systematic Association Special Volume, No. 21, (ed. K.A. Joysey and A.E. Friday), pp. 21–74 Academic Press, London.

Patterson, C. (1983). How does phylogeny differ from ontogeny?, In *Development and evolution*, British Society for Developmental Biology Symposium 6, (ed. B.C. Goodwin, N. Holder, and C. Wylie), pp. 1–31. Cambridge University Press, Cambridge.

Pennel, R.I. and Bell, P.R. (1986). Microsporogenesis in *Taxus baccata* L.: The formation of the tetrad and development of the microspores. *Bulletin of the British Museum (Natural History), Geology* **13**, 223–57.

Pocock, S.A.J. and Vasanthy, G. (1988). *Cornetipollis reticulata*, a new pollen with angiospermid features from Upper Triassic (Carnian) sediments of Arizona (U.S.A.) with notes on *Equisetosporites*. *Review of Paleobotany and Palynology* **55**, 337–56.

Rohr, R. (1977). Etude comparée de la formation de l'exine au cours de la microsporogenese chez une gymnosperme (*Taxus baccata*) et de prephanerogame (*Gingkhgo biloba*). *Cytologia* **42**, 156–67.

Rowley, J.R. and Rowley, J.S. (1986). Ontogenetic development of microspores of *Ulmus* (Ulmaceae). In *Pollen and spores: form and function*, (ed. S. Blackmore and I.K. Ferguson), pp. 19–33. Academic Press, London.

Taylor, T.N. (1973). A consideration of the morphology, ultrastructure and multicellular microgametophyte of *Cycadeoidea dacotensis* pollen. *Review of Paleobotany and Palynology* **16**, 157–64.

Taylor, T.N., Cichan, M.A., and Baldoni, A.M. (1984). The ultrastructure of Mesozoic pollen: *Pteruchus dubius* (Thomas) Townrow. *Review of Paleobotany and Palynology* **41**, 319–27.

Traverse, A. (1988). *Paleopalynology*. George Allen and Unwin, Winchester, Mass.

Waha, M. (1987). Sporoderm development of pollen tetrads in *Asimina triloba* (Annonaceae). *Pollen et Spores* **29**, 31–44.

Walker, J.W. and Doyle, J.A. (1975). The basis of angiosperm phylogeny: palynology. *Annals of the Missouri Botanical Garden* **62**, 664–723.

Watrous, L.E. and Wheeler, Q.D. (1981). The outgroup comparison method of character analysis. *Systematic Zoology* **30**, 1–11.

Wodehouse, R.P. (1935). *Pollen grains. Their structure, identification and significance in science and medicine*. McGraw-Hill, New York.

Wolpert, L. (1983). Constancy and change in the development and evolution of pattern. In *Development and evolution*, (ed. B.C. Goodwin, N. Holder, and C.C. Wylie), pp. 47–57. Cambridge University Press, Cambridge.

Zavada, M.S. (1983*a*) Pollen wall development of *Zamia floridana*. *Pollen et Spores* **25**, 287–304.

Zavada, M.S. (1983*b*). Comparative morphology of monocot pollen and trends of apertures and wall structure. *Botanical Review* **49**, 331–79.

Zavada, M.S. (1984*a*). Pollen wall development of *Austrobaileya maculata*. *Botanical Gazette* **145**, 11–21.

Zavada, M.S. (1984*b*). Angiosperm origins and evolution based on dispersed fossil pollen ultrastructure. *Annals of the Missouri Botanical Garden* **71**, 444–63.

Zavada, M.S. (1990*a*). A contribution to the pollen wall ultrastructure of orchid pollinia. *Annals of the Missouri Botanical Garden*, **77**, 785–801.

Zavada, M.S. (1990*b*). The ultrastructure of selected monosulcate pollen from the Triassic Chinle Formation, Western U.S. *Palynology* **14**, 41–51.

Zavada, M.S. and Benson, J.M. (1987). First fossil evidence for the primitive angiosperm family Lactoridaceae. *American Journal of Botany* **10**, 1590–4.

Zavada, M. S. and Crepet, W. L. (1985). Pollen wall ultrastructure of the type material of *Pteruchus africanus*, *P. dubius* and *P. papillatus*. *Pollen et Spores* **27**, 271–6.

Zavada, M. S. and Crepet, W. L. (1986). Pollen grain wall structure of *Caytonanthus arberi* (Caytoniales). *Plant Systematics and Evolution* **153**, 259–64.

Zavada, M. S. and Gabarayeva, N. I. (1991). Comparative pollen wall development of *Welwitschia mirabilis* and selected primitive angiosperms. *Bull. Torrey Botanical Club* (in press).

Zavada, M. S. and Taylor, T. N. (1986). Pollen morphology of Lactoridaceae. *Plant Systematics and Evolution* **154**, 31–9.

Zimmer, E. A., Hamby, R. K., Arnold, M. L., Leblanc, D. A., and Theriot, E. C. (1989). Ribosomal RNA phylogenies and flowering plant evolution. In *The hierarchy of life*, (ed. B. Fernholm, K. Bremer, and H. Jornvall), pp. 205–14. Elsevier, Amsterdam.

13. Patterns of development in primitive angiosperm pollen

N. I. GABARAYEVA
Komarov Botanical Institute, Professor Popova 2, Leningrad, USSR

Abstract

Three important aspects of sporoderm ontogeny in primitive angiosperms, belonging to the families Magnoliaceae, Annonaceae, and Nymphaeaceae, are discussed in relation to gymnosperm and angiosperm phylogeny. These are the formation and differentiation of the primexine matrix, the formation of endexine lamellae, and the presence of unusual configurations of endoplasmic reticulum associated with synthesis of sporopollenin precursors. Primexine matrix is considered to provide valuable information concerning the structure of the sporoderm. An endexine is apparently present in all of the eight species studied, and this has consequences for discussion of seed plant relationships.

Introduction

The sporoderm ontogeny of eight primitive angiosperm species *Michelia fuscata*, *Michelia figo*, *Manglietia tenuipes*, *Magnolia delavayi*, *Liriodendron chinensis* (Magnoliaceae); *Asimina triloba*, *Anaxagorea brevipes* (Annonaceae); and *Nymphaea coerulea* (Nymphaeaceae)—has now been investigated. All the studies have been carried out with special reference to functional morphology, and to analyse which organelles take part in forming one or another sporoderm layer and how they are involved. It is impossible to show in detail the pollen development of even a single species in this short chapter, therefore I want to underline just three important points in relation to the sporoderm ontogeny of primitive angiosperms:

Pollen and Spores (ed. S. Blackmore and S. H. Barnes), Systematics Association Special Volume No. 44, pp. 257–268, Clarendon Press, Oxford, 1991.

1. In all species investigated, during sporoderm development the tangential lamellae of the endexine, lying along the entire microspore surface, are seen very distinctly. This does not coincide with the common point of view, founded on the analysis of exines of mature pollen grains. Indeed, in mature exines of primitive angiosperms thin endexine lamellae are pressed one to another and are only weakly distinct, if visible at all.

2. The initial stages of sporoderm development, the deposition of the primexine matrix (at the beginning of the tetrad period) and young primexine, forming on that matrix (in the mid-tetrad period), provide the most valuable information on the structure of sporoderm and the degree of tapetal participation in its formation, and evidence for the phylogenetic relations of gymnosperms and angiosperms.

3. Special attention is naturally attracted by the very active role of endoplasmic reticulum (ER) in sporoderm formation in all of the investigated species, especially by the unusual aggregated ER in Magnoliaceae. Following many investigators, I conclude that ER is the principal organelle, that participates directly in the synthesis of sporopollenin (SP) precursors. This is especially evident from observations of the aggregated ER. The data of electron cytochemistry, obtained after the treatment of the material by enzymes, allow us to deduce the lipoprotein nature of the initial substances for the formation of SP precursors.

Materials and methods

The material, cultivated in Batumi Botanical Garden (Georgia) and in the greenhouses of the Komarov Botanical Institute (Leningrad), was fixed in 3 per cent glutaraldehyde for 24 hours and post-fixed in 2 per cent OsO_4, embedded in epon, or a mixture of epon and araldite. The ultra-thin sections were post-stained with uranyl acetate and lead citrate and examined in a Tesla-BS-500 and a Hitachi H-600 transmission electron microscope (TEM). The enzymatic cytochemical reactions of proteins, revealed in TEMs, were carried out by the well-known method of Dashek *et al.* (1971); the treatment of the material during or after fixation, in sections, was realized by pronase E (Serva; *Streptomyces grisens c.* 8 DMC-U liophil. research grade).

Results and discussion

In *Michelia fuscata* and *Magnolia delavayi* the primexine matrix, consisting of discrete units, or 'tufts' in the terminology of Rowley and Dahl (1977), is observed (Fig. 13.1a, e). In the majority of micrographs these unit structures look like more or less ovoid vesicles with transparent contents and dark pith; they are evidently excreted by Golgi vesicles. Being pressed one to another, these units form a hexagonal pattern (Fig. 13.1a, e). There is a striking resemblance between the primexine matrix structure in the angiosperms *Michelia fuscata* (Gabarayeva 1986*a*) and *Magnolia delavayi* (Gabarayeva 1991), on one hand, and that of the gymnosperms *Ceratozamia mexicana* and *Picea abies* (Meyer 1977), on the other. However, it seems that such a pattern in the matrix is the result of section which has not quite passed perpendicularly to the surface of the plasmalemma. Strictly perpendicular sections are probably rather rare phenomena, and in such cases the primexine matrix looks completely different; some thin, dark, radially oriented rodlets are seen in it (Fig. 13.1b). This matrix pattern confirms the model of exine structure of spiral macromolecules, organized in a three-dimensional network, which was hypothesized by Rowley (1990). However, we do not always manage to observe a primexine matrix consisting of discrete units. More often it looks fibrillar, as in *Manglietia tenuipes* (Fig. 13.1c, d), *Anaxagorea brevipes* (Fig. 13.1f), and *Nymphaea coerulea* (Fig. 13.1h); although it may be that such pictures are connected with suboptimal fixation.

The similarity of matrix structures in species from such different taxa as primitive angiosperms (*Michelia*, *Magnolia*), lower gymnosperms (*Ceratozamia*), and advanced gymnosperms (*Picea*) seems to me very significant, probably pointing to the retention of identical ancestrial characters early in the development of plants which shared common ancestors.

It is necessary to note the excretion of droplets of an osmiophylic substance from the cytoplasm outside the plasmalemma (Fig. 13.1c, d. The excretion of lipoidal droplets into the primexine matrix has only been described in *Cosmos bipinnatus* (Dickinson 1976*a*; Dickinson and Potter 1976) and in *Eleocharis palustris* (Dunbar 1973). This substance is probably the visual precursor of SP (Dickinson 1976*b*). From my observations (Gabarayeva 1987*a*), these droplets of lipoidal substance are synthesized by special, short, tubular cisternae of rough ER, are then excreted from these cisternae and, becoming wrapped into smooth ER cisternae, are transported to the plasmalemma (Fig. 13.1c), where (as a result of redifferentiation of membranes) they finally come to lie outside the plasma membrane (Fig. 13.1d). The droplets of SP precursors lie in a line along the outer border of the fibrillar matrix in *Anaxagorea brevipes* (Fig. 13.1f), although there is no clear evidence here of which organelles are responsible for their appearance (Gabarayeva, in preparation). The same osmiophyllic droplets, but larger and partly fused together (Gabarayeva 1990*a*), are seen in the fibrillar matrix of

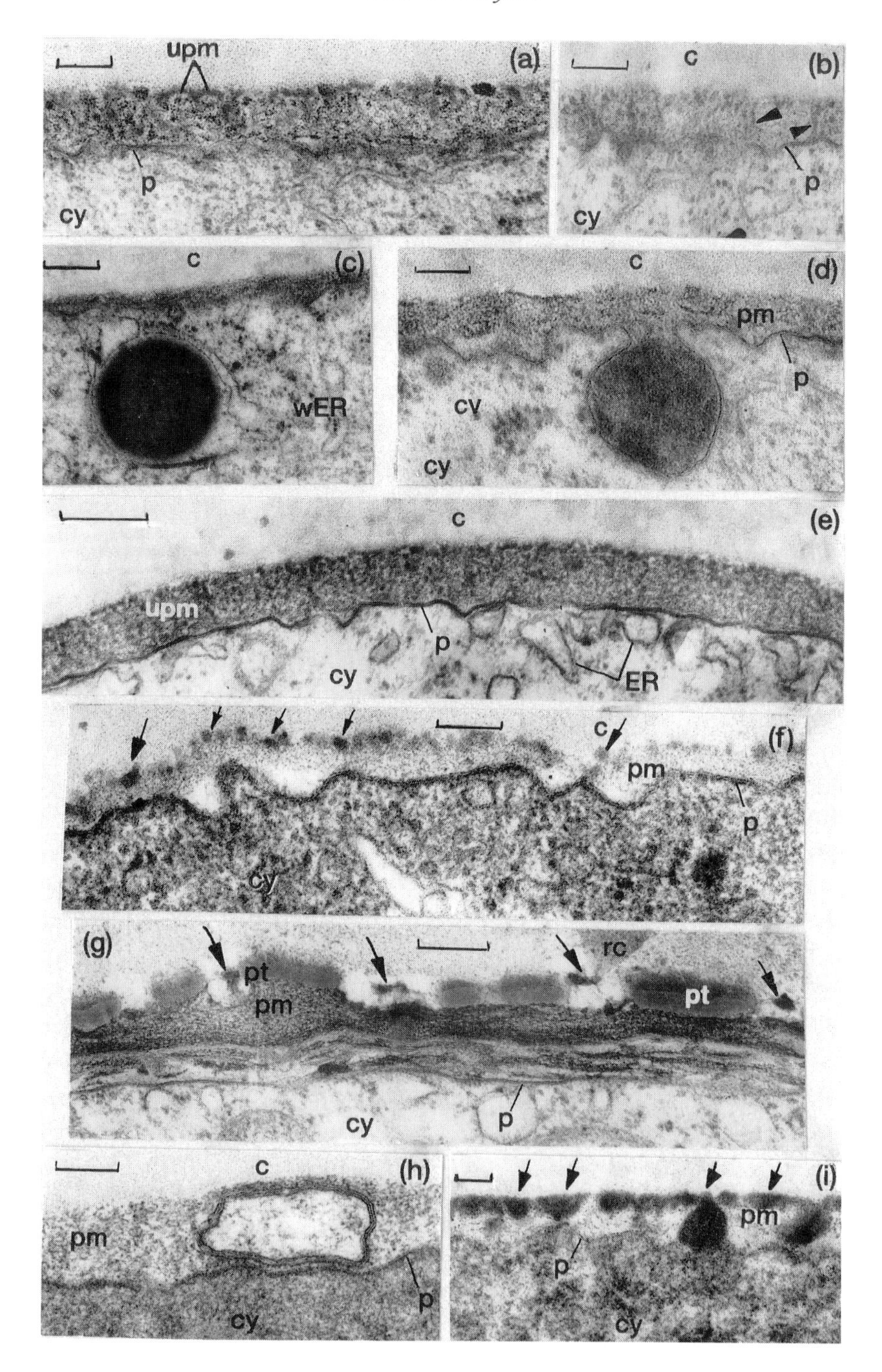
upm
(a)
p
cy
c
(b)
p
cy
c
(c)
wER
c
(d)
pm
p
cv
cy
c
(e)
upm
p
cy
ER
c
(f)
pm
p
cy
(g)
pt
pm
rc
pt
p
cy
c
(h)
pm
p
cy
(i)
pm
p
cy

Nymphaea coerulea (Fig. 13.1i), while at an earlier stage the primexine matrix of this species contains only coils of membranes (Fig. 13.1h). The initial stages of matrix development in *Asimina triloba* are very complicated and unusual (Gabarayeva 1988*a*, in press). Some masses of dark substances of tapetal origin take part in matrix formation: in places where these masses accumulate in the matrix, protoSP is never deposited (Fig. 13.1g), and at these places lacunae of the mature sporoderm form later. These observations do not coincide with an investigation of the same species by Waha (1987), but this will be discussed in a separate paper.

No special comments are required about Fig. 13.2, since lamellae of endexine in development are very clear in all the investigated species. In some species the lamellae are thin (*Michelia fuscata*, *Manglietia tenuipes*, *Liriodendron chinensis*, *Anaxagorea brevipes*; Fig. 13.2a, c, d; Gabarayeva 1986*a*, 1987*b*, 1988, 1990*b*). In other species the endexine lamellae are rather thick (*Michelia figo*, *Magnolia delavayi*, *Asimina triloba*, *Nymphaea coerulea*; Fig. 13.2b, e, f; Gabarayeva 1988*a*, 1990*a*, 1991). In all these species the endexine lamellae are seen very precisely in the process of development and appear *after* the formation of the footlayer of the ectexine. The thickness of endexine lamellae is equal in aperture and

---

Fig. 13.1 The different types of primexine matrix in primitive angiosperms. (a) Matrix of *Michelia fuscata*, consisting of discrete units; section approximately perpendicular to plasmalemma surface, some ovoid structures are seen (bar = 0.2 μm). (b) Matrix of *Michelia fuscata*, section perpendicular to plasmalemma surface; some radially oriented rod-like units (arrowheads) are seen (bar = 0.2 μm). The excretion of the droplet of osmiophyllic substance outside the plasmalemma in *Manglietia tenuipes*; the droplet of osmiophyllic substance, wrapped in cisternae of ER, is near the plasmalemma (bar = 0.25 μm). (d) As (c), but here the droplet of osmiophyllic substance with changed microstructure is already outside the plasmalemma, in its invagination) (bar = 0.2 μm). (e) Matrix of *Magnolia delavayi*, consisting of discrete units (bar = 0.3 μm). (f) Matrix of *Anaxagorea brevipes* with the discrete droplets of osmiophyllic substance along the outer border of the matrix (arrows) (bar = 0.2 μm). (g) Late matrix of *Asimina triloba*, already with some elements of primexine; the dark masses of tapetal origin are shown by arrows; protosporopollenin deposits are seen only on the matrix between them (bar = 0.25 μm). (h,i) Early matrix with coils of membranes (h), bar = 0.2 μm; and mid-matrix, containing the semi-fused droplets of osmiophyllic substance (arrows) in *Nymphaea coerulea* (i), bar = 0.25 μm. Abbreviations: upm, Units of primexine matrix; c, callose; p, plasmalemma, cy, cytoplasm of microspore; wER, the wrapper of ER; cv, coated vesicle, pm, primexine matrix; r, ribosomes; rc, the remnants of the callose envelope of the tetrad; pt, protectum.

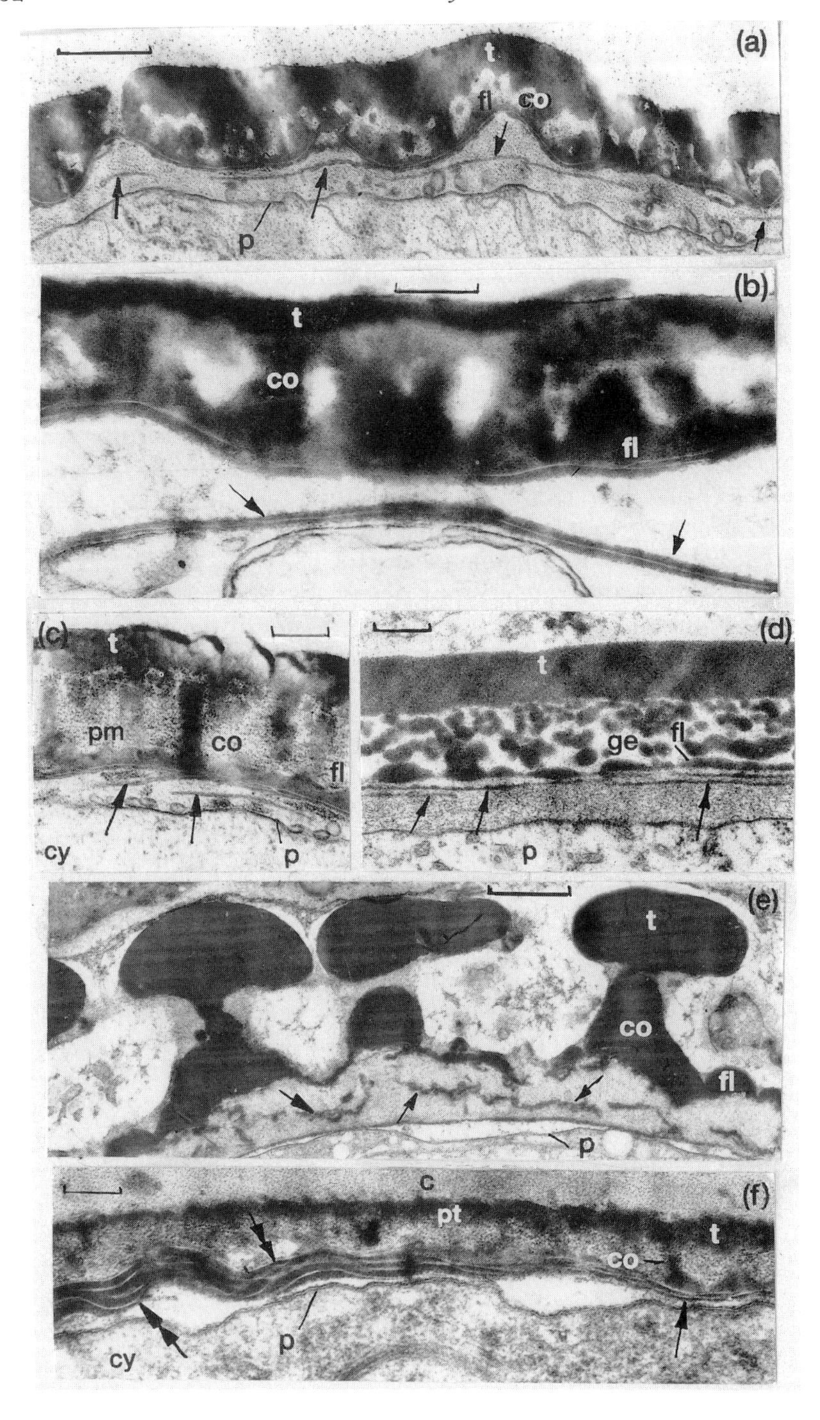
(a)
t
fl
co
p
(b)
t
co
fl
(c)
t
pm
co
fl
cy
p
(d)
t
ge
fl
p
(e)
t
co
fl
p
(f)
c
pt
t
co
cy
p

interaperture regions, with the exception of *Nymphaea coerulea*, where the endexine lamellae are thicker in the aperture region (Fig. 13.2f).

Extraordinary aggregates of ER (so called chain-mail ER) are observed in *Michelia fuscata* from the beginning of the tetrad period until the beginning of the post-tetrad period (Fig. 13.3a; Gabarayeva 1986*a*). In *Manglietia tenuipes* (Gabarayeva 1987*b*) they resemble aggregated ER observed earlier in species of *Beta* (Hoefert 1969) and in *Tradescantia bracteata* (Mepham and Lane 1970). In Fig. 13.3a it is seen that chain-mail ER, containing a grey substance, is connected directly with the plasmalemma by cisternae of rough ER branching from it and stretched out to the developing sporoderm. In *Liriodendron chinensis* from the mid-tetrad period until the mid-post-tetrad stage (Gabarayeva 1990*b*) the aggregates of another type of ER are observed: three-, four-, five-, and multi-cisternal aggregates of parallel cisternae; the outer of which carry ribosomes (Fig. 13.3b). These aggregates undulate through the cytoplasm; they often branch and are associated with lipid globules (Fig. 13.3c). It is significant that the tips of many aggregates are pressed to the plasmalemma; some of the cisternae of each aggregate contain an osmiophyllic substance, while the rest of them are electron-transparent: I have therefore called them 'zebra-aggregates'. An investigation of the chemical composition of the dark substance in cisternae of zebra-aggregates was undertaken; as a result it has been established (Gabarayeva 1990*b*) that this substance (which, I presume, is the basal one for the synthesis of the SP precursors) is probably lipoprotein.

It is worth mentioning that all the investigated species of Magnoliaceae have a very unusual intine I, consisting of large spherical granules (Gabarayeva 1986*b*, 1987*c*; data on *Magnolia* and *Liriodendron* in preparation); following the advice of L. A. Kuprijanova, I have called this 'granulina'.

In conclusion, it is necessary to return to the most confused issue of the sporoderm stratification of primitive angiosperms. For the first

Fig. 13.2 The development of endexine lamellae (arrows) in primitive angiosperms. (a) Thin lamellae of endexine in *Michelia fuscata* (bar = 0.5 μm). (b) Rather thick lamella of endexine in *Magnolia delavayi* (bar = 0.3 μm). (c) The endexine lamellae in *Manglietia tenuipes* (bar = 0.25 μm). (d) The footlayer of ectexine and thin endexine lamellae (arrowheads) in *Anaxagorea brevipes* (bar = 0.3 μm). (e) The undulate endexine lamellae in *Asimina triloba* in development (bar = 1 μm). (f) Thin lamellae of interaperture region (arrow) and thick lamellae of aperture region (double arrows) in *Nymphaea coerulea* (bar = 0.2 μm). Abbreviations: See caption for Fig. 13.1; t, tectum; co, columella; fl, footlayer of ectexine; ge, granular part of ectexine.

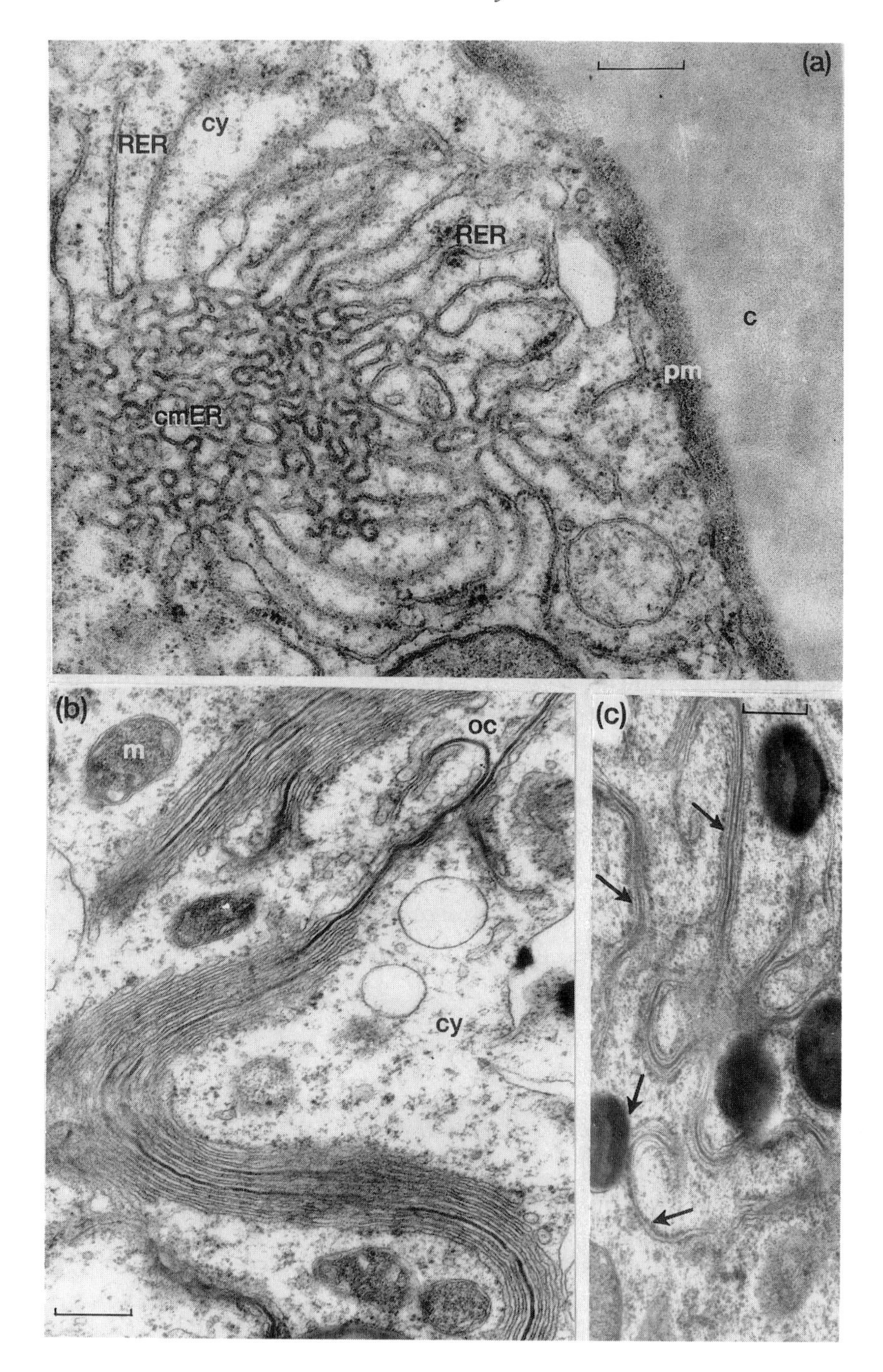
(a)
cy
RER
RER
c
pm
cmER
(b)
m
oc
cy
(c)

time in this group of plants, the existence of a lamellate endexine was demonstrated in *Asteranthe asterias* and *Hexalobus monopetalus* (Annonaceae) by Le Thomas and Lugardon (1972). However, two years later the authors changed their minds and came to conclusion that the lamellate layer was not endexine, but part of the footlayer of the ectexine, because the electron density and resistance to acetolysis were equal (Lugardon and Le Thomas 1974). The conclusion that endexine is totally absent in the majority of primitive angiosperm families, including Magnoliaceae and Annonaceae, maintains its dominant position nowadays (Walker 1976; Walker and Walker 1984; Le Thomas 1988). In searching for criteria to distinguish primitive angiosperms from gymnosperms, using the data on fossil pollen, Doyle *et al.* (1975) came to the conclusion that gymnosperms always have a lamellate endexine, and primitive angiosperms are non-lamellate, or have lamellae only under the apertures. Having relied upon these criteria (which the authors arrived at from investigations of mature pollen), Walker (1976) postulated the ancestral absence of endexine in primitive angiosperms, and, as a result, the hypothesis of the secondary evolution of endexine in angiosperms arose (Doyle *et al.* 1975; Walker 1976).

Nevertheless, endexine was shown to be present even in mature pollen of primitive angiosperms by Praglowsky (1974). In *Austrobaileya maculata* (Lauraceae), Zavada (1984) has noted the lamellate endexine under apertures and granular endexine in interapertural regions. In figs 18 and 19 of his paper one can see the thin layer of lamellae pressed against each other in interapertural regions, which become noticeable by virtue of their 'white lines'. Guédès (1982) considered that it was possible to regard pollen as lacking endexine only when the absence of tangential lamellae in the inner layer of exine had been shown ontogenetically. Ontogenetic investigations persuaded me of the presence of endexine lamellae in primitive angiosperms. Moreover, it has been demonstrated (Southworth 1974) in a whole range of plants (*Magnolia*

Fig. 13.3 The unusual aggregated ER in microspores of primitive angiosperms. (a) Chain-mail reticulum in *Michelia fuscata* in contact with the developing sporoderm by means of the cisternae of rough ER, stretched to the plasmalemma (bar = 0.5 μm). (b,c) 'zebra-aggregates' with alternating osmiophyllic and electron-transparent cisternae in *Liriodendron chinensis*, undulating in cytoplasm (b), bar = 0.3 μm; and branching 'zebra-aggregates', associated with lipid globules (c), bar = 0.3 μm). The outer cisternae carry ribosomes (arrows). Abbreviations: See caption Fig. 13.1; cmER, chain-mail aggregate of ER; m, mitochondrion; oc, osmiophyllic cisterna of aggregate; RER, rough endoplasmic reticulum.

and *Asimina* among them) that after treatment with 2-aminoethanol, only the ectexine dissolves, while the endexine remains intact. This means that these two layers can be differentiated by virtue of their chemical composition. This point is very important. One can affirm that a layer of sporoderm may be regarded as existing independently if it is characterized by its own structure, chemical composition, and ontogeny (i.e. by the time of its appearance in the course of sporoderm development). The fact is that the tangential lamellae of primitive angiosperms satisfy all the criteria for a real, independent layer, the endexine.

Having been convinced of the presence of a lamellate endexine (or a lamellate inner part of the exine in fossils) in primitive angiosperms, extant gymnosperms, fossil seed ferns (e.g. *Monoletes*; Taylor 1982), the fossil mesozoic coniferalien *Classopolis* (Taylor and Alvin 1984; Rowley and Srivastava 1986), fossil *Cyclusphaera psilata* (Taylor *et al.* 1987), and fossil *Clavatipollenites* (Walker and Walker 1984), I do not see the necessity for a complex hypothesis of the secondary origin of the endexine layer in angiosperms, because we can trace its evolutionary continuity.

To conclude, it ought to be added that all the species I have investigated (with the exception of *Anaxagorea brevipes*) have a tectate–columellate exine. This pattern of exine is evident from the very beginning of the primexine and is especially distinct at the beginning of the post-tetrad period, and begins to lose its preciseness step by step when the tectum and columellae become thicker and the pollen grains become mature.

References

Dashek, W. V., Thomas, H. R., and Rosen, W. G. (1971). Secretory cells of lily pistils. II. Electron microscope cytochemistry of canal cells. *American Journal of Botany* **58**, 909–20.

Dickinson, H. G. (1976*a*). The deposition of acetolysis-resistant polymers during the formation of pollen. *Pollen et Spores* **18**, 321–34.

Dickinson, H. G. (1976*b*). Common factors in exine deposition. In *The evolutionary significance of the exine*, (ed, I. K. Ferguson and J. Muller), pp. 67–87. Academic Press, London.

Dickinson, H. G. and Potter, U. (1976). The development of patterning in the alveolar sexine of *Cosmos bipinnatus*. *New Phytologist* **76**, 543–50.

Doyle, J. A., van Campo, M., and Lugardon, B. (1975). Observation on exine structure of *Eucommiidites* and Lower Cretaceous angiosperm pollen. *Pollen et spores* **17**, 429–86.

Dunbar, A. (1973). Pollen development in the *Eleocharis palustris* group

(*Cyperaceae*). 1. Ultrastructure and ontogeny. *Botaniska Notiser* **126**, 197–254.

Gabarayeva, N.I. (1986*a*). The development of the exine in *Michelia fuscata* (*Magnoliaceae*) in connection with the changes in cytoplasmic organelles of microspores and tapetum. *Botanischeski Zhurnal* **71**, 311–22.

Gabarayeva, N.I. (1986*b*). Ultrastructure analysis of the intine development of *Michelia fuscata* (Magnoliaceae) in connection with the changes of cytoplasmic organelles of microspores and tapetum. *Botanischeski Zhurnal* **71**, 416–28.

Gabarayeva, N.I. (1987*a*). Ultrastructure and development of sporoderm in *Manglietia tenuipes* (Magnoliaceae) during tetrad period: the primexine formation in connection with cytoplasmic organelle activity. *Botanischeski Zhurnal* **72**, 281–90.

Gabarayeva, N.I. (1987*b*). Ultrastructure and development of endexine lamellae in *Manglietia tenuipes* (Magnoliaceae) in connection with the question of endexine existence in primitive angiosperms. *Botanischeski Zhurnal* **72**, 1310–17.

Gabarayeva, N.I. (1987*c*). Ultrastructure and development of pollen grain wall in *Manglietia tenuipes* (Magnoliaceae): the formation of intine in connection with the activity of cytoplasmic organelles. *Botanischeski Zhurnal*72, 1470–7.

Gabarayeva, N.I. (1988*a*). The significance of ontogenetic investigations for the elucidation of mature sporoderm structure and phylogenetic connections on example of some Magnoliaceae and Annonaceae. In *Palynology in the USSR*, (ed. A.F. Chlonova), pp. 48–52. Novosibirsk.

Gabarayeva, N.I. (1988*b*) The unusual aggregates of endoplasmic reticulum in developing microspores of primitive angiosperms. *Theses of the XIIIth All-Union Conference on Electron Microscopy, Zvenigorod, October 1988*, p. 240.

Gabarayeva, N.I. (1990*a*). The ultrastructure investigation of main stages of development of sporoderm and changes of miscrspore cytoplasmic organelles in *Nymphaea coerulea*. *Theses of the IVth Republican conference on Electron Microscopy, Electron microscopy and contemporary technology, Kishinev, 19–20 June 1990*, pp. 48–9.

Gabarayeva, N.I. (1990*b*). About the place of synthesis of sporopollenin precursors in the developing pollen grains in fam. Magnoliaceae. *Botanischeski Zhurnal* **75**, 783–91.

Gabarayeva, N.I. (1991). The ultrastructure and development of exine and orbicules in *Magnolia delavayi* in tetrad and at the beginning of post tetrad period. *Botanischeski Zhurnal* **76**, 10–19.

Guédès, M. (1982). Exine stratification, ectexine structure and angiosperm evolution. *Grana* **21**, 161–70.

Hoefert, L.L. (1969). Ultrastructure of *Beta* pollen. I. Cytoplasmic constituents. *American Journal of Botany* **56**, 363–8.

Le Thomas, A. (1988). Les structures reproductives des *Magnoliales* africaines et malgaches: significations phylogeniques. *Monographs in Systematic Botany of the Missouri Botanical Garden* **25**, 161–74.

Le Thomas, A. and Lugardon, B. M. (1972). Sur la structure fine des tétrades de deux Annonacées (*Asteranthe asterias* et *Hexalobus monopetalus*). *Comptes Rendus de l'Academie des Sciences, Paris*, **275**, *Series D* 1747–52.

Lugardon, B. M. and Le Thomas, A. (1974). Sur la structure feuilletée de la couche basale de l'ectexine chez diverses Annonacées. *Comptes Rendus de l'Academie des Sciences, Paris, Series D* **279**, 255–8.

Mepham, R. H. and Lane, G. R. (1970). Observations on the fine structure of developing microspores of *Tradescantia bracteata. Protoplasma* **70**, 1–20.

Meyer, N. R. (1977). Comparative morphological investigations of development and ultrastructure of sporoderm of gymnosperms and angiosperms. Unpublished Ph.D. thesis, Leningrad.

Praglowski, J. (1974). *World pollen and spore flora. 3. Magnoliaceae Juss.* (Ed. S. Nilsson) Stockholm.

Rowley, J. R. (1990). The fundamental structure of the pollen exine. *Plant Systematics and Evolution, [Suppl. 5]*, 13–29.

Rowley, J. R. and Dahl, A. O. (1977). Pollen development in *Artemisia vulgaris* with special reference to glycocalyx material (1). *Pollen et Spores* **19**, 169–297.

Rowley, J. R. and Srivastava, S. K. (1986). Fine structure of *Classopollis* exines. *Canadian Journal of Botany* **64**, 3059–74.

Southworth, D. (1974). Solubility of pollen exines. *American Journal of Botany* **61**, 36–44.

Taylor, T. N. (1982). Ultrastructural studies of paleozoic seed fern pollen: sporoderm development. *Review of Palaeobotany and Palynology* **37**, 29–53.

Taylor, T. N. and Alvin, K. L. (1984). Ultrastructure and development of mesozoic pollen: *Classopolis. American Journal of Botany* **71**, 575–87.

Taylor, T. N. and Zavada, M. S., and Archangelsky, S. (1987). The ultrastructure of *Cyclusphaera psilata* from the Cretaceous of Argentina. *Grana* **26**, 74–80.

Waha, M. (1987). Sporoderm development of pollen tetrads in *Asimina triloba* (*Annonaceae*). *Pollen et Spores* **29**, 31–44.

Walker, J. W. (1976). Evolutionary significance of the exine in the pollen of primitive angiosperms. In *The evolutionary significance of the exine*, (ed. I. K. Ferguson and J. Muller), pp. 251–308, Academic Press, London.

Walker, J. W. and Walker, A. G. (1984). Ultrastructure of Lower Cretaceous angiosperm pollen and the origin and early evolution of flowering plants. *Annals of the Missouri Botanical Garden* **71**, 464–521.

Zavada, M. (1984). Pollen wall development of *Austrobaileya maculata. Botanical Gazette* **145**, 11–21.

14. A systematic analysis of pollen morphology of Acanthaceae genera with contorted corollas

ROBERT W. SCOTLAND*
Department of Botany, The Natural History Museum, Cromwell Road, London, UK

Abstract

The variation in pollen morphology is described in genera of Acanthaceae with contorted aestivation (Burkill and Clarke 1899–1900). It is argued that the pollen types of Lindau (1895) are conceptually inadequate for describing the complete range of palynological variation in the family. A parsimony analysis of pollen data is presented. The results of the analysis are radically different from classifications proposed previously. The classifications of Lindau (1895), Clarke (1885), and Burkill and Clarke (1899–1900) are compared, in terms of the new pollen data, to the results of the analysis, and are found to have relatively low consistency indices.

Introduction

Systematic studies of the family Acanthaceae have, not surprisingly, resulted in the recognition of many conflicting taxonomic groups because they use different characters or different character combinations (Nees von Esenbeck 1832, 1847; Anderson 1863, 1867; Bentham 1876; Clarke 1885; Burkill and Clarke 1899–1900; Lindau 1893, 1895; Bremekamp 1965; Balkwill and Getliffe Norris 1988). A recurring point of disagreement is the relative usefulness of pollen morphology for the

* Present address: *Department of Plant Sciences, University of Oxford, South Parks Rd, Oxford, OX1 3RB, UK*

Pollen and Spores (ed. S. Blackmore and S. H. Barnes), Systematics Association Special Volume No. 44, pp. 269–289, Clarendon Press, Oxford, 1991.

delimitation of natural groups. This is, in part, due to the great diversity of pollen morphology in the Acanthaceae (Radlkofer 1883; Lindau 1893, 1895; Raj 1961, 1973).

Several taxonomic approaches concerning the relationship between pollen morphology and recognition of groups can be identified. First, there are those that recognize taxa on the basis of other morphological characters (Nees von Esenbeck 1832, 1847; Anderson 1863, 1867; Bentham 1876; Clarke 1885; Burkill and Clarke 1899–1900), although these authors sometimes include pollen information in their descriptions. Secondly, there are those that give prominence to pollen characters to the virtual exclusion of other characters (Lindau 1893, 1895). Thirdly, there are those who accept a particular classification but amend it in the light of new pollen observations (Bremekamp 1965; Raj 1961, 1973). These different approaches have much to do with the use of the now discredited systematic procedure, *a priori* character weighting.

The pollen types of Lindau

Lindau (1895) recognized eleven pollen types, including the dustbin category, *Andere Pollenformen*, in which he placed any morphologies that did not conform to the other ten. These eleven pollen types or character concepts are still widely referred to (see Balkwill and Getliffe Norris 1988; Graham 1988; Hansen 1988; McDade 1988; Ramamoorthy and Hornelas 1988; Wasshaussen 1988; Wood 1988; Furness 1989; Hedrén 1989; Vollesen 1989; Hilsenbeck 1990). The Lindau pollen types are inadequate to account for the morphological variation of pollen in a number of respects. For example, *Glatter, runder Pollen* is supposedly shared by *Mendoncia* and *Whitfieldia* (Lindau 1895), but these genera are quite different in pollen morphology. Other types, *Spangenpollen*, *Rahmenpollen*, and *Knötchenpollen*, are, in fact, states of a single continuous morphological variable that was arbitrarily split into discrete characters. Another type, *Rippenpollen*, actually comprises at least seven discrete pollen morphologies (see below). A complete reappraisal of Lindau's pollen types is in preparation (Scotland, in preparation).

Characters from pollen morphology have been investigated as part of a broad study of character interpretation, distribution, and congruence. The results of this study have led to the view that existing terminology and character concepts are insufficient to describe the full complexity of the pollen. Therefore, before any analysis of all characters is attempted, it is important to establish the units of comparison between pollen morphologies.

Materials and methods

Taxa

As pollen in this family is so diverse, many of the morphologies are only diagnostic (autapomorphic) of individual taxa of various rank, and therefore contain no information for systematic analysis at a higher hierarchical level. Consequently, any pollen analysis of the whole family will necessarily result in a large number of polychotomies, representing unresolved relationships. This being so, it is initially expedient to analyse pollen at lower hierarchical levels within the Acanthaceae.

The analysis of pollen morphology presented here is of those genera which have a contorted aestivation pattern to the corolla, defined by Burkill and Clarke (1899–1900) as having 'no lobe wholly without or wholly within the others'. This aestivation pattern is the diagnostic character of the Ruellieae of Burkill and Clarke (1899–1900) and the Contortae of Lindau (1895). The justification for accepting the putative monophyly of the Contortae results from a parsimony analysis of all the traditional morphological characters (Scotland, in preparation). The position of the tribe Barlerieae, which was placed in the Contortae of Lindau (1895), is not considered here because most of the genera in that tribe have neither contorted nor imbricate (as that term has been understood) aestivation patterns.

Each genus was sampled until no further intrageneric variation was found. Samples were taken from, as far as possible, fully determined herbarium specimens. It is frequently possible to ascertain from the literature (Radlkofer 1883; Lindau 1893, 1895; Raj 1961, 1973) whether a genus is polymorphic for pollen morphology; if so, sampling was much more extensive. For example, 130 samples were taken from 65 species from the variable *Strobilanthes*, whereas 12 samples were taken from six species of the relatively uniform *Hygrophila*. In total, 300 samples from 150 species were examined in the study. Table 14.1 contains the complete list of genera included in the analysis.

Other genera with contorted aestivation, *Whitfieldia*, *Blechum*, *Eranthemum*, *Louteridium*, *Pseudostenosiphonium*, and *Lankesteria* were also studied. *Whitfieldia* and *Blechum* have not been included in the analysis as their respective pollen morphologies are autapomorphic for each genus. *Whitfieldia* pollen has unique tectal ornamentation, aperture type, and exine stratification. Similarly, *Blechum* has its own unique tectal ornamentation and is, apparently, the only syncolpate genus in the Acanthaceae. The other genera listed, *Eranthemum*, *Louteridium*, *Pseudostenosiphonium*, and *Lankesteria*, have yet to be described and coded and are therefore not included in the study.

Table 14.1. List of taxa and abbreviations

Acanthopale C.B. Clarke	ACAN
Aechmanthera Nees	AECH
Bravaisia DC.	BRAV
Brillantaisia Beauv.	BRIL
Chaetacanthus Nees	CHAE
Dischistocalyx T. Andres. ex Benth. and Hook.f.	DISH
Duosperma Dayton	DUOS
Dyschoriste Nees	DYSH
Echinacanthus Nees	ECHI
Epiclastopelma Lindau	EPIC
Eremomastax Lindau	EREM
Hemigraphis Nees	HEMI
Heteradelphia Lindau	HETE
Hygrophila R.Br.	HYGR
Mellera S. Moore	MELL
Mimulopsis Schweinf.	MIMU
Petalidium Nees	PETA
Phaulopsis Willd.	PHAU
Ruellia L.	RUEL
Sanchezia Ruiz and Pav.	SANC
Sautiera Decne.	SAUT
Stenosiphonium Nees	STEN
Strobilanthes Blume	STRO
Suessenguthia Merxm.	SUES
Trichanthera Kunth	TRIC

Investigation of pollen

Acetolysed pollen (Erdtman 1960), taken from herbarium specimens, was examined in a Reichert Zetopan light microscope, and a Hitachi S800 field emission scanning electron microscope, using secondary electron detection and an accelerating voltage of 8 kV.

Cladistics

Cladistics (Hennig 1966; Nelson and Platnick 1981) is a method for generating explicit systematic hypotheses, concerning relationships between organisms. No attempt at an overall review will be given here, although one aspect of cladistic procedure will be discussed briefly. Characters such as character 5 (see below) have many states and hence are referred to as multistate characters. Other characters, such as character 8, have two character states and are therefore binary characters. In either case, if the plesiomorphic (more general) condition can be determined, the character is said to be polarized. In the case of multistate characters, one has the additional problem of ordering the character states. For instance, character 5 refers to the number of apertures, i.e. 2, 3, 4, or many. Thus even if the plesiomorphic condition can be determined, it says nothing as to the relationships between the other states, unless the order of character-state change can be determined.

The ordering and polarizing of characters are, then, *a priori* constraints upon parsimony analysis. If no characters are ordered or polarized, the cladogram remains unrooted. The most widely used method to establish character polarity (and root the cladogram) is that of outgroup comparison (Watrous and Wheeler 1981; Maddison *et al.* 1984). The method relies upon the correct choice of an outgroup for determining the polarity of characters, and therefore relies upon *a priori* taxonomic knowledge. The approach adopted here is one of extreme caution as to *a priori* judgements with regard to character polarity or the order of multistate characters. This would seem to be justified in the case of Acanthaceae pollen as the characters are often unique to the Acanthaceae and would therefore be absent from any plausible outgroup. Therefore, a character such as character 2 can be polarized, that is rooted on the absence of pseudocolpi, but as pseudocolpi may be present in either of two ways, then character 2 is left unordered. However, in other cases, for example the variable character 5, the character is left unpolarized and unordered and is therefore rooted as equivocal. In one case, that of character 1, a character state is found to be present within many other genera of the Acanthaceae, and it is chosen as the plesiomorphic state.

The analysis was undertaken using the microcomputer parsimony programme Hennig86 (Farris 1988). The trees were generated using the implicit enumeration option which guarantees, for this number of taxa, to find the shortest cladograms. There are three binary and five multistate characters in the data set (see Appendix). All multistate characters were unordered as there is no justification for *a priori* assumptions regarding character state relationships. Five characters in the data set are rooted, and three are not. For example, aperture number (character 5) is rooted as equivocal (-), to represent the unknown state, because the plesiomorphic condition is unknown. The genus *Ruellia* is represented twice as it is polymorphic for aperture number, possessing either three or many apertures.

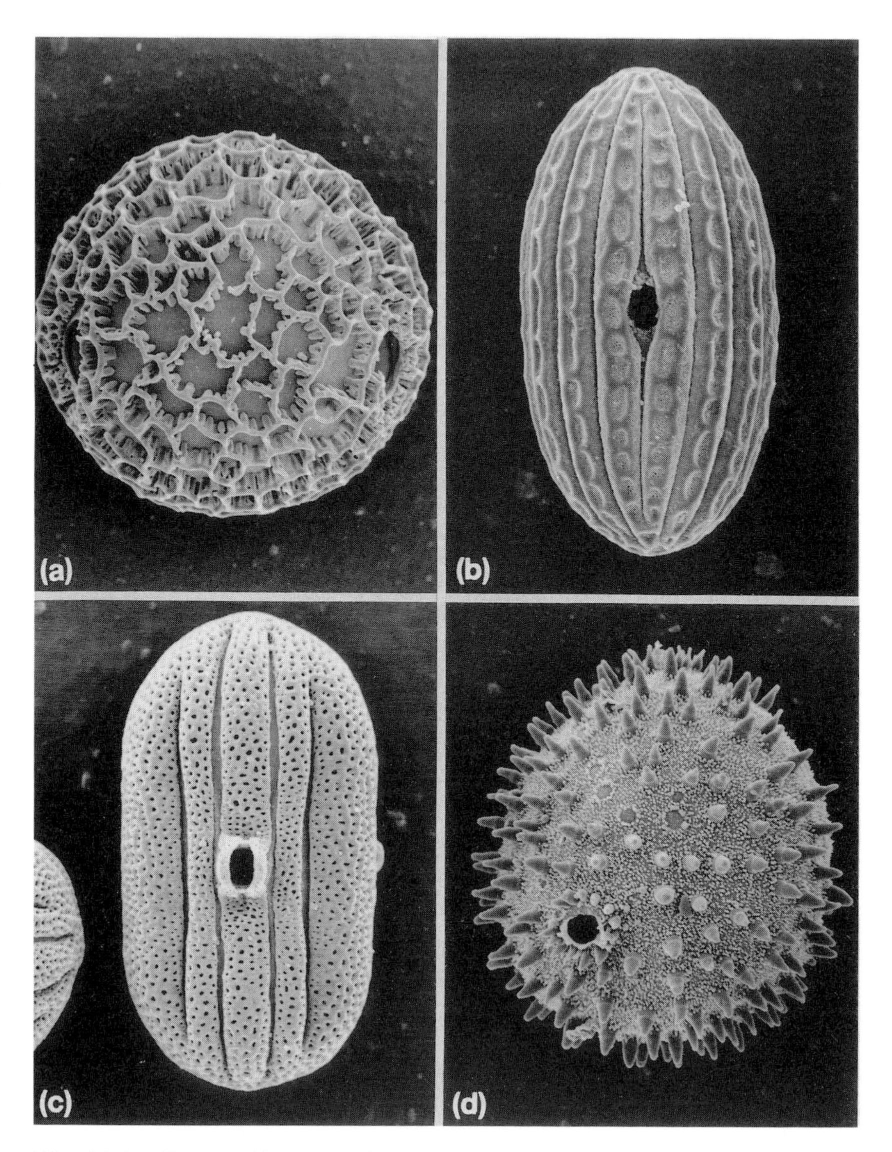

Fig. 14.1 Four pollen morphologies from the Acanthaceae. (a) *Ruellia grandiflora* (Forrsk.) Pers., × 525, Popov 7/7. (b) *Strobilanthes nutans* (Nees) T. And., × 750, Stainton 6664. (c) *Duosperma dentata* C.B. Cl., × 1125, Chase 2770. (d) *Dischistocalyx thunbergiaeflora* T. Anders., × 750, C.C. Exp. 291. See text for descriptions.

Results

Plesiomorphy, apomorphy, and autapomorphy

Figure 14.1 illustrates four pollen morphologies within the Contortae, which were known to both Burkill and Clarke (1899–1900) and Lindau (1895). These are *Wabenpollen* (open reticulum present, Fig. 14.1a), *Rippenpollen* (pseudocolpi present, Fig. 14.1b,c), and *Stachelpollen* (spines present, Fig. 14.1d).

Wabenpollen. *Ruellia* (Fig. 14.1a). Three-porate, circular, tectate columellate. Sexine composed of simple and anastomosing columellae, very open reticulum with the tectum reduced to the apices of adjacent columellae.

Rippenpollen. *Strobilanthes* (Fig. 14.1b). Three-colporate, prolate, tectate columellate. Pseudocolpi uniformly present throughout the grain, dividing the sexine into longitudinal strips. Sexine of simple columellae, bireticulate, each longitudinal sexine strip having an uneven coarse reticulum with a depressed microreticulum in the lumina.

Duosperma (Fig. 14.1c). Three-porate, prolate, tectate columellate. Four pseudocolpi associated with each aperture and no pseudocolpi in the mesocolpial region. Sexine of simple straight columellae with a fairly uniform reticulated tectum.

Stachelpollen. *Dischistocalyx* (Fig. 14.1d). Three-porate, almost circular, intectate with large, uniformly spaced spines interspersed with very small crowded columellae. Each aperture surrounded by a circle of spines.

Given these pollen morphologies, what are the taxonomic implications that result from these data? The branching diagram shown in Tree 1 (Fig. 14.2), shows the distribution of four pollen characters from three of the pollen morphologies (Fig. 14.1a,b,d). Two characters are unique to a genus and are therefore autapomorphies that cannot indicate relationships between these genera, these are atectate-pollen and pseudocolpi. All three genera are triaperturate and this character therefore cannot indicate relationships between them. The tectate character is shared by two of the genera, *Strobilanthes* and *Ruellia*. However, this character is very widely distributed in the Acanthaceae and most other angiosperms, and is best interpreted as a plesiomorphy within the Acanthaceae. In other words, the presence of tectate pollen grains is a taxonomic character diagnostic of a much higher taxonomic group. This leaves the three genera in an unresolved trichotomy, and one has

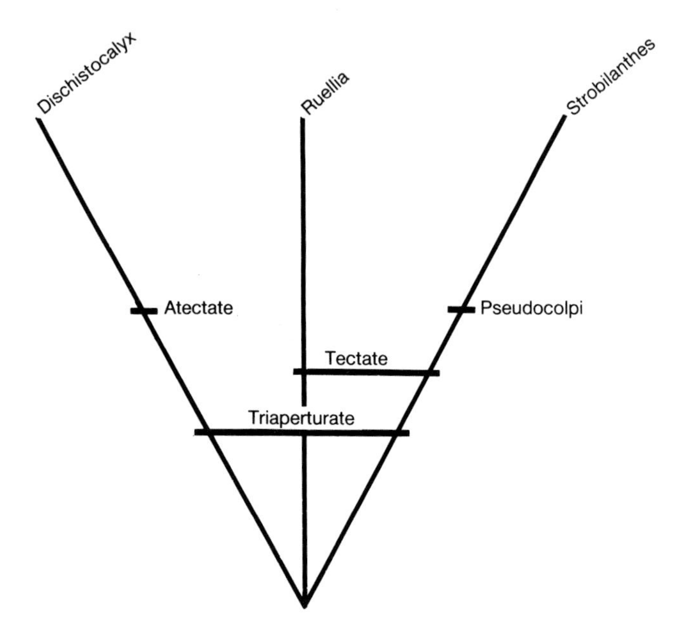

Fig. 14.2 Tree 1. The distribution of four pollen characters between three genera of the Acanthaceae.

to conclude that the pollen evidence cannot resolve the relationship between the three genera.

Within the *Rippenpollen* pollen type (based on the presence of pseudocolpi throughout the grain) of Lindau (1895) there is a great deal of morphological variation (Fig. 14.1b,c; Fig. 14.6a,b,d). This was known to Lindau, who within his concept of *Rippenpollen*, recognized other categories, namely *Trichanthereenpollen* (Fig. 14.6a), *Petalidieenpollen* (Fig. 14.6d), and *Typischer Rippenpollen* (Fig. 14.1b). However, the presence of pseudocolpi throughout the pollen grain, within Acanthaceae, is restricted to some genera within the Contortae and is not found in associated families. This information, added to the data in Tree 1 and including a fourth genus *Duosperma* (Fig. 14.1c), is shown in Tree 2 (Fig. 14.3), and illustrates the concept of homology (Patterson 1982) synapomorphy (shared derived character) of Hennig (1966). As statements of homology are subject to the test of character congruence (Patterson 1982; Rieppel 1988), these characters (the data in Trees 1 and 2) must be analysed together with all other (pollen) characters. This raises a fundamental point regarding the pollen types of Lindau (1895); that is, the very concept of *Rippenpollen* itself is a form of character weighting as one characteristic, pseudocolpi, is emphasized over all the

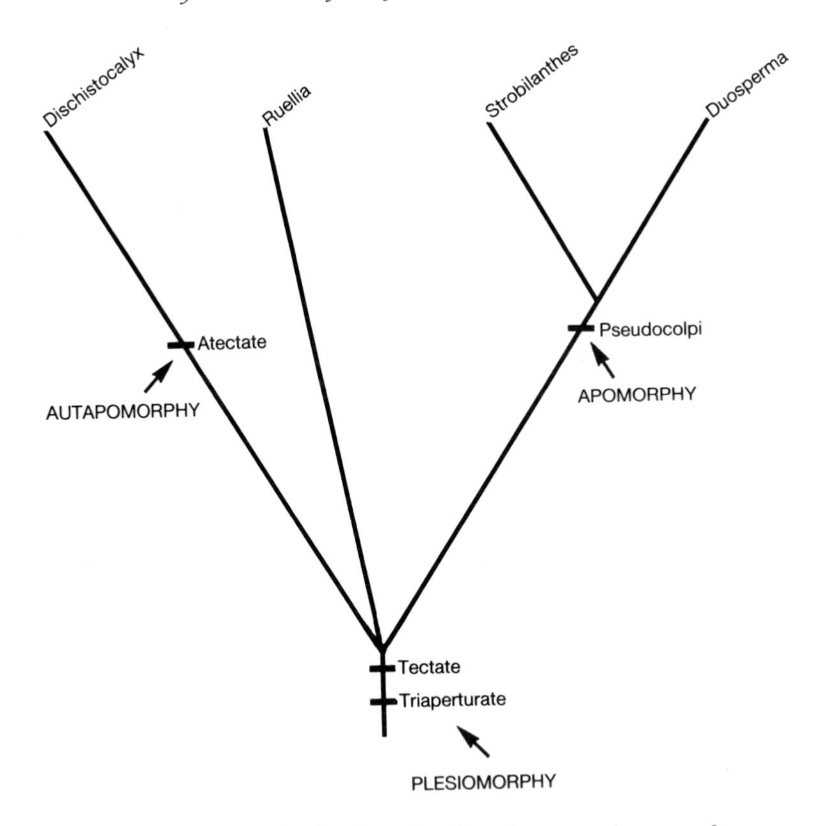

Fig. 14.3 Tree 2. The distribution of pollen characters between four genera, illustrating the concepts of apomorphy, plesiomorphy, and autapomorphy.

other characteristics. The range of variation in pollen morphology between genera is given below.

Individual pollen characters and systematic analyses

The list of eight characters (a fuller description of which is given below) used in the analyses can be found in the Appendix. Also in the Appendix is the subsequent data matrix containing the list of genera and their character states for each of the eight characters which will now be described. Dashes in the data matrix are equivalent to unknown data.

Character 1. Five different types of tectal pattern can be distinguished (Fig. 14.4). Some genera have an almost solid tectum (Fig. 14.4b), e.g. *Dyschoriste*, whereas *Ruellia* has an extremely open tectum (Fig. 14.4a). *Dischistocalyx* and *Acanthopale* are intectate (Fig. 14.4f) and of the

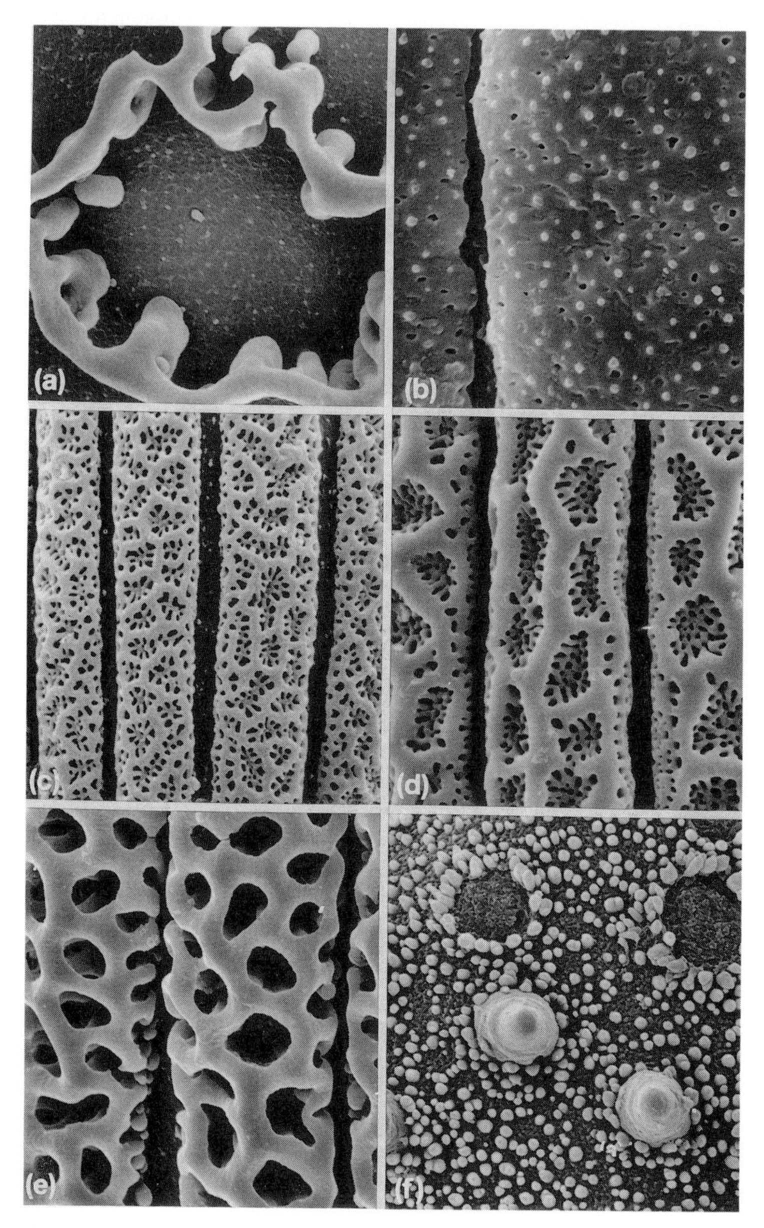

Fig. 14.4 Five tectal patterns of Acanthaceae pollen: (a) open; (b) almost closed; (c)–(d) bireticulate; (e) reticulate; (f) atectate. (a) *Ruellia grandiflora* (Forssk.) Pers., × 3000, Popov 7/7. (b) *Dyschoriste vagans* (Wight) O. Ktze, × 7500; Anon. 52. (c) *Brillantaisia nyanzarum* Burkill, × 3750, Dale 3208. (d) *Strobilanthes forrestii* Diels, × 3750, Forrest 6344. (e) *Eremomastax speciosa* (Hochst.) Cufod., × 3750, Le Testu 198. (f) *Dischistocalyx thunbergiaeflora* T. Anders., × 3750, C.C. Exp.

remainder, *Brillantaisia*, *Hygrophila*, *Strobilanthes*, and associated genera have a bireticulate tectum comprising an outer, coarse reticulum and an inner, finer one (Fig. 14.4c,d). The remaining genera have a simple reticulum (Fig. 14.4e). This was coded as an unordered multistate (five states) character and rooted on the presence of a simple reticulum (Fig. 14.4e) which, on the basis of outgroup comparison, is widely distributed throughout the family.

Character 2. Some genera have pseudocolpi distributed uniformly over the grain (Figs. 14.1b, 14.6a,b), whereas others have four pseudocolpi more closely associated with each aperture and none in the mesocolpia (Figs 14.1c, 14.6d). This was coded as an unordered multistate character (three states) and rooted on the absence of pseudocolpi.

Character 3. Some genera have distinct, raised areas of sexine associated closely with apertures, 'sexine lips' (Fig. 14.5). This was coded as a presence/absence binary character and rooted on absence.

Character 4. Of the genera with 'sexine lips to the aperture' character 3, three distinct types can be distinguished relative to the adjacent sexine and pseudocolpi (Fig. 14.5). This was coded as an unordered multistate character (three states) with the root equivocal as there is no *a priori* character state which can be judged to be the general (plesiomorphic) condition.

Character 5. The range of aperture number can be two, three, four, or more than four (Fig. 14.6). This was coded as an unordered multistate character (four states) with the root equivocal as all four character states are widely distributed in the Acanthaceae.

Character 6. The frequent palynological distinction between colpate, colporate, and porate pollen grains is not sufficient, in this context, to describe the full range of variation of aperture types present. In those genera with pseudocolpi, the pore (ora) either lies on a colpus as in *Strobilanthes* (Fig. 14.1b), or it lies between simple colpi or pseudocolpi as in *Duosperma* (Fig. 14.1c). Here it would seem that there are two distinct aperture types, although Raj (1961), surprisingly, described both as colporate. The problem with the colporate/porate distinction relative to the grains with pseudocolpi is whether the situation in *Strobilanthes* should be considered the same as a grain which is colporate without pseudocolpi present. Also, if *Duosperma* (Fig. 14.1c) is considered to be porate, then is it to be considered similar to a 'more typical' porate grain, e.g. *Ruellia* (Fig. 14.1a)? Given this situation,

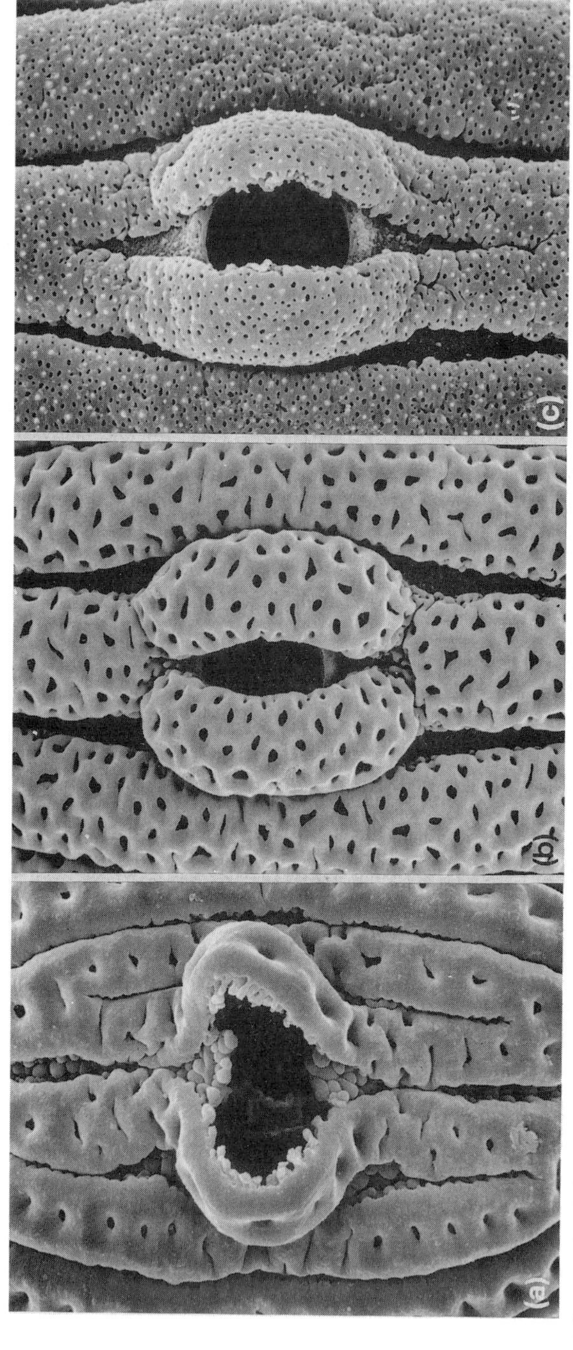

Fig. 14.5 Three different types of raised sexine 'lips' around apertures in Acanthaceae pollen. (a) *Bravaisia integerrima* (Sprengel) Standley, × 2625, *Birley 7368*. (b) *Epiclastopelma macranthium* Mildbr., × 2250. *Schlieben 2839*. (c) *Sautiera tinctorum* Decne., × 3750. Cunningham 320.

Fig. 14.6 Aperture number in Acanthaceae pollen: (a) two; (b) four: (c) many; (d) three. (a) *Bravaisia berlandieriana* (Nees) T.F. Daniel, × 1050, Cabrera 1149. (b) *Hygrophila angustifolia* R. Br, × 1400, Anon. 481b. (c) *Ruellia patula* Jacq., × 840, Popov 71/327. (d) *Petalidium glandulosum* S. Moore, × 700., Exell *et al.* 2266.

three aperture types have been recognized. For example, in genera with pseudocolpi, either the pore (ora) lies on a colpus (Fig. 14.1b) or it lies between pseudocolpi (Fig. 14.1c), with the third state being the more conventional porate condition of, for example, *Ruellia* (Fig. 14.1a). This is coded as a multistate character (three states) with the root equivocal. Another way to treat this character would be to apply the pore/colpus distinction and to code, for example *Duosperma* (Fig. 14.1c), as porate just as *Ruellia*. This does not change the topologies of the trees in any way. If this character is rooted on the colporate condition, it still makes no difference to the outcome.

Character 7. Both *Dischistocalyx* and *Acanthopale* have no tectum (Fig. 14.1d), while all other genera are tectate/columellate. This is coded as a binary character and rooted on tectate/columellate, which is the general condition within the Acanthaceae.

Character 8. The genus *Ruellia* has a tectum that is restricted to the apices of adjoining columellae (Fig. 14.1a). This is coded as a presence/absence binary character and rooted on absence.

Character 9. The genus *Petalidium* has a raised area of sexine, with columellae longer than in the surrounding areas, running longitudinally up to the apertures (Fig. 14.6d). This character is an autapomorphy, and thus consistent with any tree in which *Petalidium* is a terminal taxon, and therefore is not included in the analysis.

There are four equally parsimonious solutions for the data set (see Trees 3–6, Fig. 14.7). These trees are each 16 steps long with a consistency index of 0.93. The four trees differ in their relative placement of *Ruellia*, *Dischistocalyx*, and *Acanthopale*. *Ruellia*, *Dischistocalyx*, and *Acanthopale* either form a clade which is the sister group to all other genera (Trees 3–4); or *Ruellia*, *Dischystocalyx*, and *Acanthopale* and all other genera form a trichotomy (Trees 5–6). Also, the trees differ in placing the clade that includes *Dyschoriste*, *Sautiera*, and *Chaetacanthus* as either the sister group to the clade that includes *Mellera*, *Petalidium*, and *Mimulopsis* (Trees 4–5) or in a polychotomous sister group to the clade that includes *Trichanthera* (Trees 3 and 6). Tree 6 is also the strict consensus tree of the four trees.

Discussion

An interesting question to ask, at this stage, is how does the consensus tree based on pollen morphology (Tree 6) compare with previous classifications of these genera? One form of assessment is to compare the

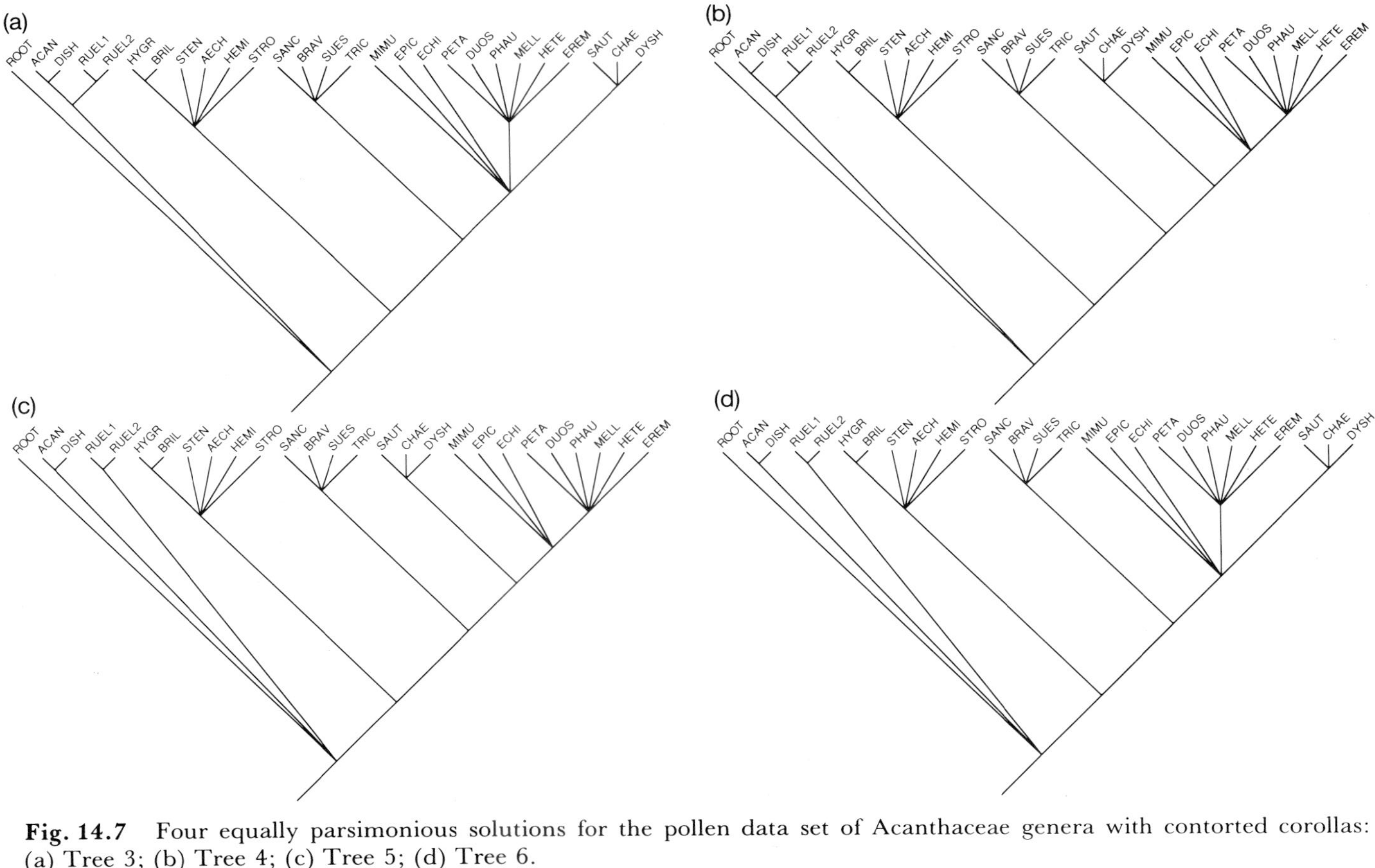

Fig. 14.7 Four equally parsimonious solutions for the pollen data set of Acanthaceae genera with contorted corollas: (a) Tree 3; (b) Tree 4; (c) Tree 5; (d) Tree 6.

fit (the better the fit, the greater the number of congruent characters) between previously proposed classifications and new pollen data. If the fit between a set of characters and a classification is 100 per cent, there will be no incongruent characters and the consistency index (the number of characters divided by the number of steps on the tree(s) that those characters produce) would be 1. These comparisons are achieved by inputting the other classifications into Hennig86 and asking the program to compute the consistency index for those classifications from the new pollen data. Therefore, the four equally parsimonious trees produced in the analysis, which have a consistency index of 0.93 (Table 14.2), have a very high degree of fit with the new pollen data. In contrast, the classifications of Lindau (1895), Clarke (1885), and Burkill and Clarke (1899–1900) have a relatively low fit with the new pollen data (Table 14.2).

The taxonomic implications from this preliminary analysis are as follows. Lindau's (1895) tribes Hygrophileae, Petalideae, and Strobilantheae are not natural groups. However, there is strong pollen evidence for his Trichanthereae being monophyletic. More specifically, to place the genera *Eremomastax* and *Mellera* in his Hygrophileae with *Hygrophila* and *Brillantaisia* is completely contrary to the pollen evidence, which places *Eremomastax* and *Mellera* in a monophyletic group with *Petalidium*, *Duosperma*, *Heteradelphia*, and *Phaulopsis*. Burkill and Clarke (1899–1900) recognized three groups within the Contortae based on the arbitrary split of the number of ovules. Not surprisingly, therefore, there is only pollen evidence for the monophyly of Hygrophileae, i.e. *Hygrophila* and *Brillantaisia*, whereas the other two subtribes, Euruellieae (equivalent to Polyspermeae and Hygrophileae of Clarke 1885) and Strobilantheae (equivalent to the Tetrandeae of Clarke 1885) are polyphyletic.

Bremekamp (1944), in his monograph of Strobilanthinae, suggested several taxonomic changes which are in general agreement with my results, namely that genera 30–36 of Lindau's (1895) Strobilantheae

Table 14.2. Comparison of the consistency indices of several classifications relative to the new pollen data

Tree	No. of steps	Consistency index
Lindau (1895)	44	0.34
Clarke (1885);		
Burkill and Clarke (1899–1900)	53	0.28
Scotland (Tree 6)	16	0.93

were 'to be transferred to the subtribe Petalidiinae' (Bremekamp 1944, p. 16). The problem, then, is to discover which other genera Bremekamp considered to lie in his Petalidiineae as he (Bremekamp) 'did not always list the genera comprising his higher taxa' (Daniel *et al.* 1990). After the transfer of these genera to the Petalidiineae (Bremekamp 1944), the Strobilanthinae contained *Strobilanthes*, *Aechmanthera*, *Hemigraphis*, *Stenosiphonium*, *Pseudostenosiphonium*, and *Lamiacanthus*. The analysis presented above suggests that, as the possession of a bireticulate tectum is restricted to Bremekamp's (1944) Strobilanthinae, and *Hygrophila* and *Brillantaisia* (i.e. part of the Hygrophileae of Lindau 1895), Bremekamp's (1944) Strobilanthinae is not resolved by the pollen data. However, the monophyletic status of a group comprising *Hygrophila* plus *Brillantaisia*, which are nested within the Strobilanthinae of Bremekamp (1944), is justified on pollen evidence by the presence of four apertures (character 5).

If, as the above analysis suggests, the appearance of pseudocolpi (character 2) occurred only once within the Contortae, then this 'pseudocolpate clade' is a previously unrecognized group, and will almost certainly contain *Strobilanthopsis* and *Ruelliopsis*, which have not been included in the study due to lack of material. Furthermore, within the 'pseudocolpate clade' other subdivisions can be identified, in particular, the group possessing raised areas of sexine around the aperture (character 3).

Conclusions

Systematics is concerned with organisms and their parts. Parts of organisms are conceptualized and described as characters. Characters vary in their complexity. Terminology can fail or succeed in conveying the morphological complexity of a character. The increased resolution of microscopes may or may not alter our perception of a character. The taxonomic hierarchy consists of groups defined by characters. The distribution of characters among organisms is often complex.

These remarks seem pertinent in the context of both historical and current systematics of Acanthaceae. Lindau's (1895) system was based on character concepts (pollen types) which are inadequate, in the light of modern microscopy, to describe the full extent of pollen morphological variation. Given that Lindau's (1895) own pollen character concepts often have no direct logical relationship to the groups he recognized, his system is open to severe criticism. Clarke (1885) and Burkill and Clarke (1899–1900) gave prominence to the number of ovules. These authors not only indulged in *a priori* weighting of ovule number above all other characters (at least within their Ruellieae), but also arbitrarily split a continuous variable into discrete units.

The relationships between genera of the Acanthaceae are poorly understood. The parsimony analysis presented above represents a first attempt at logical classification of a number of genera based on pollen morphology. Not until such an analysis of pollen is extended to include many other morphological characters will generic relationships be fully understood.

Acknowledgements

I would like to thank Stephen Blackmore and Chris Humphries for their expert and exciting supervision during this project. Also, I thank Sue Barnes for tuition in microscopy, Colin Patterson whose door was always open, Dick Brummitt for discussion of Acanthaceae and Charlie Jarvis and Fred Barrie who critically read the manuscript and put verbs in my sentences. This paper was written while the author was in receipt of a Natural History Museum Research Studentship.

References

Anderson, T. (1863). An enumeration of the species of Acanthaceae from the continent of Africa and the adjacent islands. *Journal of the Proceedings of the Linnean Society* **7**, 13–54.

Anderson, T. (1867). An enumeration of the species of the Indian Acanthaceae. *Journal of the Linnean Society of London, Botany* **9**, 425–526.

Balkwill, K. and Getliffe Norris, F. (1988). Classification of the Acanthaceae: A Southern African perspective. *Monographs in Systematic Botany from the Missouri Botanical Garden* **25**, 503–16.

Bentham, G. (1876). Acanthaceae. In *Genera plantarum*, Vol. 2, (ed. G. Bentham and J. D. Hooker), pp. 1060–122. Reeve, London.

Bremekamp, C. E. B. (1944). Materials for a monograph of the Strobilanthinae (Acanthaceae). *Verhandelingen der Nederlandsche Akademie van Wetenschappen, Afdeeling Natuurkunde* **41**, 1–305.

Bremekamp, C. E. B. (1965). Delimitation and subdivision of the Acanthaceae. *Bulletin of the Botanical Survey of India* **7**, 21–30.

Burkill, I. H. and Clarke, C. B. (1899–1900). Acanthaceae. In *Flora of tropical Africa*, Vol. 5, (ed. W. T. Thiselton-Dyer), pp. 1–262. Reeve, London.

Clarke, C. B. (1885). Acanthaceae. In *Flora of British India*, Vol. 4, (ed. J. D. Hooker), pp. 387–558. Reeve, London.

Daniel, T. F., Chuang, T. I., and Baker, M. A. (1990). Chromosome numbers of American Acanthaceae. *Systematic Botany* **15**, 13–25.

Erdtman, G. (1960). The acetolysis method – a revised description. *Svensk Botanisk Tidskrift* **54**, 561–4.

Farris, J. S. (1988). *Hennig v.* **1.5**, Hennig86 reference manual and software for MSDOS microcomputers. Published by the author.

Furness, C. A. (1989). The pollen morphology of *Ecbolium* and *Megalochlamys* (Acanthaceae). *Kew Bulletin* **44**, 681–93.

Graham, V. (1988). Delimitation and infra-generic classification of *Justicia* (Acanthaceae). *Kew Bulletin* **43**, 551–624.

Hansen, B. (1988). Revision of *Thysanostigma* (Acanthaceae). *Nordic Journal of Botany* **8**, 227–30.

Hedrén, M. (1989). Justicia Sect. Harnieria (Acanthaceae) in Tropical Africa. *Symbolae Botanicae Upsalienses* **29**, 1–141.

Hennig, W. (1966). *Phylogenetic Systematics*. University of Illinois Press, Urbana.

Hilsenbeck, R. A. (1990). Pollen morphology and systematics of *Siphonoglossa sensu lato* (Acanthaceae). *American Journal of Botany* **77**, 27–40.

Lindau, G. (1893). Beiträge zur Systematik der Acanthaceen. *Botanische Jahrbücher für Systematik, Pflazengeschichte und Pflanzengeographie* **18**, 36–64.

Lindau, G. (1895). Acanthaceae. In *Die Naturlichen Pflanzenfamilien*, Vol. 4, 3b, (ed. A. Engler and K. Prantl), pp. 274–354. Engelmann, Leipzig.

McDade, L. A. (1988). Recognition of *Aphelandra glabrata* (Acanthaceae) from Western South America, with notes on phylogenetic relationships. *Systematic Botany* **13**, 235–9.

Maddison, W. P., Donoghue, M. J., and Maddison, D. R. (1984) Outgroup analysis and parsimony. *Systematic Zoology* **33**, 83–103.

Nees von Esenbeck, C. G. (1832). Acanthaceae Indiae Orientalis. In *Plantae Asiaticae Rariores*, Vol. 3, (ed. N. Wallich), pp. 70–117. Truettel, Wurtz and Richter, London.

Nees von Esenbeck, C. G. (1847). Acanthaceae. In *Prodromus systematis naturalis regni vegetabilis*, Vol. 2. (ed. A. de Candolle), pp. 46–519. Masson, Paris.

Nelson, G. J. and Platnick, N. I. (1981). *Systematics and biogeography: cladistics and vicariance*. Columbia University Press, New York.

Patterson, C. (1982). Morphological characters and homology. In *Problems of Phylogenetic Reconstruction*, Systematics Association Special Vol. 21, (ed. K. A. Joysey and A. E. Friday), pp. 221–74. Academic Press, London.

Radlkofer, L. (1883). Ueber den systematischen Werth der Pollenbeschaffenheit bei den Acanthaceen. *Sitzunbsberichte der mathematisch-physikalischen Classe der k.b. Akademie der Wissenschaften zu München* **13**, 256–314.

Raj, B. (1961). Pollen morphological studies in the Acanthaceae. *Grana Palynologica* **3**, 3–108.

Raj, B. (1973). Further contribution to the pollen morphology of the Acanthaceae. *Journal of Palynology* **9**, 91–141.

Ramamoorthy, T. P. and Hornelas, Y. (1988). A new name and a new species in Mexican *Ruellia* (Acanthaceae). *Plant Systematics and Evolution* **151**, 161–3.

Rieppel, O. (1988). *Fundamentals of comparative biology*. Birkhauser Verlag, Berlin.

Vollesen, K. (1989). A revision of *Megalochlamys* and *Ecbolium* (Acanthaceae: Justiceae). *Kew Bulletin* **44**, 601–80.

Wasshausen, D.C. (1988). New and interesting species of Acanthaceae from Peru. *Beiträge zur Biologie der Pflanzen* **63**, 421–9.

Watrous, L.W. and Wheeler, Q.D. (1981). The outgroup comparison method of character analysis. *Systematic Zoology* **30**, 1–11.

Wood, J.R.I. (1988). Columbian Acanthaceae - some new discoveries and some reconsiderations. *Kew Bulletin* **43**, 1–51.

Appendix

List of characters and character states

1. 0, reticulate; 1, bireticulate; 2, closed; 3, open; (–), missing and not applicable.
2. 0, pseudocolpi absent; 1, pseudocolpi 12 with none in mesocolpium; 2, pseudocolpi throughout.
3. 0, absence of raised areas of sexine associated with each aperture; 1, presence of raised areas of sexine.
4. –, equivocal; 0, *Dyschoriste* type (Fig. 14.5c); 1, *Epiclastopelma* type (Fig. 14.5b); 2, *Bravaisia* type (Fig. 14.5a).
5. –, equivocal; 0, four apertures; 1, three apertures; 2, two apertures; 3, many.
6. –, equivocal; 0, pore lying on (pseudo) colpus; 1, pore lying between pseudocolpi; 2, porate.
7. 0, tectate columellate; 1, intectate.
8. 0, tectum not restricted to apices of columellae; 1, tectum restricted to apices of columellae.

Data matrix

Taxa	Characters
Root	000---00
Petalidium	01111100
Phaulopsis	01111100
Mellera	01111100
Heteradelphia	01111100
Eremomastax	01111100
Duosperma	01111100
Epiclastopelma	02111100
Echinacanthus	02111100
Mimulopsis	02111100
Trichanthera	02122000
Bravaisia	02122000
Sanchezia	02122000
Suessenguthia	02122000
Dyschoriste	22101100
Sautiera	22101100
Chaetacanthus	22101100
Hygrophila	120-0000
Brillantaisia	120-0000
Strobilanthes	120-1000
Aechmanthera	120-1000
Hemigraphis	120-1000
Stenosiphonium	120-1000
Ruellia	300-1201
Ruellia	300-3201
Dischistocalyx	-00-1210
Acanthopale	-00-3210

15. Characterizing pollen sculpture of three closely related Capparaceae species using quantitative image analysis of scanning electron micrographs

EDWARD L. VEZEY
Department of Botany and Microbiology, The University of Oklahoma, Norman, Oklahoma, USA

JOHN J. SKVARLA
Department of Botany and Microbiology and Oklahoma Biological Survey, The University of Oklahoma, Norman, Oklahoma, USA

STARIA S. VANDERPOOL
Department of Biology, University of North Dakota, Grand Forks, North Dakota, USA

Abstract

Pollen grains of three closely related species, *Cleomella longipes, Oxystylis lutea*, and *Wislizenia refracta* (Capparaceae subfamily Cleomoideae), have microreticulate sculpture, with muri surrounding scrobiculate lumina. Sculpture variation was assessed by computerized image analysis of 51 scanning electron micrographs (16–18 per species). Measurements from each grain (per cent muri coverage, lumina perforation density, and three size/shape variables) were compared by UPGMA cluster analysis, resulting in three distinct clusters corresponding to pollen grains of each species. Conversion of the five continuous characters to discrete character states using gap-coding indicates the potential utility of image analysis data for cladistic analysis.

Pollen and Spores (ed. S. Blackmore and S. H. Barnes), Systematics Association Special Volume No. 44, pp. 291–300, Clarendon Press, Oxford, 1991.

Introduction

Pollen sculpture variation provides evidence for relationships among plant species. In this study, computerized image analysis of scanning electron micrographs was used to quantify sculpture of pollen from *Cleomella longipes, Wislizenia refracta*, and *Oxystylis lutea* (Capparaceae subfamily Cleomoideae). These species are part of a monophyletic group including the six species of *Cleome* sect. *peritoma*, 10 species of *Cleomella*, and the monotypic genera *Wislizenia* and Oxystylis (Iltis 1957; Bremer and Wanntorp 1978; Vanderpool 1989). All 18 species have microreticulate pollen sculpture with muri enclosing scrobiculate lumina. This pattern has not been found in other species of *Cleome* or subfamily Cleomoideae (Al-Shehbaz 1973; Vezey and Skvarla 1990, unpublished data). Pollen sculpture thus reinforces three other synapomorphies uniting these taxa—presence of scarious stipules and hard pointed styles (Iltis 1957), and stomates on seeds (Vanderpool 1989). The three species in this investigation were selected because they form a putative monophylum within the larger group, with *Cleomella longipes* a possible sister group to *Oxystylis* and *Wislizenia* (Iltis 1957; Bremer and Wanntorp 1978).

In a preliminary investigation, Vezey and Skvarla (1990) used image analysis to examine a total of nine pollen grains of these taxa, concluding that quantitative sculpture differences were species-specific. This study expands the earlier investigation by including an average of 17 grains per species, and one additional characteristic, lumina perforation density. Cluster analysis using image analysis data was conducted to determine whether pollen grains grouped by species, and to find out which species had most similar pollen sculpture. Image analysis measurements were also converted into discrete character states using gap-coding to assess their value in cladistic analysis.

Materials and methods

Pollen was collected from herbarium sheets (Table 15.1), treated with the acetic anhydride/sulphuric acid mixture of Erdtman (1960), air dried from 95 or 100 per cent ethanol, sputter-coated with gold/palladium (60/40), and examined with a JEOL JSM 880 scanning electron microscope (SEM). A total of 51 micrographs (16–18 per species) were taken perpendicular to the equatorial region of the pollen grains, with the polar axis in a horizontal direction. The original magnification of all *Cleomella* and *Oxystylis* micrographs was × 20 000; all *Wislizenia* micrographs were taken at × 18 000.

Image analysis was conducted using Kevex Feature Analysis software (Version 2.10) on a Delta class computer (Kevex Instruments, Inc.,

Table 15.1 Taxa and collection data

Species	Collection	Location
Cleomella longipes Torr. ex. Hook.	Vanderpool 1336 OKL	Cochise Co., Arizona
Oxystylis lutea Torr. & Frem.	Vanderpool 1287 OKL	Inyo Co., California
Wislizenia refracta Engelm. ssp. *refracta*	Vanderpool 1214 OKL	Cochise Co., Arizona

San Carlos, California). Micrographs were prepared for image analysis using methods described in Vezey (1990) and Vezey and Skvarla (1990). Negatives were developed into slightly underexposed positives, then lumina were darkened with a black felt-tip pen. The resulting patterns were digitized using a television camera and an interface box by Colorado Video, Inc. (Boulder, Colorado). All micrographs were digitized at the same session, without altering lighting or distance between camera and copy stand.

An area of analysis was designated on each digitized SEM micrograph. Average polar and equatorial dimensions of the area of analysis were 4.0 × 2.4 μm for *Cleomella*, 4.1 × 2.8 μm for *Oxystylis*, and 4.7 × 3.3 μm for *Wislizenia*. Average number of whole lumina measured within each area was 15.4, 10.2, and 8.9 for *Cleomella, Oxystylis*, and *Wislizenia* micrographs, respectively.

The first variable measured in all analyses was percentage of pollen surface occupied by muri (per cent muri coverage). Next, all partial lumina along edges of the area of analysis were deleted using image analysis software. Lastly, the following quantities were determined using the remaining whole lumina on each micrograph:

1. Perforation density—perforations per square micrometer. Computed by manually counting the total number of perforations in two lumina per micrograph, then dividing by the total area of these lumina.
2. Average Wadell diameter—diameter of a circle having the same area as a lumen averaged for all whole lumina within the area of analysis.
3. Average lumen length—longest lumen dimension averaged for all whole lumina within the area of analysis.
4. Average lumen width—the mean lumen dimension perpendicular to the longest dimension averaged for all whole lumina within the area of analysis.

Table 15.2 SEM image analysis data. Measurements are averages for the midmesocolpal region of individual pollen grains

Pollen grain	Perforation density* (perforations/ μm^2)	Wadell diameter (μm)	Lumina length (μm)	Lumina width (μm)	Muri coverage (%)
C 1	78.8	0.32	0.43	0.19	78.4
C 2	69.0	0.37	0.52	0.21	77.0
C 3	79.1	0.41	0.52	0.25	75.4
C 4	-	0.41	0.52	0.26	76.7
C 5	-	0.38	0.50	0.23	75.4
C 6	80.4	0.40	0.56	0.23	77.0
C 7	78.0	0.33	0.44	0.20	81.2
C 8	74.2	0.38	0.56	0.22	73.2
C 9	-	0.33	0.50	0.17	82.7
C10	-	0.40	0.50	0.25	75.8
C11	80.5	0.37	0.47	0.23	78.8
C12	71.5	0.32	0.41	0.20	80.4
C13	-	0.36	0.57	0.18	82.1
C14	84.4	0.30	0.43	0.17	78.6
C15	78.0	0.35	0.43	0.22	80.3
C16	-	0.30	0.42	0.18	83.0
C17	78.2	0.40	0.55	0.23	76.3
C18	77.3	0.36	0.50	0.21	76.9
O 1	-	0.48	0.64	0.29	70.6
O 2	75.7	0.56	0.73	0.34	68.6
O 3	71.6	0.51	0.73	0.29	69.6
O 4	61.4	0.56	0.68	0.36	68.3
O 5	79.0	0.51	0.68	0.31	70.9
O 6	68.2	0.51	0.74	0.28	70.7
O 7	80.0	0.53	0.79	0.29	70.4
O 8	78.3	0.58	0.73	0.37	69.1
O 9	60.9	0.59	0.76	0.37	67.2
O10	60.8	0.68	0.90	0.41	65.8
O11	63.1	0.58	0.79	0.35	65.8
O12	-	0.56	0.70	0.35	69.7
O13	69.2	0.50	0.66	0.30	68.7
O14	68.8	0.56	0.73	0.34	68.4
O15	72.9	0.51	0.82	0.26	69.6
O16	70.8	0.64	0.96	0.34	62.9
O17	61.3	0.52	0.69	0.31	72.1

Table 15.2 *cont'd*

Pollen grain	Perforation density* (perforations/ μm^2)	Wadell diameter (μm)	Lumina length (μm)	Lumina width (μm)	Muri coverage (%)
W 1	-	0.69	0.98	0.39	67.6
W 2	50.4	0.64	0.82	0.40	69.8
W 3	44.8	0.82	1.05	0.51	64.9
W 4	46.8	0.61	0.73	0.40	70.0
W 5	55.5	0.68	0.87	0.43	66.4
W 6	48.6	0.81	1.08	0.48	61.8
W 7	51.5	0.67	0.86	0.41	69.2
W 8	44.8	0.65	0.81	0.41	68.1
W 9	56.1	0.76	1.08	0.43	62.5
W10	52.1	0.65	0.84	0.40	71.0
W11	56.4	0.78	1.01	0.47	65.7
W12	41.5	0.67	0.93	0.40	66.1
W13	54.9	0.74	0.89	0.50	67.4
W14	48.8	0.73	1.02	0.42	67.0
W15	44.3	0.77	1.00	0.47	69.6
W16	46.5	0.76	0.90	0.51	64.8

Abbreviations : C, *Cleomella*; O, *Oxystylis*; W, *Wislizenia*.
* The perforation density was not determined for some pollen grains because the lumina surfaces were obscured.

Variation within and among species was compared using NTSYS-pc (Version 1.50; Rohlf 1988). Data (Table 15.2) were first standardized (i.e. each character being transformed to have a mean of zero and standard deviation of one), followed by calculation of an average taxonomic distance matrix and UPGMA (unweighted pair-group using arithmetic averages) cluster analysis.

Data for each species were converted to discrete character states using simple gap-coding (Archie 1985). This involves ordering character means by size, calculating the pooled within-group standard deviation for each character, and then assigning different character states when adjacent means differ by more than this standard deviation.

Results and discussion

Data for each micrograph (Table 15.2) are summarized by species (Table 15.3) and converted into discrete character states (Table 15.4).

Table 15.3. Sculpture of each species characterized by image analysis data

Species	Perforation density (perforations/μm^2)	Wadell diameter (μm)	Lumina length (μm)	Lumina width (μm)	Muri coverage (%)
Cleomella longipes	77.5 (69.0–84.4)	0.36 (0.30–0.41)	0.49 (0.41–0.57)	0.21 (0.17–0.26)	78.3 (73.2–83.0)
Oxystylis lutea	69.4 (60.9–80.0)	0.55 (0.48–0.59)	0.75 (0.64–0.82)	0.33 (0.26–0.37)	68.7 (65.8–72.1)
Wislizenia refracta	49.5 (41.5–56.4)	0.71 (0.61–0.82)	0.93 (0.73–1.08)	0.43 (0.39–0.51)	67.0 (61.8–71.0)

Table 15.4 Data from Tables 15.2 and 15.3 converted to discrete character states using simple gap-coding; s_p is pooled within-group standard deviation

	Perforation density (perforations/ μm^2)	Wadell diameter (μm)	Lumina length (μm)	Lumina width (μm)	Muri coverage (%)
Cleomella longipes	0	0	0	0	0
Oxystylis lutea	1	1	1	1	1
Wislizenia refracta	2	2	2	2	1
s_p	5.5	0.052	0.082	0.037	2.6

UPGMA analysis (Fig. 15.2) of the multivariate data set (Table 15.2) produced three distinct groups, each group being composed of the micrographs of one species. Two of 51 grains, O10 and O16, were either aberrant or contaminants. Grains O10 and O16 were not considered when determining ranges for each character (Table 15.3). Figure 15.1 is a portion of micrograph W2 (Table 15.2), and is typical for *Wislizenia refracta*. Based on cluster analysis (Fig. 15.2) and characteristics measured in this study, the two species with most similar pollen sculpture are *Oxystylis lutea* and *Wislizenia refracta*.

Converting continuous data to discrete character states (Table 15.4) produced three ordered states for four characters, and two states for

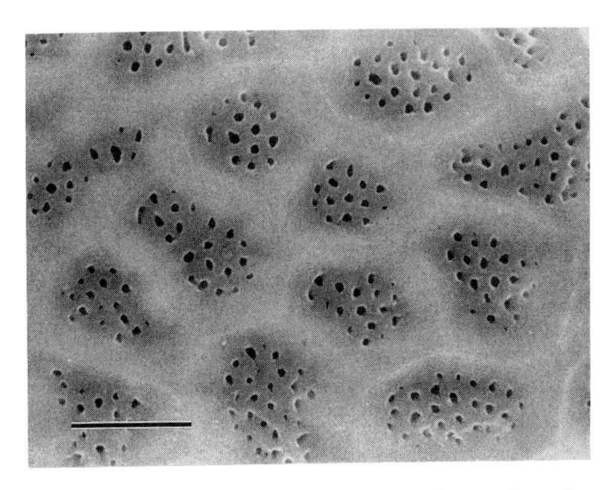

Fig. 15.1 *Wislizenia refracta*. Midmesocolpal view of pollen grain W2 (see Table 15.2). (Scale bar, 1 μm.)

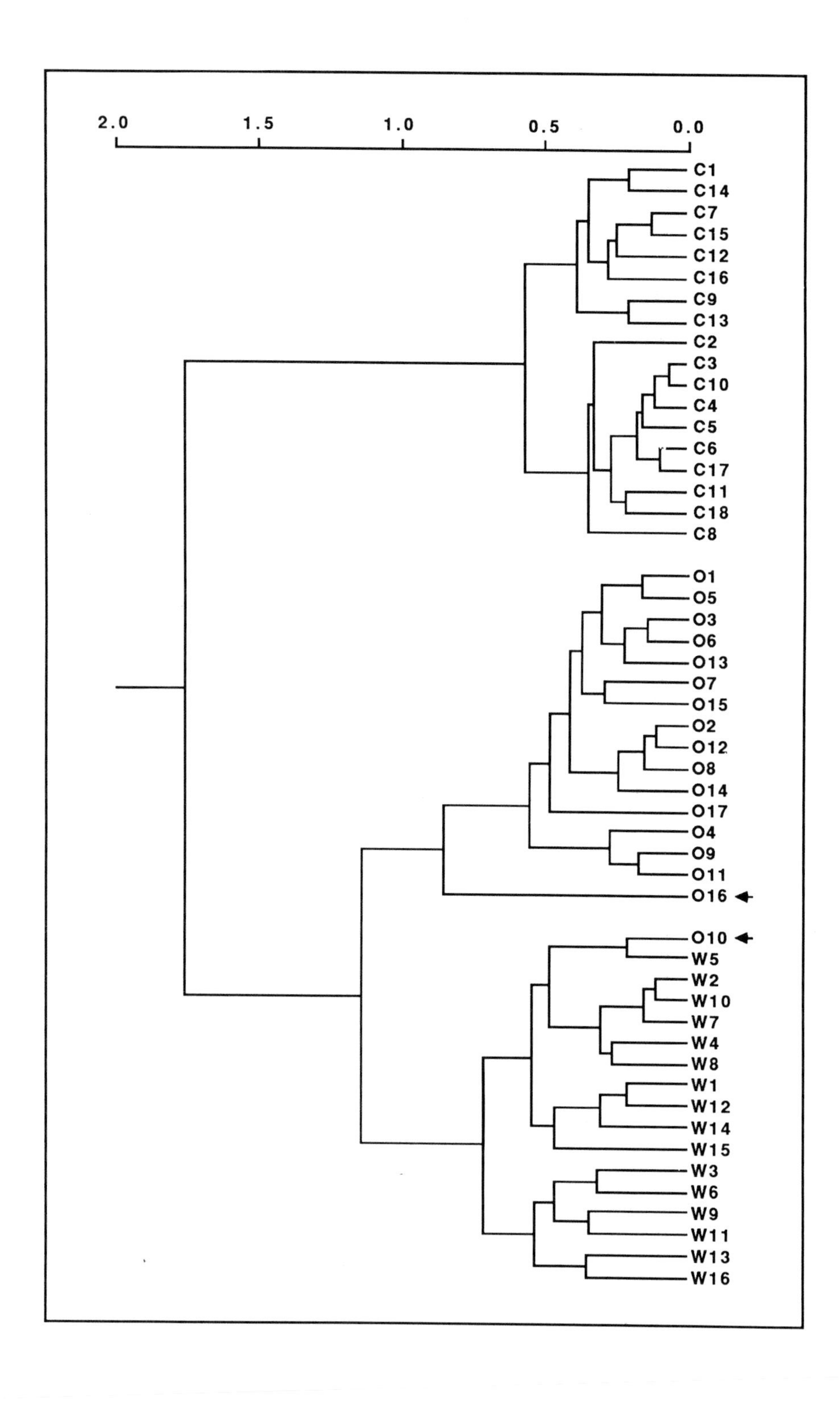
2.0
1.5
1.0
0.5
0.0
C1
C14
C7
C15
C12
C16
C9
C13
C2
C3
C10
C4
C5
C6
C17
C11
C18
C8
O1
O5
O3
O6
O13
O7
O15
O2
O12
O8
O14
O17
O4
O9
O11
O16
O10
W5
W2
W10
W7
W4
W8
W1
W12
W14
W15
W3
W6
W9
W11
W13
W16

per cent muri coverage. Polarizing these characters will be possible as the data set expands. With larger numbers of taxa, other methods for producing discrete character states, such as segment coding (Chappill 1989), may be more appropriate.

Results from Vezey and Skvarla (1990) for *Wislizenia refracta* are consistent with this study. For *Cleomella longipes* and *Oxystylis lutea*, however, eight measurements from the earlier study are considerably outside the ranges indicated in Table 15.3. Six of these (average Wadell diameter or lumina length) resulted from a greater number of very small lumina. Inconsistent results for tectum coverage (two grains of *C. longipes*) may have been caused by inclusion of grains comparable to O10 or O16 (Fig. 15.2).

Image analysis of pollen sculpture patterns, particularly those with lumina or perforations, may provide systematically useful information. As indicated above, however, results must be treated with caution until intraspecific variation is adequately sampled. Variations between populations, within populations, and on the same plant must be assessed, as well as artificial variability introduced by treatment effects. For these reasons, additional studies are necessary before the results in Table 15.3 can be said to truly characterize pollen sculpture of these species.

Acknowledgements

Our thanks to the Noble Microscopy Laboratory, University of Oklahoma, for use of facilities; to Scott D. Russell, Daniel J. Hough, and William F. Chissoe, University of Oklahoma, for technical assistance; and to James A. Doyle, University of California (Davis) for helpful comments during the symposium.

References

Al-Shehbaz, I.A. (1973). The biosystematics of the genus *Thelypodium* (Cruciferae). In *Contributions from the Gray Herbarium*, No. 204, (ed. R.C. Rollins and K. Roby), pp. 47–54. The Gray Herbarium of Harvard University.

Archie, J.W. (1985). Methods for coding variable morphological features for numerical taxonomic analysis. *Systematic Zoology* **34**, 326–45.

Bremer, K. and Wanntorp, H. (1978). Phylogenetic systematics in botany. *Taxon* **27**, 317–29.

Fig. 15.2 UPGMA cluster analysis using data from Table 15.2. O10 and O16 (arrows) are aberrant grains or contaminants.

Chappill, J. A. (1989). Quantitative characters in phylogenetic analysis. *Cladistics* **5**, 217–34.

Erdtman, G. (1960). The acetolysis method. A revised description. *Svensk Botanisk Tidskrift* **39**, 561–4.

Iltis, H.H. (1957). Studies in the Capparidaceae. III. Evolution and phylogeny of the Western North American Cleomoideae. *Annals of the Missouri Botanical Garden* **44**, 77–119.

Rohlf, F.J. (1988). NTSYS-pc. *Numerical taxonomy and multivariate analysis system.* Exeter Publishing, Setauket, New York.

Vanderpool, S.S. (1989). Micromorphology and seed-coat stomata: phylogeny of Cleomoideae (Capparaceae). Unpublished Ph.D. Dissertation, University of Oklahoma, Norman.

Vezey, E.L. (1990). Contrast enhancement of pollen perforations for image analysis. In *Proceedings of the XIIth International Congress for Electron Microscopy*, pp. 802–3. San Francisco Press, San Francisco.

Vezey, E. L. and Skvarla, J.J. (1990). Computerized feature analysis of exine sculpture patterns. *Review of Palaeobotany and Palynology*, 64, 187–196.

16. Diversification and evolution of the tapetum

E. PACINI
Dipartimento di Biologia Ambientale, Università di Siena, Via P. A. Mattioli 4, Siena, Italy

G. G. FRANCHI
Dipartimento Farmaco Chimico Tecnologico, Università di Siena, Via P. A. Mattioli 4, Siena, Italy

Abstract

Two kinds of tapeta are usually distinguished in angiosperms: the parietal or secretory type, and the amoeboid or periplasmodial type. However, the tapetum is present in all major groups of land plants and exhibits a great range of forms.

The type of tapetum present in Bryopsida, Marchantiopsida *pro parte*, and Lycopodiophyta is parietal with cell walls that persist until tapetal degeneration and passage of nutrients is relatively slow. An important modification is the intrusion of tapetal cells between the spores. In Anthocerotopsida, after intrusion the walls of tapetal cells become thicker and they are transformed into elaters. In some Marchantiopsida the tapetal cells intrude and retain a thin wall until degeneration. In Psilotophyta, Equisetophyta, Osmundales, Ophioglossales, Filicales, Salviniales, and Marsiliales, tapetal cells lose their walls and intrude forming a true periplasmodial tapetum.

Another significant modification occurs when the parietal tapetum loses its inner tangential and radial walls after the tetrad stage. This kind of tapetum is present in gymnosperms and in most angiosperms. Various types of amoeboid tapeta are derived from this latter, in which the naked protoplasts either fuse or retain their individuality. The time at which tapetal intrusion occurs may vary. For example, in *Arum italicum* it occurs during the second meiotic division, whereas in

Pollen and Spores (ed. S. Blackmore and S. H. Barnes), Systematics Association Special Volume No. 44, pp. 301–316, Clarendon Press, Oxford, 1991.

some Compositae, after a transient parietal phase, intrusion takes place just before the first haploid mitosis.

Even the parietal tapeta of spermatophytes exhibit some variations, for instance in the number of nuclei per cell, number of cell layers , and in the time and rate of degeneration. A distinctive type of parietal tapetum is present in Gramineae. The anthers are narrow and long with few grains visible in a cross-section of the loculus, all of which are in contact with the tapetum and are oriented with their apertures facing the tapetal cells.

The evolutionary trends of the tapetum lead to a more and more intimate contact between the tapetum and microspores. This is realized by means of the intrusion of tapetal cells between the spores; the loss of tapetal cell walls and, in the case of large loculi with many microspores, the hypothesized movement of microspores and pollen grains; and nutrition through direct contact of spores and pollen grains to tapetal cells.

Introduction

Land plant spores and pollen grains are produced by a variety of different taxonomic groups. Common features shared by these structures include not only the haploid genome and the fact that they are more or less dehydrated before dispersal, but also the intine and exine walls. The formation of these walls is staggered in time; the exine is always formed first. One of the functions of the walls is to accommodate changes in microspore volume (Wodehouse 1935; Pacini and Franchi 1984; Pacini 1990*a*).

Sporopollenin, the molecule of which exine is built, occurs in the walls of certain ancestors of land plants, such as unicellular or multicellular, submerged (Pickett-Heaps 1975), free subaerial (Good and Chapman 1978), and symbiont subaerial (Honegger and Brunner 1981) green algae. In green algae, sporopollenin precursors are produced by the cytoplasm of all cells rather than by specialized cells. This is inevitable because multicellular algae consist of cells of the same type. In land plants, however, these precursors are probably mainly produced by cells surrounding the spores (Dickinson 1976; Pacini *et al.* 1985). These cells, known as the tapetum, surround the spores and pollen grains during most of their development and occur in all the major groups of land plants. In addition to this principal activity, tapetal cells have many other functions, some of which are common to all plant groups while others are more restricted.

Independent of these common functions, the appearance of the tapetum varies widely from group to group. Tapetal types are better known in the angiosperms than in the lower groups, for which even morphological data are almost lacking.

Types of tapetum

The first criterion that has been used to distinguish tapetal types is the number of nuclei per cell and the way in which nuclear division occurs. Buss and Lersten (1975), in their survey of tapetal nuclear number in Leguminosae, state that binucleate or multinucleate tapetum, characteristic of Caesalpinioideae, other angiosperm families (Davis 1966), and gymnosperms (Singh 1978), is primitive, and that the uninucleate tapetum is derived. For examples of this approach to tapetal classification see the papers quoted by Buss and Lersten (1975) and D'Amato (1984). This approach to tapetal classification seems inadequate because only the angiosperms are considered and because the number of nuclei and their DNA content vary widely in reproductive structures, especially during ontogeny (D'Amato 1984).

The second criterion that has been used to recognize tapetal types is morphological, and makes a distinction between those tapeta that delimit the loculus and those that intermingle between the developing spores (Fig. 16.1). The first are referred to as parietal, glandular, or secretory; and the second as amoeboid, invasive, or periplasmodial. All of the latter terms indicate that the tapetal cytoplasm moves to envelop the meiocytes or microspores. In this system, the two main tapetal types differ essentially in the relationship between tapetal cytoplasm and the cytoplasm of the spores or pollen. This is because in the parietal type the cells maintain their individuality and position. In the amoeboid they may or may not maintain their individuality but during ontogeny they always change position and shape, enveloping the meiocytes, tetrads, or microspores. Tapetal cells lose their cell walls totally or partially before invasion (Fig. 16.1) and this process requires energy because it is an active phenomenon and there is reorganization of plasma membranes. In cases in which intrusion occurs during meiosis, there is reorganization of the tapetal cytoplasm even when the microspores are released. The involvement of the cytoskeleton during tapetal invasion was demonstrated by Tiwari and Gunning in *Tradescantia virginiana* (1986*a*), together with the fact that colchicine inhibits plasmodium formation and disrupts the pathways of sporopollenin secretion (Tiwari and Gunning 1986*b*).

In the amoeboid tapetum the plasma membrane is in close contact with the surface of the spore or pollen grain, and substances produced or transported by the tapetal protoplast are released adjacent to the sites of absorption or polymerization. Parietal tapeta, on the other

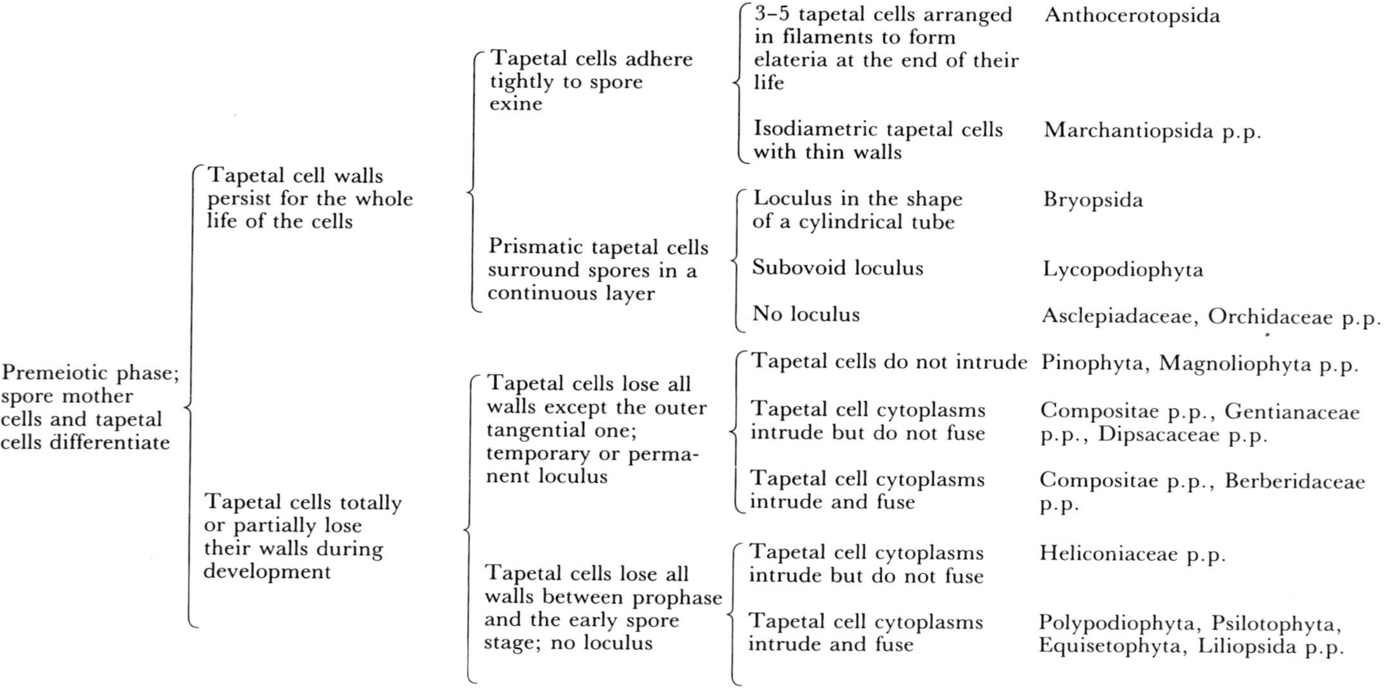

Fig. 16.1 Ontogenesis of the different tapetum types.

hand, border a cavity, the loculus, in which the spores or pollen grains develop immersed in a large mass of liquid (Fig. 16.1), and substances produced or transported by the tapetum protoplast are released far from the sites of absorption or polymerization. The shape of the loculus may vary from oval, for example in *Selaginella*, to prism-shaped, as in many angiosperms, or a hollow cylinder in the mosses (Fig. 16.1). The mass of locular fluid filling the space defined by the cells of parietal tapetum may be as much as a few cubic millimetres in large anthers such as those of *Lilium* (Willemse and Audran 1982). In the amoeboid tapetum there is much less locular fluid because the plasma membrane covers the exine microsculpture, and only increases after tapetal degeneration, when the spores or pollen grains need to remain wet until dehydration and dispersal. Figure 16.1 summarizes the occurrence of these features in the various types of tapetum. Different subtypes are recognizable in all kinds of tapetum (Pacini *et al.* 1985) on the basis of their ontogeny, cell wall persistence, cytoplasmic fusion, spatial relationships with spores, and the eventual fate of the tapetal cells (Figs 16.1, 16.2, 16.4).

The morphological criterion seems preferable because it considers tapetum ontogeny, not merely the appearance at a certain stage. This criterion also accords well with land plant phylogeny. There are two different kinds of amoeboid tapeta, that found in the Psilotophyta, Equisetophyta, and Polypodiophyta, and that found in certain monocotyledonous or dicotyledonous angiosperms. These two kinds of amoeboid tapeta can be regarded as having arisen by evolutionary convergence because their phylogenetic origin is different (Pacini *et al.* 1985).

Functions of the tapetum

The tapetum has numerous functions, since many substances are produced by the tapetum itself or conveyed by the tapetal cytoplasm from the mother plant. These have been summarized for seed plants by Echlin (1971), Pacini *et al.* (1985), and Pacini (1990*b*). The main tapetal functions in land plants are also summarized in Table 16.1, relative to the different groups. Some of these apply to all tapetum types, others are typical of a given type.

The locular fluid (or locular sap) mediates the nutrition of spores or pollen grains. All tapetally derived substances are conveyed to the locular fluid and must pass through it. The composition of this liquid changes according to the stage of development. In all tapetal types, the reabsorption or evaporation of the locular sap occurs just before sporangium or anther opens so that the contents can be dispersed

Table 16.1. Tapetal functions other than nourishment in the different groups of land plants

		Locular fluid elaboration	Lythic enzymes hydrolysing callose	Lythic enzymes hydrolysing spore pectocellulosic primary cell wall	Lythic enzymes hydrolysing tapetal cell wall	PAS-positive content of the loculus	Exine precursors	Perine precursors	Viscin thread precursors	Culture sac = peritapetal membrane	Ubisch bodies = orbicules	Elaters (adpressors)	Sporophytic proteins	Pollenkitt	Tryphine
		1	2	3	4	5	6	7	8	9	10	11	12	13	14
Bryophyta	Bryopsida	+		+			+	±							
	Marchantiopsida	+		+			+	?							
	Anthocerotopsida	+		±		+	+	?							
Psilotophyta				+	+		+						?		
Lycopodiophyta				±	+		+						?		
Equisetophyta				+	+		+					+	±		
Polypodiophyta				+	+		+	±	±†	±	±		±		
Pinophyta		+	+	+	+	±	+			±	+		±		
Magnoliophyta	Magnoliopsida	±	+	+	±*	±	+		±	±	±		+	±	±
	Liliopsida	±	+	+	±*	±	+		±	±	±		±	±	±

1, This liquid mediates almost all the other functions. 2–4, Regulatory functions concerning either the tapetal cells themselves (4) or meiocytes (2,3). 5, Associated with the function of nourishment. 6–10, Sporopollenin precursor production and polymerization sites. 11–14, Deposition of material external to the exine. 1, 5, 9, 10, Typical of parietal tapeta.
+, Present in almost all members; ±, present at least in some members; ?, no available data.
† Tryon (1985).
* Tapetal cells retain their walls in the massulate pollen of some Asclepiadaceae (Dan Dicko-Zafimahova and Audran 1981) and Orchidaceae (Sood and Mohana Rao 1986; [illegible]

(Pacini and Franchi 1984; Pacini 1990*a*, *b*). In the case of secretory tapeta, the volume of locular sap increases continuously because the loculus grows faster than the spores or pollen grains (Pacini and Franchi 1983).

The locular space probably has another function at dehiscence, that is to allow spores or pollen grains to be dispersed individually, in tetrads, or in small masses of a few monads. Dispersal also seems to be promoted by a sporopollenin layer covering the loculus surface, known as the culture sac or peritapetal membrane, produced by the tapetal cells (Heslop-Harrison 1969). When the pollen grains are dispersed in polyads or massulae, the loculus is not present at any ontogenetic stage and the tapetal cells directly surround each pollinium or massula (Kenrick and Knox 1979; Dan Dicko-Zafimahova and Audran 1981; Wolter *et al.* 1988). Pollen dispersal in these cases occurs *en masse* and particular devices must be present to promote adherence to the bodies of pollinators. In the vandoid orchids, for example, caudicles which hold pollinia are covered in elastoviscin derived from the degeneration of the tapetum (Yeung and Law 1987).

The occurrence of the loculus as a closed cavity in all major taxonomic groups, from bryophytes to angiosperms, means that the system was successful. The only exception to this arrangement in extant plants occurs in the Athocerotopsida, where the loculus is not a closed cavity and the spores do not develop more or less synchronously. Instead, the spores and tapetal cells develop inside a cylindrical cavity of annular cross-section, open at the distal pole from which ripe spores are dispersed. Along the tube all developmental stages can be observed. This system is possible because at the base of the tube there is a meristem enveloped in the gametophyte. Anthocerotopsida are the only case in which all the stages of spore development may be present at the same time in the same sporangium (Pacini and Franchi 1988*a*).

The tapetal cytoplasm also produces different enzymes which are released out of the protoplast, for example, callase for dissolving callosic walls (Stieglitz 1977). Meiospores with a callosic wall are only found in angiosperms; in lower groups there seem to be other polysaccharides (Clarke 1979). The tapetal cells also produce hydrolases responsible for the dissolution of the primary pectocellulosic wall of the microspore mother cell and the pectocellulosic wall of the tapetal cells themselves. Tapetal cells lose their walls in all types of amoeboid tapetum and in the parietal tapetum of seed plants (Figs 16.1, 16.2).

The sporopollenin precursors produced by tapetal cells can be polymerized in five different sites (see functions 6–10 listed in Table 16.1). Exine is ubiquitous around spores and pollen grains, with only a few exceptions in angiosperms, such as species with submarine pollination (Ducker and Knox 1976) and some families of monocotyledons (Kress, 1986). Perine and viscine are adjoining structures around the exine surface. The culture sac or peritapetal membrane and Ubisch bodies or orbicules, on the other hand, are formed around tapetal cells. Ubisch bodies seem to be typical of the secretory tapetum of seed plants, and have never yet been described in amoeboid tapetum. Something analogous to Ubisch bodies was described by Lugardon (1981) in Polypodiophyta, which have an amoeboid tapetum. Figures 16.1 and 16.2 list the differences between the amoeboid tapetum of pteridophytes and angiosperms; these two apparently similar types of amoeboid tapetum are of different origin (Pacini *et al.* 1985) (Fig. 16.1) and degenerate in different ways (Fig. 16.2).

The formation of elaters in *Equisetum arvense* with the intervention of the tapetum was unambiguously demonstrated by Uehara and Kúrita (1989). The production of sporophytic proteins and enzymes deposited on the exine surface does not seem peculiar to angiosperm tapeta, and can be found in some lower groups (Knox and Heslop-Harrison 1970; Pettitt 1979). At least some of these proteins are responsible for pollen stigma recognition (Heslop-Harrison 1987).

According to their physico-chemical properties, three types of substances may originate from the final dissolution of the tapetal protoplast. Pollenkitt is hydrophobic and consists mainly of lipids and carotenoids; tryphine is a mixture of hydrophobic and hydrophilic substances (Dickinson and Lewis 1973; Pacini and Casadoro 1981; Reznickova and Dickinson 1982); elastoviscin consists mainly of lipoid materials in orchid pollinarium caudicles (Schill and Wolter 1985; Yeung and Law 1987). All three degeneration products are involved either in pollen advertising or adhesion to the bodies of pollinators (Table 16.1, Fig. 16.2).

Tapetal degeneration

Tapetal degeneration could be used as another criterion for distinguishing tapetal types, but amoeboid and secretory tapeta often degenerate in the same way. Figure 16.2 lists the manner of tapetal degeneration for the different types of tapetum, in terms of persistence of cell walls, site and products of degeneration. The formation of tryphine and pollenkitt are the result of two different processes of degeneration; and, moreover, tryphine is formed by the *extra situm* degeneration of tapetal cell protoplasts. When pollenkitt is formed, the degeneration products persist in the site of the cytoplasm (*in situ* degeneration); this occurs both for parietal (Pacini and Casadoro 1981) and amoeboid

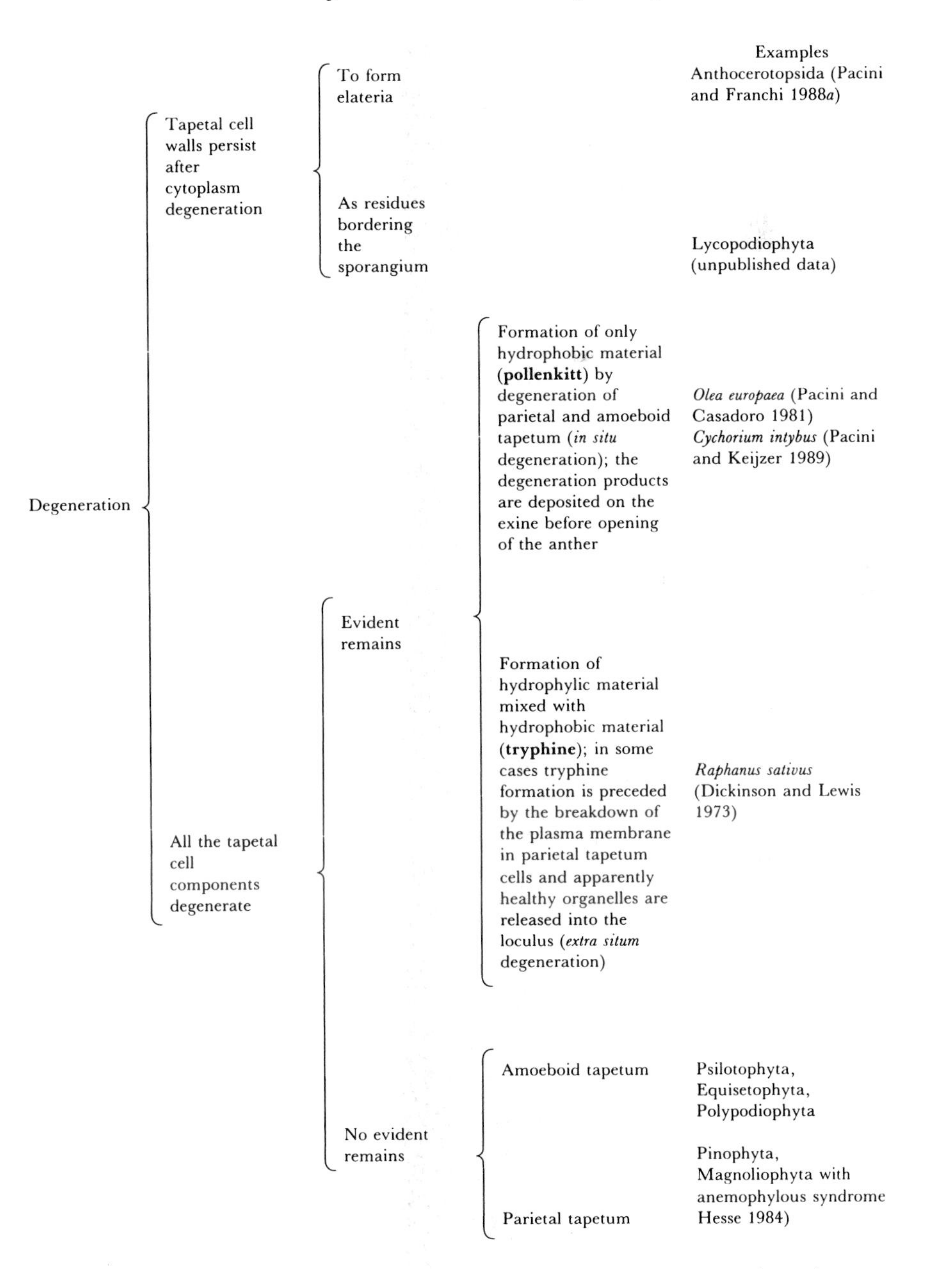

Fig. 16.2 Tapetum degeneration. Parietal and amoeboid tapeta of angiosperms sometimes degenerate in the same way

tapeta (Pacini and Keijzer 1989). The masses of pollenkitt cover the exine surface only just before anther dehiscence (Keijzer 1987). In the case of *extra situm* degeneration, first reported by Dickinson and Lewis (1973), in *Raphanus sativus*, tapetum of the secretory type starts to degenerate just after the first pollen mitosis. At first there is deposition of proteic material secreted by the tapetum inside the exine microsculpture. Then the tapetal protoplast ruptures by lysis of the plasma membrane; as a consequence still intact organelles come to lie adjacent the exine. This type of movement of tapetal components must not be regarded as invasive, because it is a passive phenomenon due to the disappearance of the plasma membrane. In contrast, the plasma membrane of amoeboid tapeta persists, separating the tapetal cytoplasm from the sporoderm up to time of complete degeneration of the organelles into lipid masses. Tapetal invasion never occurs later than the early microspore stage (Pacini *et al.* 1985, Table 5).

Trends in the evolution of the tapetum

The parietal tapetum type with walls is the oldest and the simplest type of tapetum, from which all the others originated. Subsequent variations serve to facilitate microspore nutrition and at the same time reduce competition for nourishment between spores or microgametophytes (Fig. 16.3). These variations are achieved by the loss of all or most of the cell walls (Fig. 16.3a) and/or by more intimate contact with microspores. This contact can be achieved in various ways. The simplest is the formation of septa decreasing the number of microspores per loculus and increasing the inner surface of the parietal tapetum (Fig. 16.3b). This feature occurs in some angiosperm and pteridophyte families (Lersten 1971; Tobe and Raven 1986). Another adaptation is present in Gramineae, where the anthers are narrow and elongated with few microspores in cross-section, each having a single pore adhering to the parietal tapetum (Pacini *et al.* 1985). Nutrition is direct, without any intermediary; the reduced loculus has lost its function of mediating the release of nourishing and regulatory substances. The most common way of achieving close

Fig. 16.3 Tapetum, pollen grain types in angiosperms and competition for nourishment between the pollen grains of a loculus. This kind of competition depends only on the relative position of each grain; any genetic competition is incidental.

	Tapetum, pollen grain types	Comments
(a) 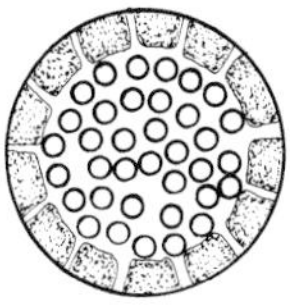	Parietal tapetum with many microspores per loculus e.g. many families of dicots, monocots, and almost all gymnosperms	Competition for nourishment is high if the system does not move (cf. Pacini and Franchi 1988*b*); reduced polarization for the entry of nutrients according to pore presence and number
(b) 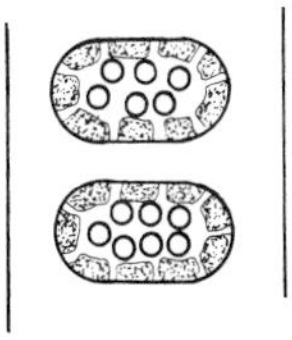	Parietal tapetum in multilocular anthers with few microspores per loculus, e.g. Balsaminaceae	Competition for nourishment is low, even if the system does not move; reduced polarization for the entry of nutrients according to pore presence and number
(c) 	Parietal tapetum with few microspores in a cross-section of the anther; a single pore per grain adhering to tapetum, e.g. Gramineae	No competition for nourishment; polarization for the entry of nutrients
(d) 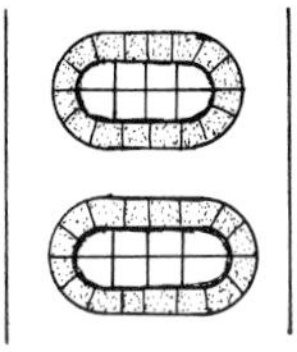	Parietal tapetum with persisting cell walls; polyads, e.g. Mimosaceae (Kenrick and Knox (1979) found amoeboid tapetum in three species of *Acacia*)	No competition for nourishment; polarization for the entry of nutrients
(e) 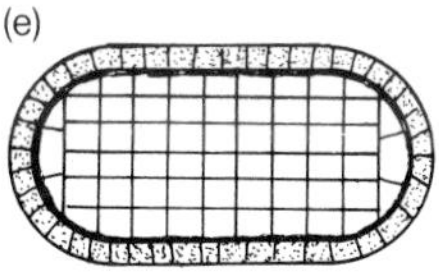	Parietal tapetum with persisting cell walls; massulae, e.g. Asclepiadaceae, Orchidaceae	Devices to reduce competition for nourishment, e.g. exine only at the periphery of the massula; pollen grains joined by cytomictic channels; strong polarization for the entry of nutrients
(f) 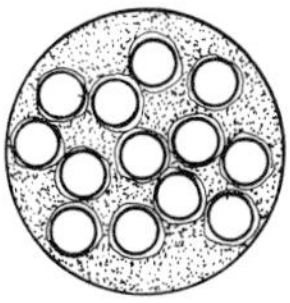	Amoeboid 'syncytial' tapetum forming a periplasmodium, a limited number of microspores per loculus, e.g. Compositae p.p., Araceae	Little or no competition for nourishment; no polarization for the entry of nutrients
(g) 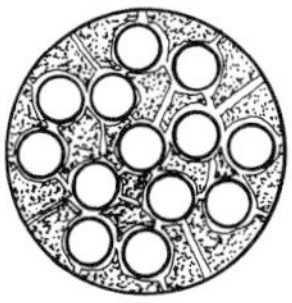	Non-syncytial amoeboid tapetum, tapetal cells fill the loculus and surround the microspores; a limited number of microspores per loculus, e.g. Compositae p.p., Cannaceae	

tapetal cytoplasm

			Examples	Authority
Macro- and microsporangia			*Selaginella* sp.	Buchen and Sievers 1981
Anthers	Tapetal cells having different functions in individuals of the same species	Heterantherous and heterostylous species	*Primula* sp.	Stevens and Murray 1981
		Heterantherous species with feeding and pollinating anthers in the same flower	*Lagerstroemia indica*	Pacini and Bellani 1986
	Two kinds of tapetal cells in the same loculus		Massulatae orchids with elastoviscin	Wolter *et al.* 1988

Fig. 16.4 Examples of the occurrence of different tapetum types in plants of the same species.

contact between the tapetum and spores or pollen grains is the formation of an amoeboid tapetum (Fig. 16.3f, g). Amoeboid tapeta may originate at different stages of development from prophase up to early microspore stage; directly or after a transient parietal phase; and by fusion or otherwise of the tapetal cell cytoplasms (Pacini *et al.* 1985). The amoeboid tapetum has arisen many times in land plants, but we do not know the stimulus or *raison d'être* of these transformations from secretory to amoeboid tapeta. It is necessary to recall that both kinds of tapeta have been reported in at least 12 families, both primitive and advanced, e.g. Caprifoliaceae, Lauraceae, and Zingiberaceae (Pacini *et al.* 1985).

When the tapetum surrounds compound pollen grains in massulae or pollinia, the loculus is absent and cell walls persist up to tapetum degeneration (Fig. 16.3d, e).

Pollen morphology and tapetal types

From the above discussion it seems that the tapetum varies according to pollen grain features (Fig. 16.4). However, it should be noted that when more than one type of pollen grain is present in a species the tapetum varies too. This occurs in ferns (Buchen and Sievers 1981); in heterantherous heterostylous species (Stevens and Murray 1981); and in heterantherous species with feeding and pollinating anthers in the same flower (Pacini and Bellani 1986). Two kinds of tapetal cells may coexist even in the same anther, for example in massulatae orchids, where some tapetal cells have the typical nourishing and regulatory functions, and others produce elastoviscin (Wolter *et al.* 1988).

Most pollen grains are more or less spherical, with a mean diameter ranging from 30 to 100 μm. Nevertheless, others have unusual shapes and sizes. For example, the pumpkin *Cucurbita maxima* has unusually large pollen grains (500 μm in diameter), and the loculus contains only two rows of grains in cross-section. Another case is that of submerged marine monocotyledons having thread-shaped pollen grains (10–30 μm thick and up to 5 mm long): here the tapetum is amoeboid (Pettitt *et al.* 1981) because this is the only kind of tapetum able to ensure uniform nutrition of such unusual pollen. These are just two examples of how tapetum and pollen features are correlated in order to ensure a large number of viable pollen grains. In lower plants, in contrast, it is difficult to demonstrate how tapetum and spores are reciprocally influenced.

References

Buchen, B. and Sievers, A. (1981). Sporogenesis and pollen grain formation. In *Cytomorphogenesis in plants*, (ed. O. Kiermayer), pp. 349–76. Springer Verlag, Wien.

Buss, P.A. and Lersten, N.R. (1975). Survey of tapetal nuclear number as a taxonomic character in Leguminosae. *Botanical Gazette* **136**, 388–95.

Clarke, G.C.S. (1979). Spore morphology and bryophyte systematics. In *Bryophyte systematics*, (ed. G.C.S. Clarke and J.G. Duckett), pp. 234–316. Academic Press, London.

D'Amato, F. (1984). Role of polyploidy in reproductive organ tissues. In *Embryology of angiosperms*, (ed. B.M. Johri), pp. 519–66. Springer Verlag, Berlin.

Dan Dicko-Zafimahova, L. and Audran, J.C. (1981). Étude ontogénique de la pollinie de *Calotropis procera* (Asclepiadaceae). Apports préliminaires de la microscopie électronique. *Grana* **20**, 81–99.

Davis, G.L. (1966). *Systematic embryology of the angiosperms*. Wiley and Sons, New York.

Dickinson, H.G. (1976). Common factors in exine deposition. In *The evolutionary significance of exine*, (ed. I.K. Ferguson and J. Muller), pp. 67–89. Academic Press, London.

Dickinson, H.G. and Lewis, D. (1973). The formation of tryphine coating the pollen grains of *Raphanus* and its properties relating to the self-incompatibility system. *Proceedings of the Royal Society of London* **B184**, 149–65.

Ducker, S.C. and Knox, R.B. (1976). Submarine pollination in seagrasses. *Nature* **263**, 705–6.

Echlin, P. (1971). The role of tapetum during microsporogenesis of angiosperms. In *Pollen: development and physiology*, (ed. J. Heslop-Harrison), pp. 41–61. Butterworths, London.

Good, B.H. and Chapman, R.H. (1978). The ultrastructures of *Phycopeltis* (Chroolepidiaceae: Chlorophyta). I. Sporopollenin in the cell walls. *American Journal of Botany* **65**, 27–33.

Heslop-Harrison, J. (1969). An acetolysis-resistant membrane investing tapetum and sporogenous tissue in the anthers of certain Compositae. *Canadian Journal of Botany* **47**, 541–2.

Heslop-Harrison, J. (1987). Pollen germination and pollen-tube growth. *International Reviews in Plant Physiology* **107**, 1–78.

Hesse, M. (1984). Pollenkitt is lacking in Gnetatae: *Ephedra* and *Welwitschia*; further proof for its restriction to the Angiosperms. *Plant Systematics and Evolution* **149**, 155–85.

Honegger, R. and Brunner, U. (1981). Sporopolenin in the cell walls of *Coccomixa* and *Myrmecia* phycobionts of various lichens: an ultrastructural investigation. *Canadian Journal of Botany* **59**, 2713–34.

Keijzer, C.J. (1987). The process of anther dehiscence and pollen dispersal.

2. The formation and the transfer mechanism of pollenkitt, cell-wall development of the loculus tissues and a function of the orbicles in pollen dispersal. *New Phytologist* **105**, 499–507.

Kenrick, J. and Knox, R. B. (1979). Pollen development and cytochemistry in some Australian species of *Acacia. Australian Journal of Botany* **27**, 413–27.

Knox R. B. and Heslop-Harrison, J. (1970). Pollen wall proteins: localization and enzymatic activity. *Journal of Cell Science* **6**, 1–27.

Kress, W.J. (1986). Exineless pollen structure and pollination systems of tropical Heliconia (Heliconiaceae). In *Pollen and spores: Form and function* (eds S. Blackmore and I.K. Ferguson), pp. 329–45. Academic Press, London.

Lersten, N. R. (1971). A review of septate microsporangia in vascular plants. *Iowa State Journal of Science* **45**, 487–97.

Lugardon, B. (1981). Les globules des Filicinées, homologues des corps d'Ubisch des Spermatophytes. *Pollen et Spores* **23**, 93–124.

Pacini, E. (1990*a*). Harmomegathic characters of Pteridophyta spores and Spermatophyta pollen. *Plant Systematics and Evolution*, (Suppl. 5), 53–69.

Pacini, E. (1990*b*). Tapetum and microspore function. In *Microspores: evolution and ontogeny* (ed. S. Blackmore and R. B. Knox) pp. 213–37. Academic Press, London.

Pacini, E. and Bellani, L. M. (1986). *Lagerstroemia indica* L. pollen: form and function. In *Pollen and spores: Form and function*, (ed. S. Blackmore and I. K. Ferguson), pp. 347–57. Academic Press, London.

Pacini, E. and Casadoro, G. (1981). Tapetum plastids of *Olea europaea. Protoplasma* **106**, 289–97.

Pacini, E. and Franchi, G. G. (1983). Pollen grain development in *Smilax aspera* L. and possible functions of the loculus. In *Pollen: Biology and implications for plant breeding*, (ed. D. L. Mulcahy and E. Ottaviano), pp. 183–90. Elsevier, New York.

Pacini, E. and Franchi, G. G. (1984). 'Harmomegathy': un problema aperto e misconosciuto. *Giornale Botanico Italiano* **118**, 271–82.

Pacini, E. and Franchi, G. G. (1988*a*). Sporophyte development in *Phaecoceros laevis* (L.) Prosk. *Giornale Botanico Italiano* **122**, (Suppl. 1), 51.

Pacini, E. and Franchi, G. G. (1988*b*). Do microspores/pollen grains move inside the loculus? *Abstracts of the 7th International Palynological Congress Brisbane*, p. 126.

Pacini, E. and Keijzer, C.J. (1989). Ontogeny of intruding non-periplasmodial tapetum in the wild chicory, *Cichorium intybus* (Compositae). *Plant Systematics and Evolution* **167**, 149–64.

Pacini, E., Franchi, G. G., and Hesse, M. (1985). The tapetum: its form, function, and possible phylogeny in Embryophyta. *Plant Systematics and Evolution* **149**, 155–85.

Pettitt, J. M. (1979). Ultrastructure and cytochemistry of spore wall morphogenesis. In *The Experimental biology of ferns*, (ed. A. F. Dyer), pp. 213–52. Academic Press, London.

Pettitt, J. M., Ducker, S. C., and Knox, R. B. (1981). Submarine pollination. *Scientific American* **244**, 134–43.

Pickett-Heaps, J. D. (1975). *Green algae: structure reproduction and evolution in selected genera*. Sinauer Associates, Sunderland, Massachusetts.

Reznickova, S. A. and Dickinson, H. G. (1982). Ultrastructural aspects of storage lipid mobilization in the tapetum of *Lilium hybrida* var. Enchantement. *Planta* **155**, 400–8.

Schill, R. and Wolter, M. (1985). Ontogeny of elastoviscin in the Orchidaceae. *Nordic Journal of Botany* **5**, 575–80.

Singh, H. (1978). Embryology of gymnosperms. In *Encyclopedia of Plant Anatomy*, (ed. W. Zimmermann *et al.*) Vol. X/2. Gebruder Borntraeger, Berlin.

Sood, S. K. and Mohana Rao, P. R. (1986). Development of male and female gametophytes in *Herminium angustifolium* (Orchidaceae). *Phytomorphology* **36**, 11–15.

Stevens, V. A. and Murray, B. G. (1981). Studies on heteromorphic self-incompatibility systems. The cytochemistry and ultrastructure of the tapetum of *Primula obconica*. *Journal of Cell Science* **50**, 419–31.

Stieglitz, H. (1977). Role of β-1-3-glucan in postmeiotic microspore release. *Developmental Biology* **57**, 87–97.

Tiwari, S. C. and Gunning, B. E. S. (1986*a*). Cytoskeleton, cell surface and the development of invasive plasmodial tapetum in *Tradescantia virginiana* L. *Protoplasma* **133**, 89–99.

Tiwari, S. C. and Gunning, B. E. S. (1986*b*). Colchicine inhibits plasmodium formation and disrupts pathways of sporopollenin secretion in the anther tapetum of *Tradescantia virginiana* L. *Protoplasma* **133**, 115–28.

Tobe, H. and Raven, P. H. (1986). Evolution of polysporangiate anthers in Onagraceae. *American Journal of Botany* **73**, 475–88.

Tryon, A. F. (1985). Spores of myrmecophylous ferns. *Proceedings of the Royal Society of Edinburgh* **86B**, 105–10.

Uehara, K. and Kurita, S. (1989). An ultrastructural study of spore wall morphogenesis in *Equisetum arvense*. *American Journal of Botany* **76**, 939–51.

Willemse, M. T. M. and Audran, J. C. (1982). Transfer of developing bean pollen in lily anthers. *Acta Botanica Neerlandica* **31**, 221–5.

Wodehouse, R. P. (1935). Pollen grains. Their structure, identification and significance in science and medicine. McGraw-Hill, New York.

Wolter. M. and Schill, R. (1986). Ontogenie von pollen, massulae und pollinien bei den Orchideen. *Tropische und subtropische Pflanzenweld* **56**, 1–93.

Wolter, M., Seuffer, C., and Schill, R. (1988). The ontogeny of pollinia and elastoviscin in the anther of *Doritis pulcherrima* (Orchidaceae). *Nordic Journal of Botany* **8**, 77–88.

Yeung, E. C. and Law, S. K. (1987). The formation of hyaline caudicle in two vandoid orchids. *Canadian Journal of Botany* **65**, 1459–64.

17. Development of tapetum and orbicules of *Catharanthus roseus* (Apocynaceae)

G.A. EL-GHAZALY
S. NILSSON
Naturhistoriska Riksmuseet, Palynologiska Laboratoriet, Stockholm, Sweden

Abstract

The development of the tapetum and orbicules in *Catharanthus roseus* (L.) G. Don has been examined using transmission electron microscopy (TEM). Precursors of orbicules are formed as spheroidal vesicles within the tapetal cytoplasm and are associated with the endoplasmic reticulum. The pro-orbicules are extruded through the cell membrane of the tapetal cell and this coincides with the dissolution of the tapetal cell wall and the callosic wall of the microspores. The orbicules are developed by irregular deposition of sporopollenin on the pro-orbicules. A thin membranous layer is developed between the orbicules, forming the orbicular membrane. At free microspore stage globular bodies are seen in the cytoplasm of the tapetal cells. Some of these bodies are large and coated with a thin electron-dense layer, others are small bodies, electron-dense and gathered in groups. At vacuolate stage the tapetal cells extend into the anther locule, acting as a periplasmodial tapetum. At late vacuolate stage the tapetal cells retreat from the anther locule and possess an organized and apparently functional structure. The tapetal cells start to degenerate just before anthesis of the pollen grains. The diversity in the structure of orbicules and tapetum may suggest different functions at different stages of development and may have arisen by remarkable processes of adaptation.

Introduction

This is the second in a series of three papers on the ontogeny and histochemistry of the pollen wall, orbicules, and tapetum of *Catharanthus*

Pollen and Spores (ed. S. Blackmore and S.H. Barnes), Systematics Association Special Volume No. 44, pp. 317–329, Clarendon Press, Oxford, 1991.

roseus. The first paper concerned the development of the pollen wall (El-Ghazaly 1990), while the present study describes diversity in the ultrastructure and development of tapetal cells and orbicules, from the early tetrad stage through to mature pollen grains.

Numerous reports refer to the morphology of the tapetum based on light microscopic observations, for example those of Schnarf (1923), Kosmath (1927), and Ubisch (1927). Since the advent of electron microscopy and development of new techniques, the structure and ontogeny of tapetum and orbicules (Ubisch bodies) have been investigated in many families of angiosperms. Two major types of tapetum have been described: glandular or secretory and amoeboid or plasmodial. The first type has been observed in species of 175 angiosperm families (Albertini *et al.* 1987) and species investigated in detail include *Poa annua* (Rowley 1963), *Lilium longiflorum* (Heslop-Harrison 1968), *Helleborus foetidus* (Echlin and Godwin 1968), *Olea europaea*, (Pacini and Juniper 1979), *Lilium hybrida* (Reznickova and Dickinson 1982), and *Triticum aestivum* (El-Ghazaly and Jensen 1986). Amoeboid or plasmodial tapetum has been observed in species of 32 angiosperm families (Albertini *et al.* 1987), for example *Tradescantia bracteata* (Mepham and Lane 1969), *Rhoeo discolor* (Albertini 1975), and *Gentiana acaulis* (Lombardo and Carraro 1976).

Materials and methods

Anthers of *Catharanthus roseus*, containing stages from early microspore tetrads to fully mature pollen grains, were collected from greenhouse-grown plants. The anthers were fixed in 2 per cent glutaraldehyde in 0.025 M cacodylate buffer (pH 6.8, room temperature, 24 h), rinsed in the buffer, and placed in 1 per cent osmium tetroxide in 0.025 M cacodylate buffer (pH 6.8, room temperature, *c.* 18 h). Following dehydration in acetone series, the anthers were embedded in epon-araldite resin. Sections were stained with uranyl acetate and lead citrate. Stained sections were examined with a Zeiss EM-10A.

Results

Tetrad stage

The microspore tetrads are surrounded by one layer of tapetal cells. The tapetal cytoplasm stains densely in contrast with microspore cytoplasm (Fig. 17.1), the nuclei of the tapetal cells are large with conspicuous nucleoli, and the cell walls are traversed by plasmodesmata. The cells are characterized by an abundance of rough endoplasmic reticulum

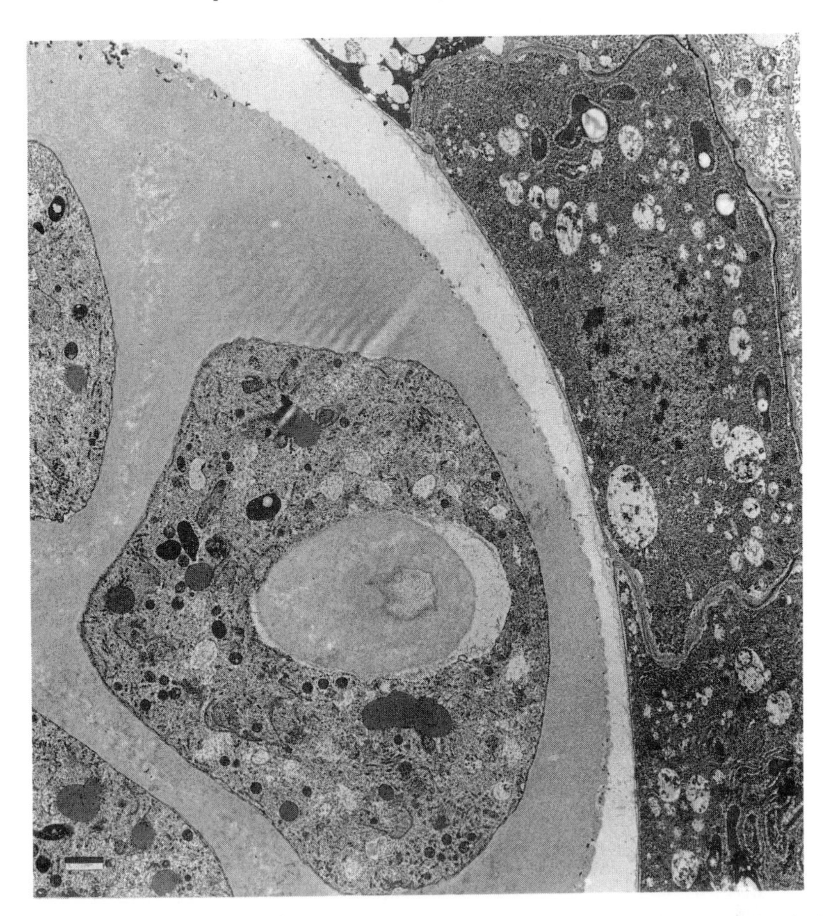

Fig. 17.1 Early tetrad and tapetal cells. Tapetal cytoplasm stains densely in contrast with the microspore cytoplasm; × 5000.

profiles, mitochondria, plastids, and dictyosomes (Fig. 17.2a).

By the late tetrad stage, the tapetal cells have enlarged considerably. Small globular bodies (pro-orbicules) of varying size (50–100 nm in diameter) are formed within the tapetal cells and pass through the plasma membrane (Fig. 17.2b). The pro-orbicules stain densely and appear to have a membrane surrounding them (Fig. 17.3a). They are observed closely associated with the endoplasmic reticulum (Fig. 17.3b, arrowheads). Where the pro-orbicules come into contact with the plasma membrane, their membrane becomes incorporated into the plasma membrane, with a consequent discharge of the pro-orbicules to the outside of the cell so that they lie between the plasma membrane and the cell wall (Figs 17.2b, 17.3a, b). Lysis of the tapetal cell walls occurs

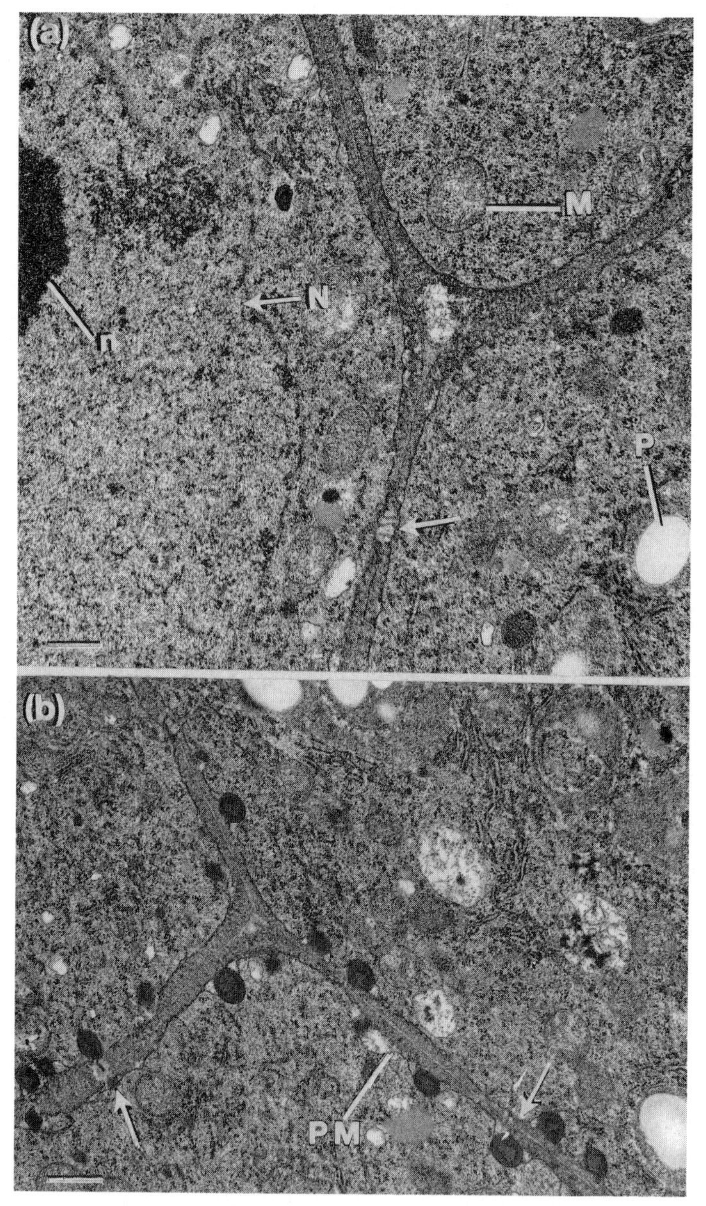

Fig. 17.2 (a) Tapetal cells of a young tetrad; note the plasmodesmata (arrow) that connect the tapetal cells, mitochondria (M), plastids with starch grains (P), and a part of the nucleus (N) and an electron-dense nucleolus (n); × 7200. (b) Tapetal cells of tetrad stage; plasmodesmata (arrows) connecting the cells, numerous electron-dense bodies (orbicules) situated on the plasma membrane (PM); × 7200.

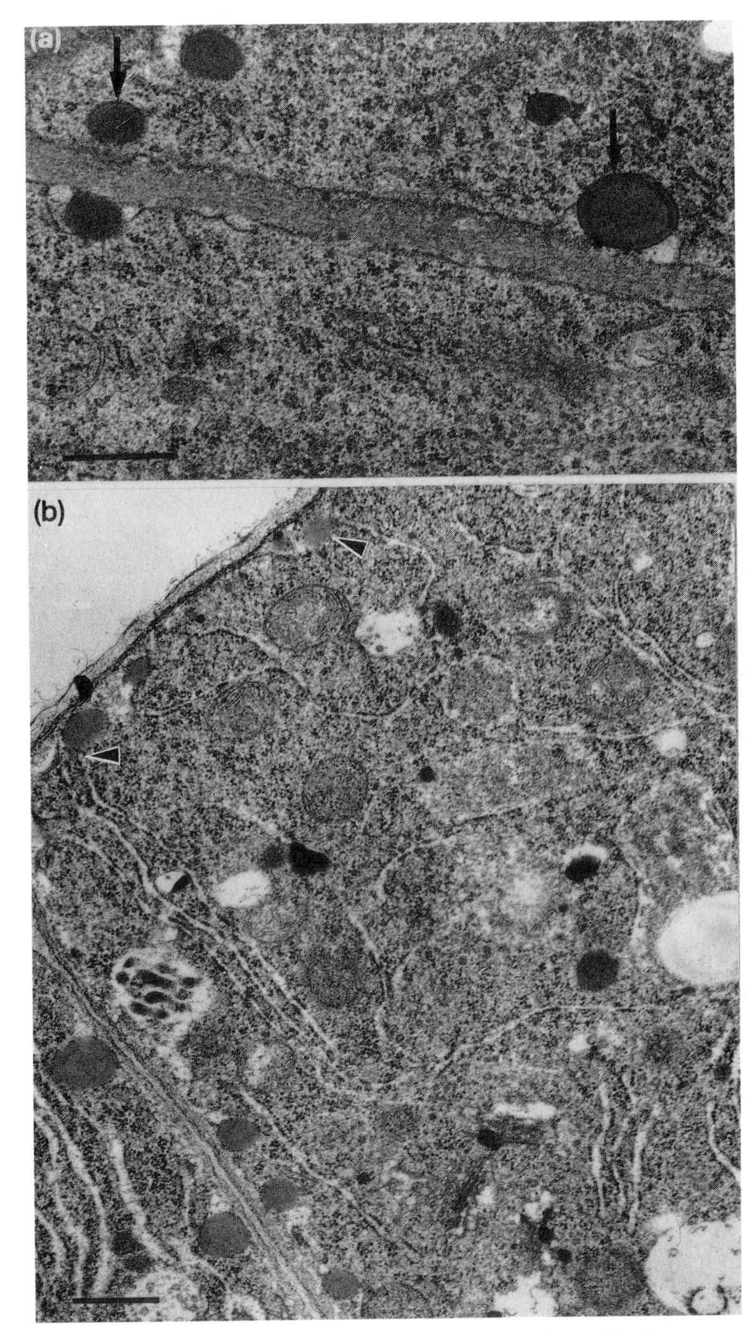

Fig. 17.3 (a) Magnified part of Fig. 17.2(b); note the orbicules surrounded by a thin membrane (arrows); × 15 200. (b) Pro-orbicules are associated with the endoplasmic reticulum (arrowheads); note the dissolution of the cell wall; × 11 875.

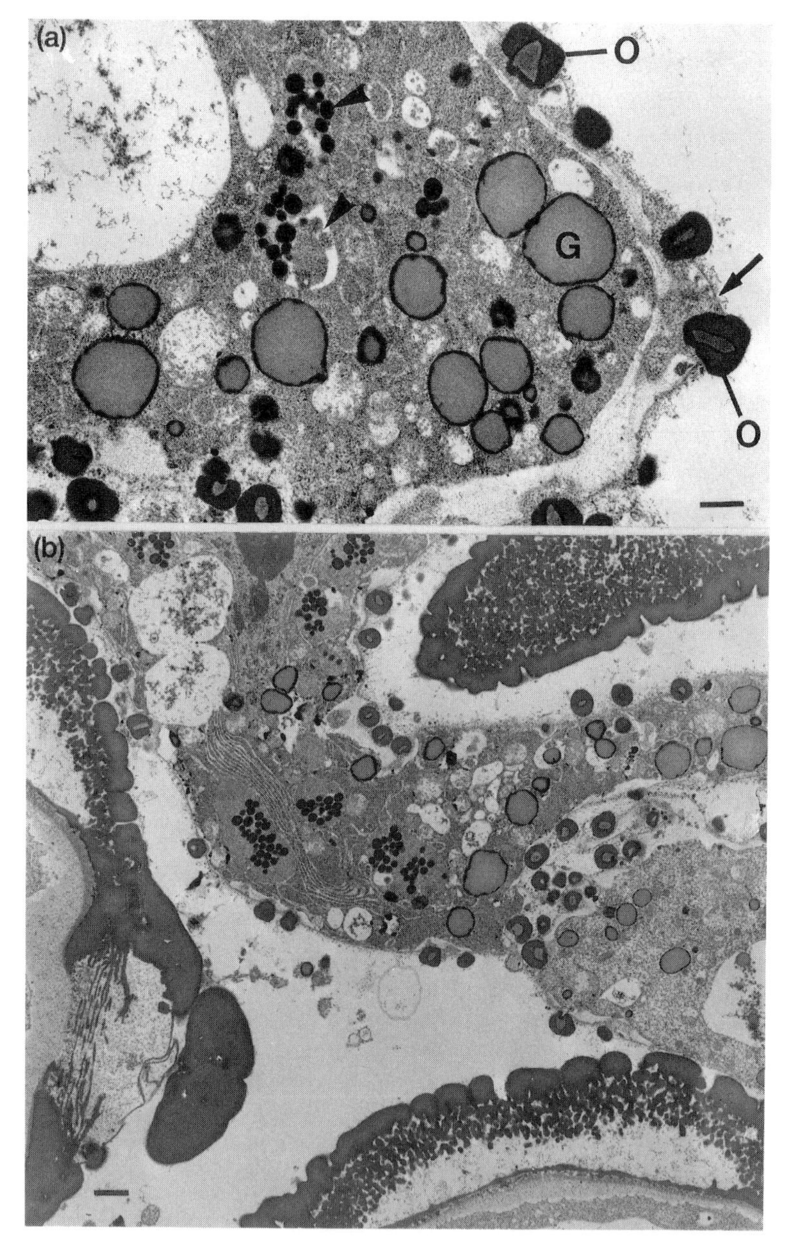

Fig. 17.4 (a) Tapetal cell of the free microspore stage; the plasma membrane is coated with orbicular membrane (arrow), which connects the orbicules (0); the cytoplasm contains large globules (G) and electron-dense small globules (arrowheads); × 5850. (b) Microspores in the vacuolate stage; note tapetal cells that are expanded and surround the microspores; × 4500.

at this stage. The pro-orbicules become irregularly coated with electron-dense material, probably sporopollenin.

Free microspore stage

Deposition of sporopollenin continues on the orbicules and the wall units of the developing microspores. The remnants of the wall of the tapetal cells disappear and the plasma membrane becomes coated with a fibrillar layer, which connects the orbicules. This fibrillar layer becomes compact and thin, forming the orbicular membrane (Fig. 17.4a). The cytoplasm characteristically has small vacuoles, electron-dense small globules arranged in groups inside plastids, and large globules with a thin electron-dense coat (Fig. 17.4a).

Vacuolate microspore stage

The plasma membrane of the tapetal cells remains intact and is covered by orbicular membranes and orbicules. The tapetum becomes disorganized and is characterized by an increase of vesiculation within the cytoplasm (Fig. 17.4b). Tapetal cells expand into the anther locules and surround the microspores (Fig. 17.4b). By the end of this stage of development the tapetal cells retreat. Their cytoplasm no longer contains the two types of globules observed before (Fig. 17.5a) and mitochondria, dictyosomes, and endoplasmic reticulum are now more apparent. During this stage the microspore wall attains a maximum rate of development. The orbicule central core becomes more electron-dense than before and is delimited by an even denser thin coat, that is followed by a thick sporopollenin wall (Fig. 17.5a).

Pollen-grain stage

At young pollen-grain stage, the tapetum possesses an organized and apparently functional structure. As development proceeds the cells maintain their integrity and become highly active and secretory. At this stage small lipoidal globules are enclosed within plastids and numerous mitochondria, dictyosomes, and endoplasmic reticulum are observed (Fig. 17.5b). The tapetum maintains a high degree of organization until just prior to anthesis, when the cytoplasm frequently contains small electron-dense lipoidal groups and occasionally less electron-dense large globules. The orbicular membrane is still in contact with the tapetal cells and is sprinkled with orbicules (Fig. 17.6a). At anthesis the pollen grain cytoplasm includes lipid and amyloplasts, while the tapetal cytoplasm is completely disorganized and the orbicules rest irregularly upon the remnant of the tapetum (Fig. 17.6b). When pollen grains were mounted in a drop of water and observed with a light

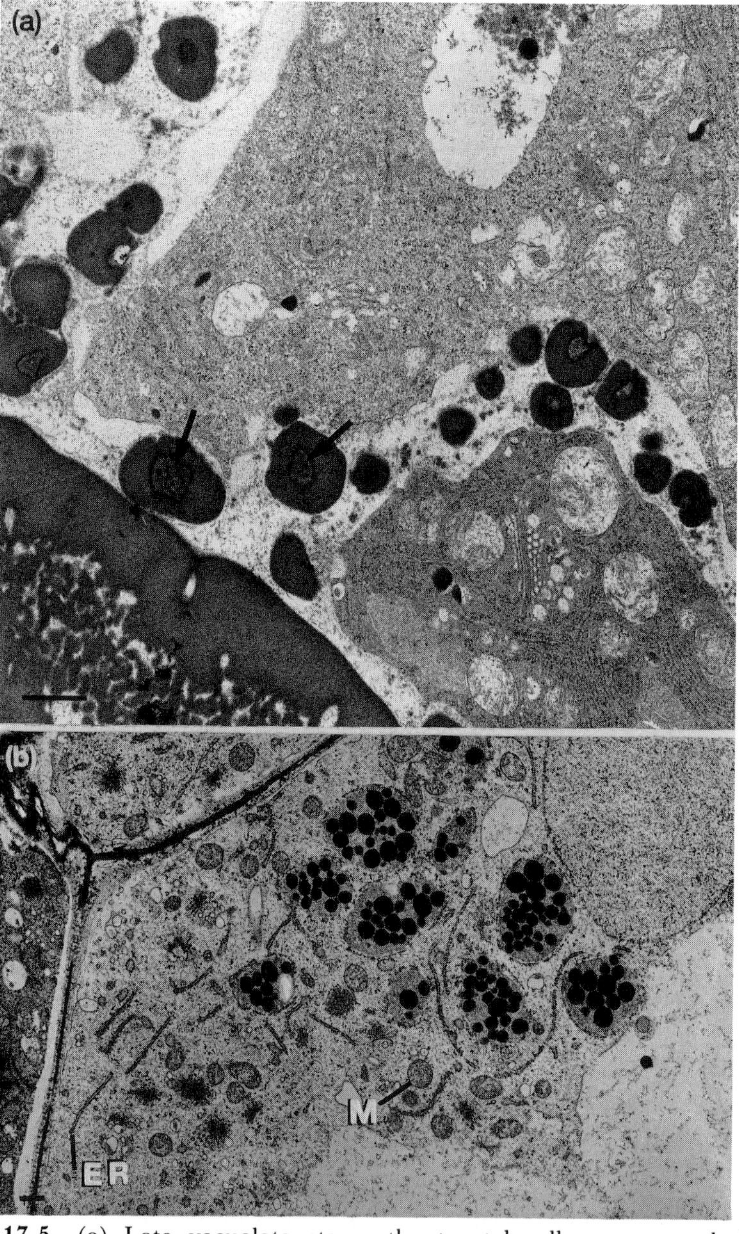

Fig. 17.5 (a) Late vacuolate stage; the tapetal cells retreat and their cytoplasm does not contain the globules observed before (cf. Fig. 17.4a,b); note the electron-dense membrane surrounding the central core of orbicules (arrows); × 9000. (b) Tapetal cells of a young pollen grain; note the numerous mitochondria (M), endoplasmic reticulum fragments (ER), and groups of small electron-dense globules; × 3600.

microscope, numerous orbicules were seen covering the aperture of the pollen (Fig. 17.6c).

Discussion

The development of the tapetum in *Catharanthus roseus* passes through successive stages of metabolism, disorganization, reorganization and, finally, ending in dissolution. Diversity in orbicules and tapetum of *Catharanthus* at different developmental stages provides some functional aspects. At the tetrad stage, it seems that the tapetum does not contribute, or contributes only slightly, to the primexine while the microspores are invested with the callose layer. At this stage the tapetum has not entered the phase of decline, associated with possible release of metabolites into the anther cavity. Southworth (1971) has shown that both phenylalanine and leucine would not penetrate the callose wall. However, Echlin (1971) suggested that the possibility of a tapetal contribution to the primexine at this stage cannot be excluded, and presumed that if any material passes from the tapetum to the microspore it would be a precursor, or unpolymerized form, of sporopollenin.

During the late tetrad stage, precursors of orbicules are formed as spheroidal vesicles (pro-orbicules) within the tapetal cytoplasm, in association with the endoplasmic reticulum. The ontogeny of orbicules in *Catharanthus* is similar to that of *Triticum* (El-Ghazaly and Jensen 1986). It is also similar, to a certain extent, to that of *Helleborus foetidus* (Echlin and Godwin 1968) and *Cannabis sativa* (Heslop-Harrison 1962, 1963), where pro-orbicules developed in the cytoplasm and sporopollenin was deposited on their surface as soon as they had extruded outside the plasma membrane of the tapetal cells.

In *Catharanthus*, the ribosomes and mitochondria are not associated with pro-orbicule formation, and this distinguishes *Catharanthus* from *Helleborus* (Echlin and Godwin 1968) and *Cannabis* (Heslop-Harrison 1963). Lysis of the tapetal cell wall occurs after the pro-orbicules have extruded outside the plasma membrane. Mepham and Lane (1969) postulated that dissolution of the tapetal cell wall in *Tradescantia* is a result of the release of hydrolytic enzymes from dictyosome-derived vesicles. The orbicules in *Catharanthus* may maintain a similar function.

At the free microspore stage the metabolism of the tapetal cells appears to be associated primarily with the synthesis of lipid material. Lipid globules, which are electron-dense, appear with the plastids, besides which there are electron-dense, large, free globules observed in the tapetal cytoplasm. Similar globules were described by Cousin (1979) as paralipoidal globules enclosed within a chromoplast-like structure, and large globules. After secretion of the globules in the

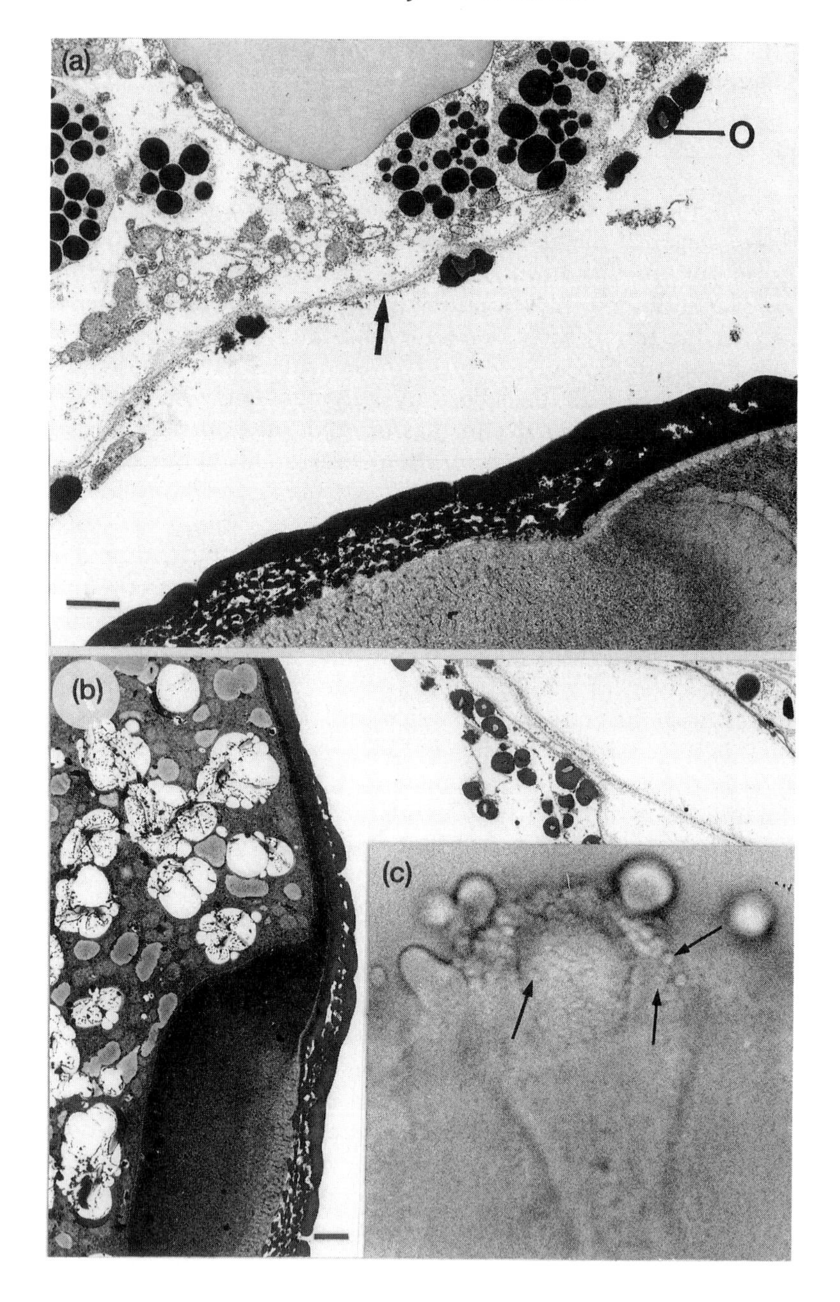
(a)
O
(b)
(c)

tapetal cytoplasm of *Catharanthus*, the cells dissociate from one another, become amoeboid, extend into anther locules, and are interdispersed among the microspores.

The formation of orbicules in *Catharanthus* and the development of a membrane that connects them are characteristics of a secretory tapetum; while the early loss of the tapetal cell walls and the intrusion of the tapetal protoplasts among the developing microspores of *Catharanthus* are characteristic of a periplasmodial tapetum. These observations of features of secretory and plasmodial tapetum at different stages of development in *Catharanthus* distinguish this species from the general trend where the tapetum is secretory (glandular) or amoeboid (periplasmodial). At the end of the vacuolate stage the tapetal cells retreat, and their cytoplasm does not contain the globules observed earlier. This observation indicates that the globules secreted in the tapetal cytoplasm may be incorporated into the developing microspores, and that tapetum plays an important role in the development of microspore wall, particularly during the vacuolate stage, when the wall proceeds towards a maximum rate of development.

Orbicules appear to have another role during this stage, when they become more electron-dense than before, with a central core delimited by a dark membrane. These observations suggest that orbicules are in an active state and are probably taking part in microspore development. Vithanage and Knox (1976) showed that tapetal esterase reaches maximum activity during the mid-vacuolate stage. At the vacuolate stage the orbicule walls, in *Catharanthus*, increase in thickness. It is presumed that the material passing from the tapetum to the microspores would be nutrients and precursors of sporopollenin. Some of this material could be deposited on the orbicule wall to avoid high concentrations of material outside the loculus of sporogenous tissue, consequently preventing osmosis and collapse of the developing microspores.

The tapetum in *Catharanthus* shows some similarities to that of *Tradescantia* (Mepham and Lane 1969), in that the changes that occur in tapetal cells after the vacuolate stage appear to be a reorganization rather than a degeneration. *Catharanthus* is distinguished by retreat of the tapetal cells after vacuolation, and the tapetal membranes do not

Fig. 17.6 (a) Part of a tapetal cell and pollen wall of a mature pollen grain; the small lipoidal groups are frequent; note the orbicular membrane (arrow) connecting the orbicules (O); × 7200. (b) A mature pollen grain before anthesis; note the pollen grain cytoplasm, including lipid and amyloplast; the tapetum degenerate and orbicules and orbicular membranes are irregularly arranged; × 4500. (c) Light micrograph of a pollen grain before anthesis, showing accumulation of orbicules on the aperture (arrows); × 900.

penetrate through the exine of developing pollen grains, as in the case of *Tradescantia*.

The tapetum tissue of *Catharanthus* maintains a high degree of organization until just prior to anthesis. Nutrients for pollen growth and, in particular, intine development may be derived from the tapetum at this stage. The reappearance of amyloplast and lipid in the cytoplasm of pollen grains at about the time of pollen tube germination may support the contention that the tapetum in *Catharanthus* stays functionally active during maturation of pollen grains.

The presence of an orbicular membrane, connecting the orbicules after degeneration of tapetal cells, has been described in secretory tapetum (Banerjee 1967; Christensen *et al.* 1972; Cerceau-Larrival and Roland-Heydacker 1976; El-Ghazaly and Jensen 1986).

In *Catharanthus* the orbicular membrane and orbicules cover the mature pollen grains and frequently fill the aperture regions at anthesis. This may suggest an additional function of orbicules as pollen-borne substances involved in the sporophytic incompatibility system between pollen and stigma.

References

Albertini, L. (1975). Étude autoradiographique de l'incorporation du trytophane ^{3}H dans les microsporocytes et dans le tapis du *Rhoeo discolor* Hanse. *Caryologia* **28**, 445–58.

Albertini, L., Souvré, A., and Audran, J.C. (1987). Le tapis de l'anthère et ses relations avec les microsporocytes et les grains de pollen. *Revue de Cytologie et de Biologie Végetales, Botanique* **10**, 211–42.

Banerjee, U.C. (1967). Ultrastructure of the tapetal membranes in grasses. *Grana Palynologica* **7**, 365–77.

Cerceau-Larrival, M.-Th. and Roland-Heydacker, F. (1976). Ontogénie et ultrastructure de pollens d'Ombellifères. Tapis et Corps d'Ubisch. *Comptes Rendus Hebdomadaires des Seances de l'Academic des Sciences, Paris* **283**, 29–32.

Christensen, J.E., Horner, H.T. Jr, and Lersten, N.R. (1972). Pollen wall and tapetal orbicular wall development in *Sorghum bicolor* (Gramineae). *American Journal of Botany* **59**, 43–58.

Cousin, M.T. (1979). Tapetum and pollen grains of *Vinka rosea* (Apocynaceae). Ultrastructure and investigation with scanning electron microscope. *Grana* **18**, 115–28.

Echlin, P. (1971). The role of the tapetum during microsporogenesis of angiosperms. In *Pollen: development and physiology* (ed. J. Heslop-Harrison), pp. 41–61. Butterworths, London.

Echlin, P. and Godwin, H. (1968). The ultrastructure and ontogeny of pollen

in *Helleborus foetidus* L. I. The development of the tapetum and Ubisch bodies. *Journal of Cell Science* **3**, 161–74.
El-Ghazaly, G. (1990). Development of pollen grains of *Catharanthus roseus* (Apocynaceae). *Review of Palaeobotany and Palynolgy*, **64**.
El-Ghazaly, G. and Jensen, W. A. (1986). Studies of the development of wheat (*Triticum aestivum*) pollen. I. Formation of the pollen wall and Ubisch bodies. *Grana* **25**, 1–29.
Heslop-Harrison, J. (1962). Origin of exine. *Nature* **195**, 1069–71.
Heslop-Harrison, J. (1963). Ultrastructural aspects of differentiation in sporogenous tissue. *Symposia of the Society for Experimental Biology* **17**, 315–40.
Heslop-Harrison, J. (1968). Tapetal origin of pollen-coat substances in *Lilium*. *New Phytologist* **67**, 779–86.
Kosmath, L. (1927). Studien über das Antherentapetum. *Österreichische Botanische Zeitschrift* **76**, 235–41.
Lombardo, G. and Carraro, L. (1976). Tapetal ultrastructural changes during pollen development. III. Studies on *Gentiana acaulis*. *Caryologia* **29**, 347–9.
Mepham, R. H. and Lane, G. R. (1969). Formation and development of the tapetal periplasmodium in *Tradescantia bracteata*. *Protoplasma* **68**, 175–92.
Pacini, E. and Juniper, B. E. (1979). The ultrastructure of pollen grain development in the olive (*Olea europaea*). II. Secretion by the tapetal cells. *New Phytologist* **83**, 165–74.
Reznickova, S. A. and Dickinson, H. G. (1982). Ultrastructural aspects of storage lipid mobilization in the tapetum of *Lilium hybrida* var. Enchantment. *Planta* **155**, 400–8.
Rowley, J. R. (1963). Ubisch body development in *Poa annua*. *Grana Palynologica* **4**, 25–36.
Schnarf, K. (1923). Kleine Beiträge zur Entwicklungsgeschichte der Angiospermen. IV. Uber das Verhalten des Antheren-Tapetums einiger Pflanzen. *Österreichische Botanische Zeitschrift* **72**, 242–5.
Southworth, D. (1971). Incorporation of radioactive precursors into developing pollen walls. In *Pollen: development and physiology*, (ed. J. Heslop-Harrison), pp. 115–20. Butterworths, London.
Ubisch, G. von (1927). Zur Entwicklungsgeschichte der Antheren. *Planta* **3**, 490–5.
Vithanage, H. I. and Knox, R. B. (1976). Pollen wall proteins; quantitative cytochemistry of the origins of intine and exine enzymes in *Brassica oleracea*. *Journal of Cell Science* **21**, 423–35.

18. Structural and functional variation in pollen intines

J. HESLOP-HARRISON
Y. HESLOP-HARRISON
Cell Physiology Unit, Welsh Plant Breeding Station, University College of Wales, Plas Gogerddan, nr. Aberystwyth SY23 3EB, UK

Abstract

The basic three-layered organization of the angiosperm intine is related primarily to its function during pollen hydration, activation, germination and early tube growth. Structural variation amongst families, genera, and species is attributable mainly to two circumstances: that the intine necessarily tends to conform with the architecture of the exine, and that the additional adaptations concerned with regulating the water economy of the grain and controlling germination rate are present in some groups.

Exine morphogenesis is the dominant factor in the early deposition of the wall layers, and the framework so created affects the features of the later-forming intine. However, it is not a matter of passive conformation of a plastic matrix to a template; the cellulosic fibrils of the inner stratum of the intine are often polarized towards exine features, revealing a more specific control, apparently associated in some instances with microtubule disposition in the vegetative cell.

The stratification of the intine is generally well developed at the aperture sites, but much reduced elsewhere. Some aspects of the function of the central, protein-bearing, layer at the apertures during germination have long been appreciated, but new features are now coming to light. The stratum is the source not only of the proteins discharged into the stigma during early hydration, but evidently also of the enzymes concerned in slackening the fabric of the intine itself preparatory to the emergence of the tube. Following the early ingress of the water, the vegetative cell cytoplasm is activated and Ca^{2+} gradients directed

Pollen and Spores (ed. S. Blackmore and S. H. Barnes), Systematics Association Special Volume No. 44, pp. 331–343, Clarendon Press, Oxford, 1991.

towards the germination sites are established; at these sites short actin fibrils accumulate, together with polysaccharide-containing tube-wall precursor bodies, as the tube tip is formed. The chapter includes a review of the ways in which these events are regulated by structural and chemical adaptations of the intine, notably in pollens subject to environmental stress.

In this brief review we discuss the principal structural characteristics of the intine layers of angiosperm pollens, illustrating aspects of the variation by reference to examples of different degrees of complexity drawn from a range of taxonomic groups. With the intine, as with other components of the pollen wall, the mere study of structure can have little significance for pollen biology in general unless what is observed can be linked to function (J. Heslop-Harrison 1979*a*), and it is with function that we are primarily concerned in what follows. Furthermore, while the comparisons we make may well have some broad evolutionary significance, it is obvious enough that they cannot be taken to offer a basis for phylogenetic speculation: this indulgence must await the accumulation of a more extensive body of data than is currently available.

Three main roles are fulfilled by the pollen intine:

(1) as a key part of the germination mechanism;

(2) as a repository of enzymic and other proteins involved in the interaction with the stigma; and

(3) as the progenitor of the pollen tube wall.

These functions represent an irreducible minimum, in the sense that each plays an essential part in the release of the male gametophyte and in its interactions with the stigma. The basic requirements are met by broadly similar adaptations throughout the angiosperms: but the structural details vary, and numerous additional morphological and physiological specializations associated with various secondary functions exist. Intines do not carry the kinds of taxonomic specificity so often associated with pollen exines; yet their location means that they must necessarily conform with some of the principal architectural features of the overlying exine—notably the form and disposition of apertures—and this can bestow a degree of morphological individuality. Furthermore, in some groups the intines possess additional features related to pollen ecology, including especially adaptations associated with stress tolerance.

The intine is laid down after the major features of exine structure have been defined. The early polysaccharide depositions are primarily of pectic substances, and these are followed by the accretion of a layer with a dominant cellulosic component. These designations have functional significance, but they cannot be taken as implying precise chemical identification. Thus the term 'pectin' covers a wide spectrum of compounds, definable in general terms as galacturonans or rhamnogalacturans, but variable in their properties—including solubility—according to the degree of esterification and cross-linking and the composition of side-chains. As well as playing a significant part in determining physical behaviour, which we shall shortly consider, the heterogeneity of this component of pollen intines is often seen in the different responses to so-called pectin stains, such as ruthenium red and alcian blue, which sometimes reveal stratification, the chemical significance of which remains obscure.

An apparently universal feature of the pectic part of pollen intines is the presence of protein inclusions. It is not excluded that proteoglycans occur in a dispersed fashion throughout all or part of the pectic zone, but the bulk of the protein conveyed in the intine to be released at the time of germination is held in discrete inclusions, readily distinguishable by electron microscopy. These protein inclusions are incorporated at the time of deposition of the intine in the form of leaflets, tubules, or more massive columns (Knox and Heslop-Harrison 1970; Knox 1971; J. Heslop-Harrison 1975), and they occupy a distinct zone between an outer more homogeneous pectin layer immediately underlying the exine and the inner cellulosic layer.

Microfibrillar pectins are associated with the pollen tube, but no such substructure can be discerned in the pectic layers of the intine. In contrast, the cellulosic layer adjacent to the plasmalemma of the vegetative cell does have a microfibrillar component. The designation as 'cellulosic' is based upon staining properties and resistance to alkaline digestion, and analogy with the cellulosic component of the pollen tube wall suggests that chemically it is likely to be a glucan with both β-1,3 and β-1,4 linkages, the latter predominating (Herth *et al.* 1974). The resistance of this component to exine- and pectin-solubilizing agents make it possible to isolate the cellulosic layer to produce 'intine ghosts', a useful technique for demonstrating its role during germination and early tube growth (J. Heslop-Harrison 1979*b*). Electron microscopy shows that the isolated microfibrils have apparent diameters in the range 5–15 nm (Y. Heslop-Harrison and J. Heslop-Harrison 1982), but they are usually associated to form ribbons, or aggregated in strands or cables of dimensions great enough to be resolved with the optical microscope. The simplest intine organization, common to most monocotyledons, is summarized in Fig. 18.1a. In omniaperturate pollens such as those of some families of the Englerian order Scitamineae, this type of stratification is present over the whole

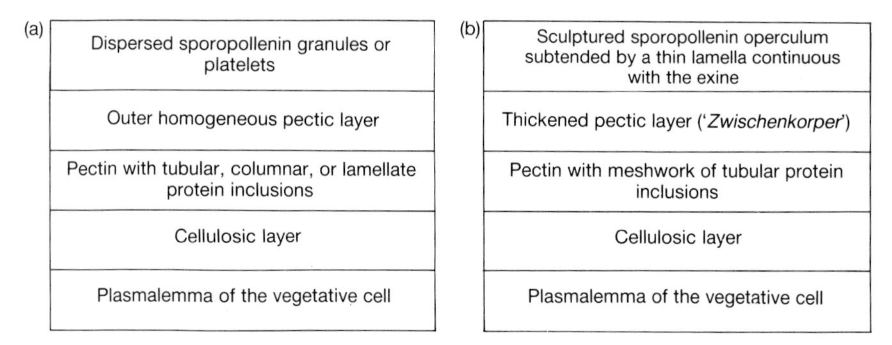

Fig. 18.1 (a) Summary of the intine stratification common to many inaperturate and omniaperturate species, and characteristic of the colpial region of many monocotyledons. (b) Summary of intine stratification in the Gramineae.

surface of the grain, and Skvarla and Rowley (1970), in their study of *Canna*, were the first to appreciate that it corresponds precisely with the stratification at the apertures in other monocotyledons where the sites are structurally demarcated. Indeed, some genera reveal intermediate states where the functional aperture sites are dispersed or less well defined (Kress *et al.* 1978; Kress and Stone 1982).

The functions of the three basic strata become clear during germination (J. Heslop-Harrison 1975; Y. Heslop-Harrison 1977). The pollen grain of *Iris reticulata* provides a model. Here the sporopollenin of the exine is dispersed over the whole surface in islets or small granules (Fig. 18.2a), exposing large areas of the outer continuous pectic layer of the intine (Fig. 18.2b). Isolated intine ghosts show that the inner cellulosic layer forms a continuous sheath, without any differentiation indicative of aperture sites (Fig. 18.2c). During hydration *in vitro*, dissolution of the outer layers leads first to the release of solubilized

---

Fig. 18.2 (a)–(g) Pollen of *Iris reticulata*, x *c.* 1450. (a) Distribution of sporopollenin of the exine, auramine O staining. (b) outer pectic layer of the intine as revealed by alcian blue staining. (c) Isolated intine, showing absence of any defined apertural zone. Fluorescence micrograph, calcofluor white staining for cellulosic glucans. (d) Living pollen grain, emission of pectin, alcian blue staining. (e) Living pollen grain, emission of protein; Coomassie blue staining. (f) As (c), intine isolated from a germinating grain with single emergent tube. (g) As (f), intine with one emergent tube but with demarcation of a second potential germination site. (h) Whole pollen grain, calcofluor white staining, with two contiguous potential germination sites, one beginning to form a tube tip, × *c.* 2300. (i) As (c), intine with two emergent tube tips, × *c.* 1450.

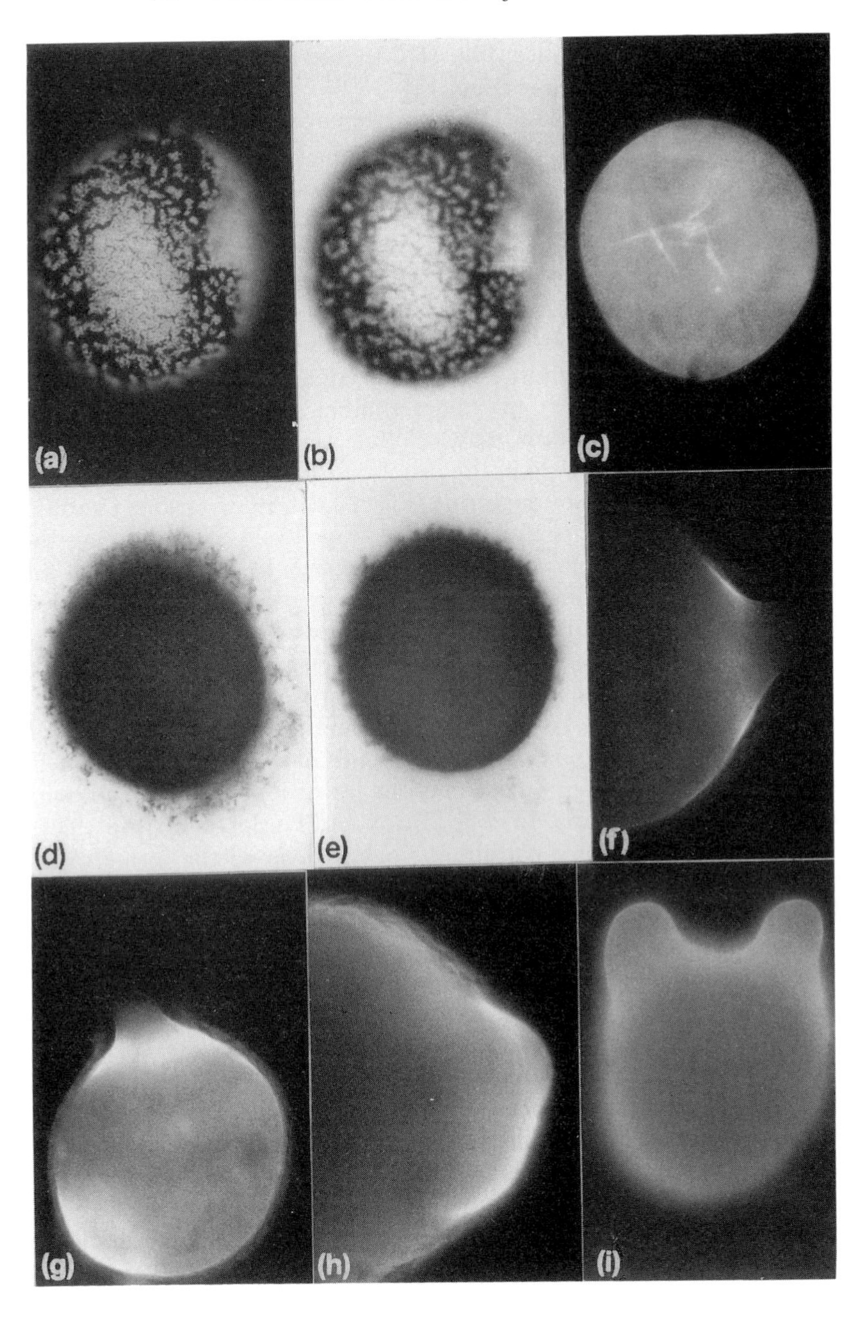
(a)
(b)
(c)
(d)
(e)
(f)
(g)
(h)
(i)

pectin over the whole surface (Fig. 18.2d), and then of protein from the inclusions of the inner pectic zone (Fig. 18.2e). However, as in the case of *Crocus*, described in detail in an earlier paper (Y. Heslop-Harrison 1977), the emissions from pollen grains on the stigma are polarized towards the contact surface with the contiguous papilla. It is through this interface that water passes from the stigma into the grain (J. Heslop-Harrison 1979*b*), and the contact site also marks where the pollen tube tip will emerge. This illustrates a significant principle, since it shows how polarity is established in a grain which initially has no evident intrinsic asymmetries. The importance of internal gradients, sometimes exogenously induced, in polarizing the grain is mentioned further below, but here we may note another suggestive phenomenon associated with germination *in vitro*. While single tubes are often produced, the walls originating in the usual manner as an extension of the inner cellulosic stratum of the intine (Fig. 18.2f), many grains establish more than one site for tube emergence (Figs 18.2g–i) reinforcing the conclusion that in pollens like that of *I. reticulata* the positioning of the tube is a matter of indifference.

Among the Iridaceae transitional states may be seen between pollens with no well-defined apertures to those with clear dorsiventrality and a demarcated colpus. The latter, a familiar feature of the pollens of most Liliiflorae, is associated with pre-determination of the germination sites, the preferred positions being at the ends of the colpial slit. In these sites the three-layered organization of the intine is well developed, with a conspicuous central protein zone (Knox and Heslop-Harrison 1970).

Hydration of the outer continuous pectic layer of the intine in the pollen types just discussed accelerates the rupturing of the exine in genera such as *Crocus*, and this function of the corresponding stratum is even more clearly seen in the Gramineae. The organization of the intine at the single aperture in pollens of this family is summarized in Fig. 18.1b. The key difference lies in the greater development of the outer layer at the aperture, where it forms the *Zwischenkörper* of various earlier authors. This body, which can be distinguished from the inner strata by its staining properties, overlies and seals off the middle, protein-containing part of the intine. It undergoes rapid gelation and swelling during hydration, and is responsible for lifting the operculum in advance of the release of the apertural proteins. We have described details of the mechanism in earlier papers (J. Heslop-Harrison 1979*b*; J. Heslop-Harrison and Y. Heslop-Harrison 1980), and we need only note here that the properties of the material of the outer layer show that while chemically it can be referred to the general class of pectins, its physical behaviour, including its propensity for very rapid gelation, reflects specialization for a critical early function in germination.

'Pectins' with similar roles are found in a very wide variety of pollen types, from those of gymnosperms, like *Taxus*, to those of advanced angiosperm families.

Still greater specializations are found among dicotyledons. *Corylus avellana* provides an example which probably has parallels in many other genera with tricolpate and tricolporate grains (Y. Heslop-Harrison *et al.* 1986). Here we see the intine functioning in osmoregulation; and it is fitting to acknowledge the fine contribution of Schoch-Bodmer (1936), who more than half a century ago reported in detail on this function in hazel pollen. The key structural characteristics of the apertural zone of *C. avellana* pollen are summarized in Fig. 18.3a, and details are shown in Fig. 18.4a–i, which are self-explanatory. Fig. 18.4a–c show how the shape of the grain and of the part of the intine commonly referred to as the oncus changes during initial hydration, and Fig. 18.4d–h illustrates features of the overall stratification of the apertural intine as revealed cytochemically. The aperture of the grain of *C. avellana* is unusual in that the tiny sporopollenin operculum (Fig. 18.4d, e) covers a zone of protein (Fig. 18.4h). This is continuous with protein held in the bacular layer of the exine, and is evidently of sporophytic origin. The pectin of the oncus itself is differentiated into three zones (Fig. 18.4f). The outer is protein-free, while the inner carries dispersed protein, staining poorly with general protein stains (Fig. 18.4h) but showing intense esterase activity (Fig. 18.4i). During hydration *in vitro*, the protein held immediately beneath the operculum disperses first (Fig. 18.4j), and

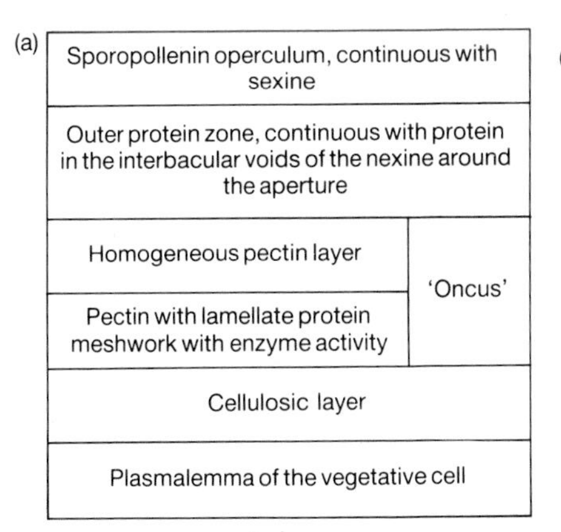

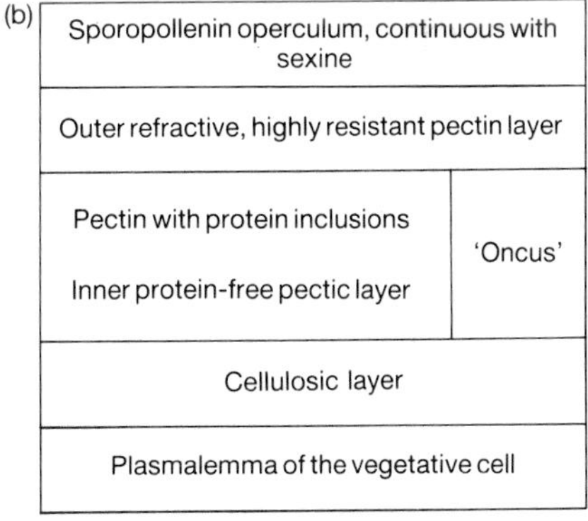

Fig. 18.3 (a) Summary of intine stratification at the apertural site of *Corylus avellana* pollen (data from Y. Heslop-Harrison *et al.* 1986). (b) Summary of intine stratification at the apertural site of *Eucalyptus* pollen (data from J. Heslop-Harrison and Y. Heslop-Harrison 1985).

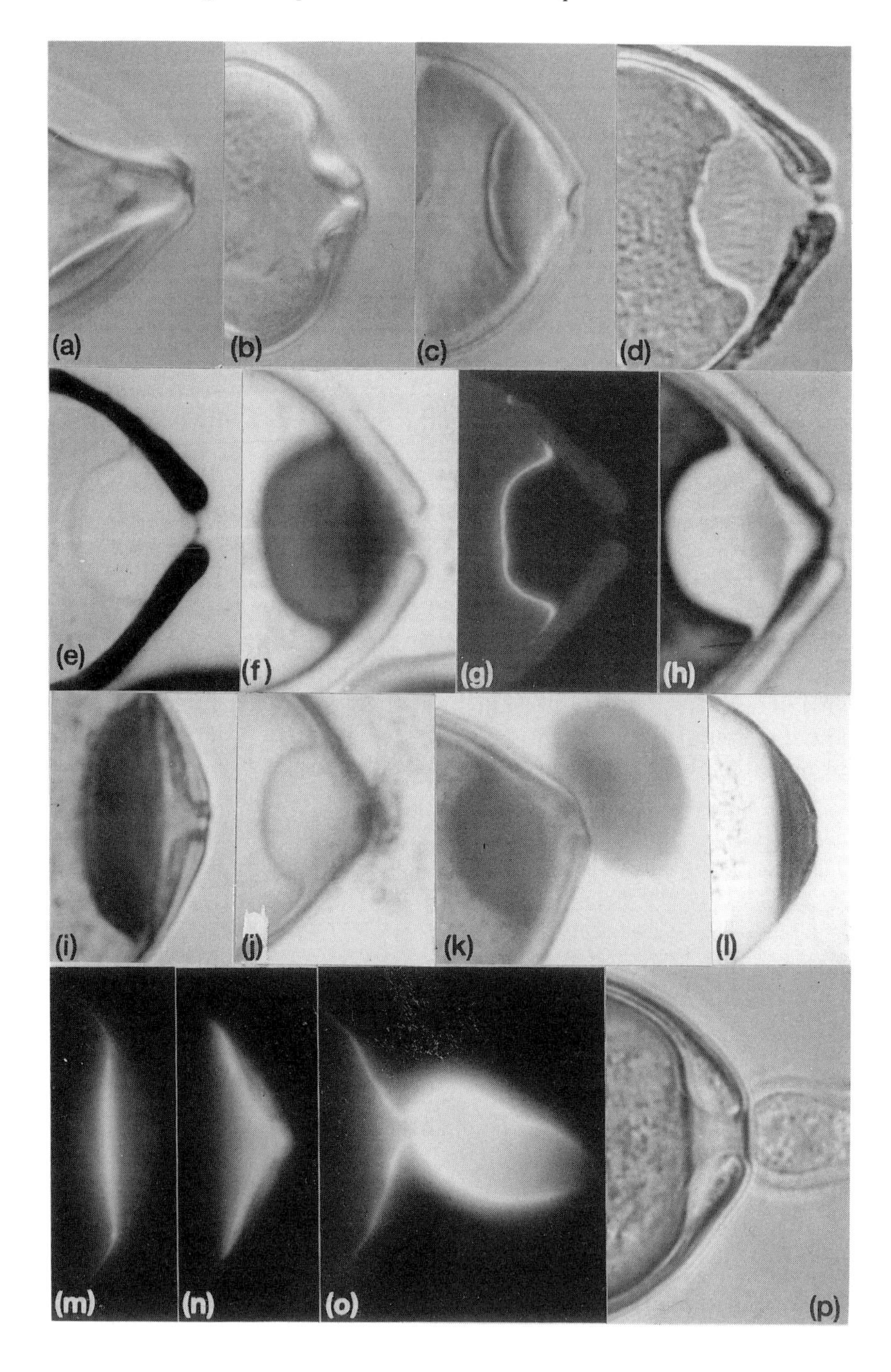
(a)
(b)
(c)
(d)
(e)
(f)
(g)
(h)
(i)
(j)
(k)
(l)
(m)
(n)
(o)
(p)

then the oncus undergoes dissolution with the emission of a cloud of gelatinous pectin (Fig. 18.4k), eventually flattening (Fig. 18.4l). As this proceeds, the inner cellulosic layer (Fig. 18.4g) flattens (Fig. 18.4m), and ultimately the tube tip is formed as a prominence directed towards the aperture (Fig. 18.4n). The emerging tube penetrates the residual oncus pectin through a narrow shaft, and its diameter increases once this is traversed (Fig. 18.4o, p).

Hazel pollen illustrates some of the complexities of intine function in an essentially mesophytic species, showing how the various requirements are associated with structural and chemical differentiations at the apertural sites. The genus *Eucalyptus*, with a pollen of essentially similar morphology, reveals yet other types of adaptation, associated in this instance with resistance to heat and desiccation (J. Heslop-Harrison and Y. Heslop-Harrison 1985). The stratification at the apertures of the tricolporate grain is summarized in Fig. 18.3b. In this instance the outer zone of the intine immediately beneath the thin sporopollenin layer overlying the aperture is apparently composed of yet another type of pectic polysaccharide, notable in this case for its resistance to dissolution during hydration of the grain. It is displaced in a body as the pectins of the underlying oncus gradually gelatinize, eventually itself undergoing solubilization. There can be little doubt that this is an adaptation associated with stress tolerance. The result of the slow and sometimes erratic operation of the apertural mechanism is to produce a wide dispersal of germination times. The desiccated pollen is physically robust, and quite remarkably resistant to heat: thus a small proportion

---

Fig. 18.4 Apertural regions of pollen grains of *Corylus avellana*, × *c*. 1800. Preparation details given in Y. Heslop-Harrison *et al.* (1986). (a)–(c) Living grain in three stage of hydration, showing progressive swelling of the oncus. Differential interference micrographs. (d)–(h) Resin-embedded grains sectioned at 1–1.5 μm; (d) Differential interference contrast; (e) acid fuchsin staining of the exine showing the small operculum; (f) alcian blue staining of the pectin of the oncus and the outer stratum of the intine extending around the grain; (g) inner cellulosic layer of the intine, fluorescence micrograph, calcofluor white staining; (h) protein distribution, Coomassie blue staining; (i), esterase localization. (j) Living grain, protein emission *in vitro*; Coomassie blue staining. (k) Living grain, pectin emission *in vitro*; alcian blue staining. (l) As (k); pectin mostly lost from the oncus. (m) As (l); fluorescence micrograph, calcofluor white staining, showing the flattening of the inner cellulosic layer of the intine conforming to the reduced profile of the oncus pectin. (n) As (m); later stage with early initiation of a pollen tube tip. (o,p) Grain with emergent pollen tube; (o) fluorescence micrograph, calcofluor white staining for cellulosic glucans of the intine and the pollen tube wall; (p) differential interference contrast.

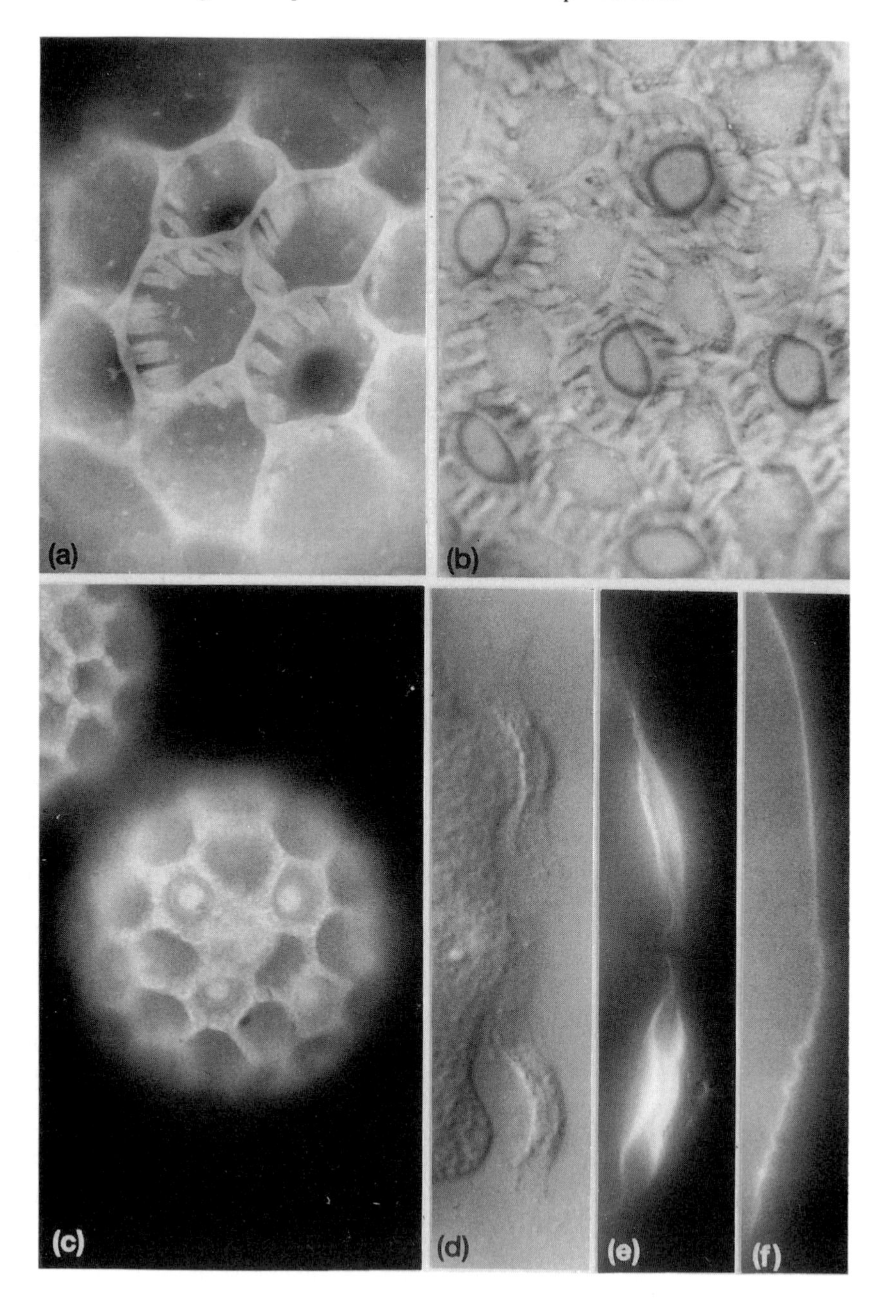
(a)
(b)
(c)
(d)
(e)
(f)

of *E. rhodantha* pollen actually survived exposure for 24 h to a temperature of 70°C (J. Heslop-Harrison and Y. Heslop-Harrison 1985).

The examples discussed in the foregoing by no means span the full range of intine variation in the angiosperms. In some instances additional layers are added at the apertural sites, the pectin strata alternating with more than one cellulosic layer. A second, outer, cellulosic stratum has been observed in *Linum grandiflorum* by Dulberger (1989), and sometimes such additional deposits form a separate elaborately patterned shield, as in *Abutilon hybridum* (Y. Heslop-Harrison and J. Heslop-Harrison 1982). As a concluding example, we may consider the remarkable situation in *Cobaea scandens* of the Polemoniaceae. Fig. 18.5a–f summarizes some of the features of this extraordinary grain, where at each aperture the intine carries three cellulosic layers, the outer pair forming a kind of capsule enclosing a small protein charge. As yet nothing can be said of the functions of such a mechanism, but we may be sure that they include the three principal roles outlined in the introductory paragraphs, and possibly others too.

An intriguing feature of many of the more complex apertural mechanisms associated with osmoregulation and germination is that they function even in dead pollen. They are in fact 'pre-set', the various components being laid down in the anther during the later stages of pollen development (J. Heslop-Harrison 1979*b*). In all cases we have examined, however, the processes cease at the point where one would expect the pollen tube tip to be differentiated in the inner cellulosic layer of the intine: this evidently requires the participation of a living vegetative cell. Recent findings, the details of which would take us outside of the range of this review, provide clues to the key events involved. During hydration, complex circulatory movements begin in the protoplast as the storage actin is transformed into a system of fibrils.

Fig. 18.5 Pollen of *Cobaea scandens*. (a) Detail of exine structure, showing distribution of apertures. Fluorescence micrograph, auramine 0 staining for sporopollenin, × c. 2000. (b) As (a) alcian blue staining for pectins; the aperture sites are defined by annuli, heavily staining with this preparation, which subtend the 'capsules' seen in (c), (d) and (e), × c. 2000. (c) Whole pollen grain, 1-anilinonaphthalene-8-sulphonic acid staining for protein; fluorescence micrograph showing the protein held in discrete capsules at the apertures, × c. 900. (d)–(f), all × c. 3750. (d) Partly digested pollen grain with two protein-containing capsules separated from the surface of the vegetative cell, differential interference micrograph; (e) as (d), fluorescence micrograph, calcofluor white staining for cellulose, the capsule protein is seen to be closely invested by cellulosic plates; (f), as (e), cellulosic inner layer of the intine left at the surface of the vegetative cell after the detachment of the capsules.

Eventually, the apertures become the focus of fibrillar actin networks (J. Heslop-Harrison *et al.* 1986) and accumulations of wall-precursor materials, and there is evidence from some species that this may be related to the development of Ca^{2+} gradients in the cytoplasm. In species such as *I. reticulata* (Fig. 18.2g), thickening of the inner cellulosic layer begins at this time, and the first deposition of callose takes place in the form of an annulus around the site of the prospective tube tip. In the appropriate osmotic conditions, the tip itself gradually emerges as a papilla, which then begins cylindrical growth to form the tube. Circumstantial evidence from polyphonious pollens suggests that while these events require a living protoplast, they do not depend upon gene transcription, also being 'pre-programmed' during pollen development.

References

Dulberger, R. (1989). The apertural wall in pollen in *Linum gradiflorum*. *Annals of Botany* **83**, 421–31.

Herth, W., Franke, W.W., Bittiger, H., Kuppel, A., and Keilich, G. (1974). Alkali-resistant fibrils of β-1,3- and β-1,4-glucans: structural polysaccharides in the pollen-tube wall of *Lilium longiflorum*. *Cytobiologie* **8**, 344–67.

Heslop-Harrison, J. (1975). The physiology of the pollen grain surface. *Proceedings of the Royal Society of London* **B199**, 275–99.

Heslop-Harrison, J. (1979*a*). Pollen walls as adaptive systems. *Annals of the Missouri Botanical Garden* **66**, 813–29.

Heslop-Harrison, J. (1979*b*). Aspects of the structure, cytochemistry and germination of the pollen of rye (*Secale cereale* L.). *Annals of Botany* **44**, (Suppl. 1), 1–47.

Heslop-Harrison, J. and Heslop-Harrison, Y. (1980). Cytochemistry and function of the *Zwischenkörper* in grass pollens. *Pollen et Spores* **22**, 5–10.

Heslop-Harrison, J. and Heslop-Harrison, Y. (1985). Germination of stress tolerant *Eucalyptus* pollen. *Journal of Cell Science* **73**, 135–57.

Heslop-Harrison, J., Heslop-Harrison, Y., Cresti, M., Tiezzi, A., and Ciampolini, F. (1986). Actin during pollen germination. *Journal of Cell Science* **86**, 1–8.

Heslop-Harrison, Y. (1977). The pollen–stigma interaction. Pollen tube penetration in *Crocus*. *Annals of Botany* **41**, 913–22.

Heslop-Harrison, Y. and Heslop-Harrison, J. (1982). The microfibrillar component of the pollen intine: some structural features. *Annals of Botany* **50**, 831–42.

Heslop-Harrison, Y., Heslop-Harrison, J.S., and Heslop-Harrison, J. (1986). Germination of *Corylus avellana* L. (hazel) pollen: hydration and the function of the oncus. *Acta Botanica Neerlandica* **35**, 265–84.

Knox, R.B. (1971). Pollen-wall proteins: localisation and enzymic activity

during development in *Gladiolus*. *Journal of Cell Science* **9**, 209–37.
Knox, R. B. and Heslop-Harrison, J. (1970). Pollen-wall proteins: localisation and enzymic activity. *Journal of Cell Science* **6**, 1–27.
Kress, J. and Stone, D. E. (1982). Nature of the sporoderm in monocotyledons, with special reference to the pollen grains of *Canna* and *Heliconia*. *Grana* **21**, 129–48.
Kress, J., Stone, D. E., and Sellers, S. C. (1978). Ultrastructure of exine-less pollen: *Heliconia* (Heliconiaceae). *American Journal of Botany* **65**, 1064–76.
Schoch-Bodmer, H. (1936). Zur Physiologie der Pollenkeimung bei *Corylus avellana*: pollen- und Narbensaugkrafte, Quellungserscheinungen der Kolloide des Pollens. *Protoplasma* **15**, 337–71.
Skvarla, J. and Rowley, J. (1970). The pollen wall of *Canna* and its similarity to the germinal aperture of other pollen. *American Journal of Botany* **57**, 519–29.

19. The evolution of gametes—from motility to double fertilization

R. BRUCE KNOX
SOPHIE C. DUCKER
School of Botany, University of Melbourne, Parkville, Victoria, Australia

Abstract

Considerable advances have occurred in the understanding of the structure and evolution of the male gamete of plants. Motile sperm were animalcules in the seventeenth century, and the flagellated sperms of plants were not recognized until the end of the eighteenth century. With the advent of high resolution electron microscopy in the twentieth century, together with molecular approaches including monoclonal antibody technology, analysis of their structural and heritable characteristics is a reality. Motile male gametes are most complex in early gymnosperms, the cycads and *Ginkgo*. What cytological changes have accompanied the evolution of motile sperm in these systems?

The patterns of diversification of angiosperm pollen are well established, but what of the male gamete? The sperm cells are held within the pollen grains or tubes, and—we believe—passively transferred for fertilization. There are now more than 30 genera of both dicots and monocots for which evidence has been obtained for the organization of the reproductive cells and vegetative nucleus to form *male germ units*. These are single transmitting units implicated in transfer of the sperm cells and vegetative nucleus, that is all the DNA of male heredity, within the pollen tube. The hypothesis has been proposed that it is not necessarily a matter of chance which of the pair of sperm cells in each pollen tube fuses with the egg cell at fertilization, and which fuses with the central cell. The very definition of a *gamete* puts the situation in perspective: a *gamete* is a reproductive cell involved in embryo formation, so the question arises as to which of the *pair of sperm cells* in each angiosperm pollen tube is the male gamete? The present, very limited, evidence suggests

Pollen and Spores (ed. S. Blackmore and S. H. Barnes), Systematics Association Special Volume No. 44, pp. 345–361, Clarendon Press, Oxford, 1991.

that each species has either its own version of a male germ unit, or none has been detected so far. Between the male gamete and the associated sperm, significant differences can occur in cellular organization and size, nuclear control, and the number and proportions of plastids to mitochondria. If these findings are generalized, it becomes necessary to redescribe double fertilization, and the way is open to isolate sperm-specific genes expressed at fertilization and associated with sperm–egg recognition.

Introduction

The function of sperm cells is inevitably associated with motility in order for the sperm to meet the egg and fertilization to occur. Sperms were first recognized in the middle of the seventeenth century by Leeuwenhoek (1678). He observed animalcules in a sample of urine/semen from a patient infected with gonorrhoea. Leeuwenhoek described them as small 'bodies with a long motile tail'. Animalcule was a term first used to denote small animals in 1599, and later infusoria visible with the newly invented microscope.

Ability to move is a character of the kingdom Animalia (Fig. 19.1), as suggested by Linnaeus (1735). But what occurs in the kingdom Plantae? We owe recognition of maleness in plants to Nehemiah Grew (1682), who found that the 'attire' (or stamens) contained 'sperme' or 'spermatick globulets' (pollen grains) and were associated with fertility. Linnaeus (1753) differentiated between the male and female organs of the plant, which were basic characters in his classification of flowering plants. He considered that the cryptogams had no sexual reproduction.

The first description of the male organs of plants are those of Schmidel (1762), who illustrated the male organs of a liverwort, *Jungermannia*. He found that the contents of the masculine globules (the antheridia) showed movement, and were considered to represent 'masculine globules', 'semen'. Because Müller (1773) observed movement in *Gonium*, a plate-like green alga, he classified it as an animal, a worm, in the *Vermes*. Amazingly, in spite of the primitiveness of microscopes in the eighteenth century, Müller (1786) saw the movement of *Ceratium* (now regarded as an alga) with '*undique ciliis minutissimus*'. This then is the discovery of the now universal agent of motility of the sperms found in all animals and the majority of plants.

The fundamental occurrence of cilia on motile infusoria or other algae was later confirmed by Ehrenberg (1832) classifying them all as animals. Nees von Esenbeck (1822) considered that *Sphagnum* moss gave birth to animals and these monads, the flagellated gametes, floated away. Unger (1839) discovered the spiral filaments in mosses and liverworts, calling them *Spirillum bryozoon*. He emphasized his observations that the animalcules of bryophytes are closer to the sperms of animals

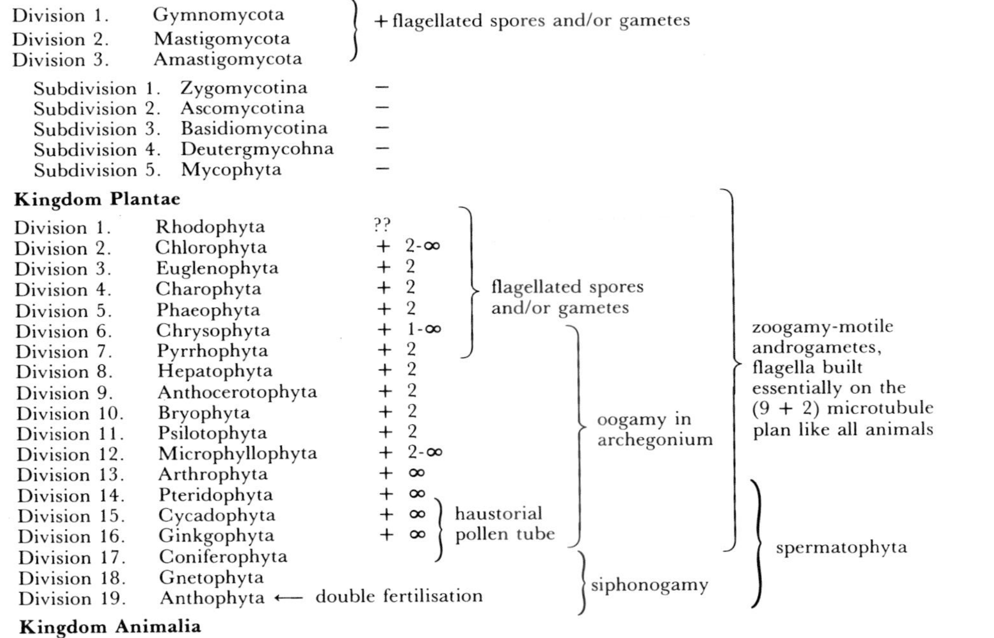

Fig. 19.1 Types of gametes according to Bold *et al.*'s (1987) classification of organisms. +, Indicates gametes are flagellate, and the number is given on the right; –, Indicates non-flagellate types. The brackets show the types of sperm-delivery systems.

than to *Spirillum*. In 1843, he published the book *Die Pflanze im Moment der Thierwerdung* (The plant turns animal). He summed up the earlier literature, and illustrated *Vaucheria* flagella. This *plant transforming into animal* concept was firmly embedded in the literature of the time. Even the famous French writer Alexandre Dumas wrote: '*Ainsi, donc, a certaines époques, dans certains organes, la plante se fait animal*' (the plant decides to become an animal at a certain stage of its life) (Thuret 1843).

The discovery of sperm cells

Amici (1824) and Brogniart (1828) described how pollen grains germinated, producing the pollen tube, which entered the stigma. In the same decade, Brown (1831) showed that the pollen tubes moved deeply within the ovary. The knowledge of pollen tube growth in flowering plants was summed up by Bischoff (1836), with the particles in the tubes showing Brownian movement while the tube is living, but ceasing when dead. Bischoff, too, postulated that among the cryptogams, ferns and mosses may have sexual reproduction.

Fritsche (1837), working with *Chara*, showed that the male organs, the antheridia, are filled with pollenstrings, containing in the spiral elements the unrecognized male gametes. But by the 1840s, Thuret had described the biflagellate gametes of *Chara*: '*l'animalcule, un fils roulé en tire-bouchon, agite les tentacules sans cesse avec une grande rapidité*' (Thuret 1840). He compared the different zoospores of the algae, but still believed in the transformation of the plant into an animal (Thuret 1843). He found that in certain brown algae, the two flagella were sited anteriorly and posteriorly. By the 1850s, he had observed motile zoospores in ferns (Thuret 1850). The French group surrounding Thuret (Leszczye-Suminsky 1849) found that the gametes of the ferns were multiflagellate and the spindle-shaped sperms showed spiral movement while proceeding into the archegonium.

Isogamy was discovered (Fig. 19.1), the fusion of identical flagellated gametes or zoospores giving rise to a new plant. In contrast, oogamy involved the movement of a motile spermatozoid, or androgamete, only to a motionless egg. This female gamete can be attached or free-floating. Every fertilization that involves a motile sperm is termed zoogamy, and it occurs in all divisions of the algae, bryophytes, pteridophytes, and the lower gymnosperms. There are no motile androgametes in the higher gymnosperms and flowering plants. So far, also, they have not been demonstrated in the red algae, which have an amoeboid male gamete.

Hofmeister (1851) finally summed up the universal role of the androgametes, showing that they provided a similar mode of fertilization. By the 1870s, sexual reproduction was firmly recognized with the description of the fusion of gamete nuclei in sea-urchins (Hertwig 1876). Schmitz (1879), working with *Spirogyra*, showed that this was also the essence of sexual reproduction in plants. As early as 1888, Ballowitz described the motile system of the sperm of bullfinches—showing that each flagellum breaks down into 11 threads. This is an unexpected premonition of the (9 + 2) system of microtubules later to be shown by electron microscopy to be standard components of the flagella in the sperm of animals and plants alike.

Advances in our understanding of the cytology of flowering plant sperms came in the last two decades of the nineteenth century, with the work of Guignard (1891) who considered that fertilization involved an interaction between the egg and one sperm only. Nawaschin (1898) discovered double fertilization in angiosperms, and thus the initiation of the endosperm by the second sperm. Strasburger, who had previously shown that fertilization in flowering plants involved the fusion of the sperm and egg cells, discussed the functions of the larger vegetative and smaller generative cells in pollen (1892).

The sperms of seed plants are reproductive protoplasts

In angiosperms, the work of the European cytologists at the end of the nineteenth century showed that the pair of sperm cells are the product of a single mitotic division of the generative cell. In the first half of the twentieth century, cytologists using the light microscope did little to increase understanding of the sperm cell. Sperm cells in some systems were found to be linked together, held by a common wall, for example in *Asclepias* (Finn 1925).

The picture that emerged is of spherical reduced cells that contained all the male genetic information. They were regarded as being passively transferred within the pollen tube, and their only likeness to the sperms of lower plants was found just before fertilization. In *Fritillaria*, *Lilium*, and *Monotropa* the sperm nuclei were seen to become spindle-shaped, and spirally coiled, reminiscent of the ferns (see references in Yuasa 1954). Maheshwari (1950) concluded that there was even doubt whether the sperms were cells, and not simply nuclei, as suggested by the still widely used classification of pollen into bi- and trinucleate types.

This state of ignorance ended with the application of electron microscopy, with its vastly increased resolution of these small cells. While most ultrastructural interest was focused on the intricately patterned wall of pollen grains, Larson (1965) confirmed that the sperms of *Zea* are cellular. Jensen and Fisher (1968) went on to establish the nature of their cellular organelles and structure, showing that the sperms are protoplasts without a discernible cell wall.

The male germ unit forms a transmitting vehicle within the pollen tube

A new era in understanding of plant sperms began with the finding of an association between the tube nucleus and the pair of sperm cells which are fused together in germinating pollen of *Plumbago* (Russell and Cass 1981). The tube nucleus had developed a long connective that linked it with the leading sperm cell of the pair. Meanwhile, in two other systems, *Spinacia* (Wilms and Aelst 1983) and *Brassica* (Dumas *et al.* 1984), similar associations had been found in the mature tricellular pollen. From these data, the concept of the male germ unit arose (Dumas *et al.* 1984).

In these genera the pair of sperm cells are not only linked together, but one is closely associated with the vegetative nucleus to form a male germ unit. This provides a single transmitting unit during passage through the pollen tube for all the DNA of male heredity. Some 30 genera of angiosperms have been shown to possess a male germ unit (Knox and Singh 1987); Knox *et al.* 1988). In the tricellular pollen of *Plumbago* (Russell 1985), the larger sperm, associated with the vegetative nucleus, is plastid-rich but mitochondria-poor. The smaller sperm, attached only to its partner, is the reciprocal.

Recently, evidence from quantitative microscopy indicates that the pair of sperm cells in the tricellular pollen grains of *Brassica campestris* and *B. oleracea* are dimorphic in terms of cell volume and organelle content (McConchie *et al.* 1987). Both cells contain a normal complement of organelles, including mitochondria, endoplasmic reticulum, ribosomes, microtubules, Golgi bodies, vesicles, and a prominent nucleus. However, both lack plastids, and the mitochondria are small and ellipsoid with simple morphologies.

In both species, the sperm cell associated with the vegetative nucleus is approximately 20 per cent larger in cell volume than its partner, and has a significantly greater number of mitochondria (up to 34 in *B. campestris*, and up to 15 in *B. oleracea*). The unattached sperm in both species has the lowest number of mitochondria reported in a sperm cell of a flowering plant (four in *B. campestris*, six in *B. oleracea*). These data indicate that the possibility of male cytoplasmic inheritance in these species is greatly reduced. The formation of the male germ unit during pollen development in *Brassica* has recently been studied (Murgia *et al.*(1989), but its role in sperm transfer in the pollen tube remains to be explored.

In the bicellular system, *Rhododendron*, the generative cell in mature pollen has an unusual spindle-shaped structure, with long terminal extensions that wrap around the body of the cells, and one is connected with an intine wall ingrowth. There is also a close association between generative cell and vegetative or tube nucleus in *Rhododendron* pollen tubes (Kaul *et al.* 1987) forming a male germ unit (Fig. 19.2a). Twenty-four hours after incubation of the pollen in germination medium, the generative cell had passed into the pollen tube, usually behind the vegetative or tube nucleus. The generative cell and its extensions are closely associated with the tube nucleus (Fig. 19.2b). By 48 h of incubation, the generative cell has divided to produce two sperm cells, which remain attached to each other, and extensions are in close proximity ($\simeq$ 55 nm) to the tube nucleus (Fig. 19.2c).

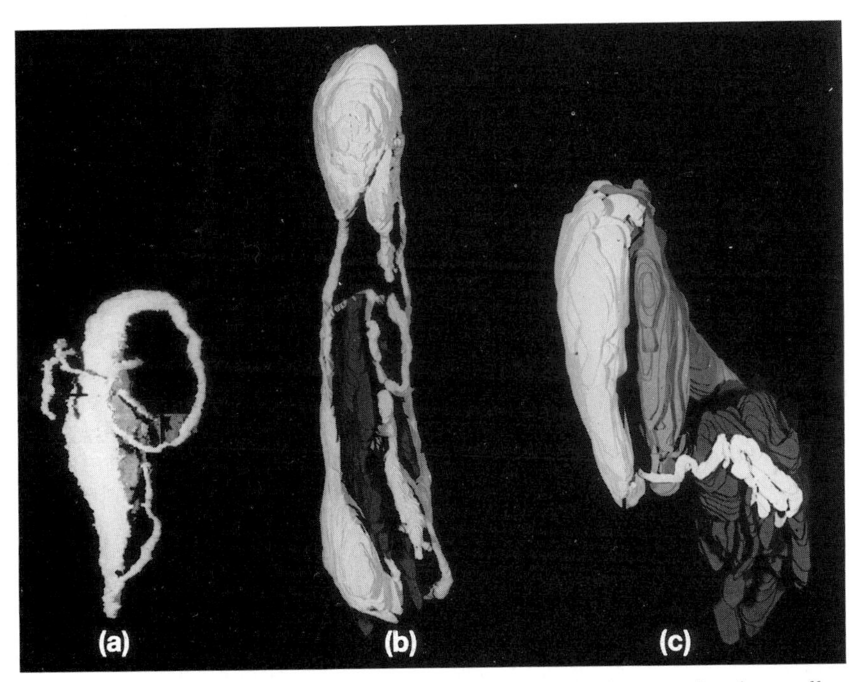

Fig. 19.2 The development of the male germ unit in germinating pollen grain and tubes of *Rhododendron laetum* as shown by three-dimensional reconstructions of serial thin sections (manual method). (a) generative cell in mature grain (vegetative or tube nucleus not illustrated). (b) generative cell and tube nucleus after culture *in vitro* for 24 h. (c) Sperm cells and tube nucleus after culture for 48 h. The tube nucleus is dark grey; the generative cell is white; sperm cells are white and light grey.

A feature of this model is that both sperm cells have extensions clasping the tube nucleus. One of the pair is generally larger than the other, averaging 99 v. 65 μm^2. No significant differences were evident in sperm cell volume, averaging 43 v. 33 μm^3 nor in organelle content: means of 94 v. 61 mitochondria and 20 v. 12 plastids. Because both sperm cells are attached to the tube nucleus, the pairs are classified according to the larger v. smaller sperm in terms of surface area. This may or may not be a meaningful relationship in biological terms, as it is evident that the larger sperm generally has the larger number of mitochondria and plastids.

Key features of the male germ unit in *Rhododendron* are:

(1) association commences between generative cell and tube nucleus within the pollen tube;

(2) association is based on both sperms being associated with the tube nucleus.

Sperm–egg recognition is pre-determined for each pair of sperm cells

The definition of a *gamete* creates an impression of inequality when viewing each pair of sperm cells. According to the Oxford English Dictionary, a gamete is a reproductive cell involved in embryo formation. Thus, the egg is the female gamete. But only one of the two sperm cells in each angiosperm pollen tube is the male gamete. An explanation for these differences comes from two hypotheses proposed to explain double fertilization. First, there is the chance hypothesis: both sperm cells are identical, and it is merely the first past the post that fuses with the egg. Secondly, there is the specific recognition hypothesis (Knox *et al.* 1988). This states that it is not just a matter of chance which of the pair of sperm cells in each pollen tube fuses with the egg at fertilization, and which fuses with the central cell (Knox and Singh 1987; Knox *et al.* 1988). The sperm cell that can fuse with the egg is predetermined at or soon after generative cell division and possesses specific recognition receptors.

If the specific recognition hypothesis is generalized, it becomes necessary to change our concept of double fertilization. The mitosis that cuts off the vegetative and generative cells arises when the microspore becomes bicellular. During subsequent development in bicellular pollen types, these cells then pass into the pollen tube at its germination, and the sperm cell division does not occur until the tube has penetrated into the style. In tricellular systems, there is a further division of the generative cell within the grain to yield the sperm cells. The eventual result is the same: a *male gamete* to fertilize the egg cell and an *associate cell* destined to fuse with the polar nuclei to form the primary endosperm cell. Between the male gamete and the associate cell, there is scope for differences in cellular organization and size, nuclear control, and the number and proportions of plastids to mitochondria.

Such differences may explain the difficulties of interspecies crossing and even why in some cases a cross, but not its reciprocal might succeed. A further and important consequence is that any genetic modification induced in the generative cell *before* it divides would have unpredictable consequences, and any genetic modification applied after this division would only *directly* affect the zygote if it were applied to the male gamete (Knox and Singh 1990). If the associate cell were affected, possible modifications would be to the endosperm and this could *indirectly* affect the zygote.

Assuming for the moment that this distinction between male gamete and the associate cell proves to be justified, there is the fascinating prospect that the surface recognition molecules, that develop and redistribute in animal sperms during their maturation, might have different counterparts on both the male gamete and the associate cell surfaces (Hough *et al.* 1986; Knox *et al.* 1988). Any emerging reproductive technology that sought to influence events at or around fertilization would have to take account of these new concepts.

Evidence for or against this hypothesis is being obtained in several ways. In *Plumbago*, Russell (1985) demonstrated the existence of preferential fertilization by TEM studies of fertilized zygotes. The plastid-rich sperm has a high probability of fusing with the egg. This is the first cytological evidence for preferential fertilization. In the pollen tube, the sperms are held in pairs within the plasma membrane of the inner pollen tube. In *Plumbago*, this membrane is shed shortly before tube entry into the embryo sac (Russell 1985). When released for fertilization, the sperm cells are surrounded only by their own plasma membrane. This is to date the only report concerning such membrane changes.

Several laboratories have isolated sperm cells, and are exploring their cell biology. Sperm cells from mature pollen of maize and wheat have been isolated (Cass and Fabi 1988; Matthys-Rochon *et al.* 1987), and a characteristic feature is that the elongate sperm cells round off when isolated, appearing as spherical protoplasts or gametoplasts.

The potential exists to detect differences in surface antigens of the pairs of sperm cells using monoclonal antibody technology, as has been achieved for *Brassica* (Hough *et al.* 1986) and *Plumbago* (Pennell *et al.* 1987). Given further improvements in recombinant DNA technology, it should be possible in future to detect sperm-specific gene expression associated with fertilization. We will now review the potential applications of these new technologies.

Sperm-specific proteins

Fluorescent probes can be used to visualize the cellular profiles of the sperm cells to complement quantitative electron microscopic studies that have so far been used to define male germ units. DNA fluorochromes reveal the sperm cells by the fluorescence of the nuclei (Hough *et al.* 1986). Video-image processing techniques can be used to both enhance the cell profiles, and to reveal the pattern of the fluorescent probe binding to these very small cells. In *Rhododendron*, the sperm cells are approximately 3×7 μm in diameter (Taylor *et al.* 1989).

Microtubules in sperm cells have been observed in ultrastructural studies of a few mature tricellular pollen types, and tubulin has recently been detected by immunofluorescence combined with video-image processing in *Rhododendron*. Ultrastructural analyses have not detected microfilaments in generative or sperm cells, whereas our immunofluorescence studies of actin indicate that there are abundant actin microfilaments in both generative and sperm cells (Taylor *et al.* 1989). Just as with the flagellar proteins of lower plant sperms, so the sperms of flowering plants have abundant tubulin and actin proteins expressed.

The sperm cells are natural protoplasts, without detectable cell wall polysaccharides in most systems (Knox *et al.* 1988). This means that their plasma membrane proteins are accessible. There are a number of probes for labelling of plasma membrane proteins, such as radioactive iodination, biotin, fluorescent probes, and monoclonal antibodies (MAbs). Some of these methods, however, are either hazardous or involve time-consuming procedures. Also, the antibody approach detects only those antigens which are immunodominant in mice, and the less immunogenic surface proteins are more difficult to detect with antibodies.

Recently, biotin has been reported as a safe and effective reagent for labelling of surface proteins. The membranes of intact animal cells and plant protoplasts are impermeable to biotin esters. The high binding affinity of streptavidin with biotin allows the detection of the biotin-bound proteins after electrophoresis on SDS-PAGE. Results of surface-labelling of membrane proteins can be compared with labelling of disrupted cells, in which all the cytoplasmic proteins will be labelled.

With the isolation of pollen protoplasts (see references in Blomstedt *et al.* 1990) and sperm cells from a number of angiosperms (Hough *et al.* 1986; Russell 1986; Matthys-Rochon *et al.* 1987; Cass and Fabi 1988; Knox *et al.* 1988; Southworth and Knox 1989; Taylor *et al.* 1989) the study of the reproductive membranes involved in fertilization has become feasible. Electrophoretic techniques have been used to characterize the proteins in sperm-enriched fractions from mature pollen (Knox *et al.* 1988). We have used biotin-labelling to compare the membrane surface proteins of intact leaf and pollen protoplasts of *Brassica*, *Gladiolus*, and *Lilium* (Blomstedt *et al.* 1990). We have identified common and unique membrane surface proteins of intact somatic and reproductive protoplasts in the dicot *Brassica* and the monocot *Lilium*. The sperm cell protoplasts possess at least one unique polypeptide.

Isolation of pre-fertilization sperm cells

The *in vivo/in vitro* method has been developed to provide for the first time the potential to isolate for analysis large quantities of sperm cells that are at various stages leading up to fertilization (Shivanna *et al.* 1988). Briefly, pollen is germinated *in vivo* on the stigma; style segments are excised at various times (and lengths down the style) afterwards and the pollen tubes allowed to grow out *in vitro* into growth medium. The pollen tubes burst to release the intact sperms after treatment with cell wall digesting enzymes or osmotic shock.

The isolated sperms are small ellipsoidal protoplasts, e.g. *c.* 2 × 3 μm in *Gladiolus*, 8 × 10 μm in lily, and 3 × 4 μm in *Rhododendron*. After isolation, a proportion of the sperms occur in pairs, linked at one end by finger-like connections. The pairs of sperm cells are dimorphic in terms of surface area and volume (Shivanna *et al.* 1988).

Sperm surface protein receptors

Surface antigens from a number of somatic protoplasts have been detected by monoclonal antibody (MAb) technology. These have been produced by using whole protoplasts or plasma membrane preparations as immunogens. Although MAbs can detect protoplast surface antigens, the majority of the antibodies made to date appear to be directed to carbohydrates, and this appears to be the case for sperm cells (Pennell *et al.* 1988). Because of cross-reactivity of the glycosyl components of glycoproteins, such MAbs are not useful specific probes to identify surface proteins. Despite these potential problems, we have used sperm-enriched fractions of *Brassica* pollen and pollen tubes during 1989–90 to prepare a library of MAbs. These have been screened as positive using immunofluorescence methods, either using sperm cells bound to Nuclepore filters (Hough *et al.* 1986) or cryostat sections. Sperm cells are simultaneously probed with a DNA fluorochrome for microscopic detection. Selected hybridoma cell lines have been cloned, resulting in

MAbs specific, as judged by immunofluorescence microscopy, for the pollen plasma membrane; certain cytoplasmic organelles of pollen and pollen tube; and the sperm cells. Immunogold-labelling is in progress to determine the precise cellular sites of MAb binding.

What kind of sperm antigens do we expect to detect? First, there are surface proteins and glycoproteins of the plasma membranes—the peripheral and integral proteins. In animal sperms, freeze-fracture microscopy has demonstrated discrete membrane domains that correlate with the location of specific sperm proteins. Intramembrane particle patterns have been detected by these methods in generative cells (Emons *et al.* 1988), but not in sperm cells. Secondly, there are cytoplasmic and nuclear antigens, which may be useful in providing sperm-specific probes in future studies of sperm–egg interactions.

Potential for molecular cloning

An advantage of sperm cells is that they will show haploid gene expression, and that the genes involved are perhaps numbered in hundreds rather than the thousands that have been detected in a pollen grain. In our laboratory, we have prepared several cDNA libraries from mRNA isolated from tricellular pollen, e.g. *Brassica campestris, B. napus* (Theerakulpisut *et al.* 1989), and *Lolium perenne* (Knox *et al.* 1989). The potential exists to apply the techniques of molecular cloning to sperm cells. This will have the advantage of identifying sperm-sepcific genes and providing new cellular probes for sperm cells. These could be expressed in the sperm of mature pollen in tricellular systems, or show period specificity, and not be expressed perhaps until the pollen tube is approaching the embryo sac prior to fertilization. For this purpose, we have developed the *in vivo/in vitro* technique for sperm isolation (Shivanna *et al.* 1988). However, it is likely that the amount of mRNA in the sperm cells may be too low to make it possible to prepare a cDNA library using classic methods.

The sperm-specific proteins identified may also be of low abundance. Under these conditions, it may not be possible to pick up cDNA clones by screening the library. We suggest the use of the polymerase chain reaction (PCR, Doherty *et al.* 1989), which can be used to amplify cDNAs. We have used PCR amplification in our laboratory to pick up full-length clones of some pollen cDNAs, for example allergenic proteins of *Lolium*.

Conclusions and future perspectives

The sperm cell as the male gamete must contain all the necessary genetic information to produce the next sporophytic generation of plants. But for the little known paper by Yuasa (1954), seldom before has it been possible to make a comparison of sperms of lower and higher plants, and to speculate about their differences.

1. Non-flagellated sperms are found in only four out of 19 divisions of plants. These include the red algae, the higher gymnosperms, and the flowering plants. These are among the most successful of plant groups, the red algae in the marine, the others in the terrestrial environment.

2. With the exception of these four divisions, motility in sperms is based on the presence of flagella and is dependent on water as a medium.

3. In the red algae, sperm motility is amoeboid. Dingle (1970) has shown that when amoebae transform from the amoeboid to flagellate states, a temperature shock increases the number of flagella in *Naegleria gruberi*. It is tempting to speculate that some basic switch of this kind could have operated during sperm evolution in plants. In the majority of groups, the active sperm state is flagellate, while in the four divisions with non-flagellate sperm, we postulate that the active state is amoeboid. In amoebae, motility is based on an actomyosin system. Recent immunofluorescence experiments with pollen tubes suggest that the actomyosin system may be responsible for sperm movement within the pollen tubes of flowering plants (Heslop-Harrison and Heslop-Harrison 1989). The three-dimensional models of generative and sperm cells and the associated tube nuclei in *Rhododendron* (Taylor *et al.* 1989) certainly suggest movement of this kind.

4. It has been shown repeatedly, including the classic work of Nawaschin (1909), that the sperm cells of flowering plants assume a spiral configuration at the time of fertilization. This change preserves a relict mode of penetration of the embryo sac. The aqueous environment has been replaced by this unit, containing the female cells which are ready to receive the male unit.

So many questions are posed by these findings. Answers can now be obtained, using new approaches of molecular biology, to such questions as:

(1) Does any trace of the gene sequences encoding the flagellar protein remain in the genome of the groups with non-flagellate sperm cells?

(2) Are there sperm-specific genes expressed during double fertilization that determine the fate and function of the male germ unit?

(3) What is the significance of the finding that 18 out of 19 plant groups have a single fertilization to produce an embryo?

(4) What is the molecular basis of double fertilization in flowering plants?

Acknowledgements

We thank Drs Paul C. Silva, Wm. J. Woelkerling, Gary Nelson, Tim Entwistle, G. I. MacFadden, and John Pettitt for discussions and Dr Steve Blackmore for his courtesy in delivering this lecture. We are grateful to the Librarians of the New York Botanical Garden, the Farlow Herbarium, the Melbourne National Herbarium, and the Ewart Library, Melbourne University for bibliographic assistance and Chris O'Brien for photographic help. We thank the Australian Research Council for financial support to RBK.

References

Amici, J. B. (1824). Observations microscopiques sur diverses éspèces des plantes. *Annales des Sciences Naturelles, Botanique* **2**, 41–70, 211–48.

Ballowitz, E. (1988). Untersuchugen über die Struktur der Spermatozoen. *Archiv für mikroskopische Anatomie* **32**, 401–73.

Bischoff, G. W. (1836). *Lehrbuch der Botanik. Naturgeschichte der drei Reiche*, Vols 4 and 5. Schweizerbart's Verlagshandlung, Stuttgart.

Blomstedt, C., Xu. H., Singh, M. B., and Knox, R. B. (1990). Biotin-labeling of somatic and reproductive protoplasts of *Gladiolus*, lily and *Brassica*. Protoplast '91 (ed. K. Glimelius) Uppsala (in press).

Bold, H. C., Alexopoulos, C. J., and Delevoryas, T. (1987). *Morphology of plants and fungi.* Harper and Row, New York.

Brongniart, A. T. (1828). Nouvelles recherches sur le pollen. *Annales des Sciences Naturelles, Botanique* **15**, 381–401.

Brown, R. (1831). Observations on the organs and mode of fecundation in Orchideae and Asclepiadeae. *Transactions of the Linnean Society. London* **16**, 685–745.

Cass, D. D. and Fabi, G. C. (1988). Structure and properties of sperm cells isolated from the pollen of *Zea mays. Canadian Journal of Botany* **66**, 819–25.

Cass *et al.* 1986

Charzynska *et al* 1989.

Decaisne, J. and Thuret, G. (1845). Recherche sur les anthérides et les spores de quelques *Fucus. Annales des Sciences Naturelles, Botanique, Sér. 3,* **3**, 1–15.

Dingle, A. D. (1970). Control of flagellum number in *Naegleria. Journal of Cell Science* **7**, 463–81.

Doherty, P. J., Huesca-Contreras, M., Dosch, H. M., and Pan, S. (1989). Rapid amplification of complementary DNA from small amounts of unfractionated RNA. *Analytical Biochemistry* **177**, 7–10.

Dumas, C., Knox, R. B., McConchie, C. A., and Russell, S. D. (1984).

Emerging physiological concepts in fertilization. *What's New in Plant Physiology* **15**, 17–20.

Ehrenberg, C.G. (1832). Über die Entwicklung und Lebensdauer der Infusionsthiere; nebst ferneren Beiträgen zu einer Vergleichung ihrer organischen Systeme. *Abhandlungen der Königlichen Akademie der Wissenschaften, Berlin* **1831**, 1–154.

Emons, A.M.C., Kroh, M., Knuiman, B., and Platel, T. (1988). Intramembrane particle pattern in vegetative and generative plasma membranes of lily pollen grain and pollen tube. In *Plant sperm cells as tools for biotechnology*, (ed. H.J. Wilm, and C.J. Keijzer), 41–8. Pudoc, Wageningen.

Finn, W.W. (1925). Male cells in angiosperms. 1. Spermatogenesis and fertilisation in *Asclepias cornuti. Botanical Gazette* **80**, 1–25.

Fritsche, J. (1837). Über den Pollen. *Mémoires presentés à l'académie impériale des sciences de St.-Pétersbourg par divers savants* **3**, 649–72.

Grew, N. (1682). *The anatomy of plants*, Sources of Science, No. 11, Johnson Reprint Corporation, New York (1965).

Guignard, L. (1891). Nouvelles études sur la fécundation comparaison des phénoménes morphologiques observés chez les plantes et chez les animaux. *Annales des Sciences Naturelles, Botanique, Sér 7* **14**, 163–296.

Hertwig, O. (1876). Beiträge zur Kenntniss, der Bildung, Befruchtung und Theilung des thierischen Eies. *Morphologische Jahrbücher* **1**, 347–434.

Heslop-Harrison, J. and Heslop-Harrison, Y. (1989). Actomyosin and movement in the angiosperm pollen tube: an interpretation of recent results. *Sexual Plant Reproduction* **2**, 199–207.

Hofmeister, W.F. (1851). *Vergleichende Untersuchungen der Keimung, Entfaltung und Fruchtbildung höherer Kryptogamen.* F. Hofmeister, Leipzig.

Hough, T., Sing, M.B., Smart, I., and Knox, R.B. (1986). Immunofluorescent screening of monoclonal antibodies to surface antigens of animal and plant cells bound to polycarbonate membranes. *Journal of Immunological Methods* **92**, 103–7.

Jensen, W.A. and Fisher, D.B. (1968). Cotton embryogenesis: The sperm. *Protoplasma* **65**, 277–86.

Kaul, V., Theunis, C., Palser, B., Knox, R.B., and Williams, E.G. (1987). Association of the generative cell and vegetative nucleus in pollen tubes of *Rhododendron laetum. Annals of Botany* **59**, 227–35.

Knox, R.B. and Singh, M.B. (1987). New perspectives in pollen biology and fertilization. In *Annals of Botany Centenary Volume* **60**, 15–37.

Knox, R.B. and Singh, M.B. (1990). Reproduction and recognition phenomena in the Poaceae. In *Reproductive versatility in the grasses* (ed G.P. Chapman). pp. 220–39. Cambridge University Press.

Knox, R.B., Southworth, D., and Singh, M. (1988). Sperm cell determinants and control of fertilization in plants. In *Eukaryote cell recognition*, (ed. G.C. Chapman), pp. 175–93. Cambridge University Press.

Knox, R. B., Singh, M. B., Hough, T., and Theerakulpisut, P. (1989). The rye-grass pollen allergen, *Lol p1*. In *Allergy and molecular biology*, (ed. T. Merrett). USA. *Advances in Biosciences* **74**, 161–71. Pergamon Press, Oxford.

Larson, D. A. (1965). Fine structural changes in the cytoplasm of germinating pollen. *American Journal of Botany* **52**, 139–54.

Leeuwenhoek, A. van (1677). Concerning little Animals by him observed in Rain- Well- Sea- and Snow water. *Philosophical Transactions* **12**, (133), 821–31.

Leeuwenhoek, A. van (1678). Observationes D. Anthonii Lewenhoek, de natis e semini genitali Animaliculis. *Philosophical Transactions* **12**, (142), 1040–3.

Leszczye-Suminsky, M. H. (1849). Sur le développement des fougères. *Annales des Sciences Naturelles, Botanique, Sér 3* **11**, 114–26.

Linnaeus, C. (1735). *Systema naturae*. T. Haak, Leiden.

Linnaeus, C. (1753). *Species plantarum*, 2 Vols. Laurentii Salvii, Stockholm Repr. 1957–1959. Ray Society, London.

McConchie, C. A., Russell, S. D., Dumas, C., Tuohy, M., and Knox, R. B. (1987). Quantitative cytology of the sperm cells of *Brassica campestris* and *B. oleracea*. *Planta* **170**, 446–52.

Maheshwari, P. (1950). *Embryology of angiosperms*. McGraw-Hill, New York.

Matthys-Rochon, E., Vergne, P., Detchepare, S., and Dumas, C. (1987). Male germ unit isolation from three tricellular pollen species: *Brassica oleracea, Zea mays* and *Triticum aestivum*. *Plant Physiology* **83**, 464–6.

Müller, O. F. (1773). *Vermium terrestrium et fluviatilum seu animalium infusorum, helminthicorum et testaceorum, non marinorum secincta historia*, Vol. 1. Martin Hallager, Hauniae et Lipsiae.

Müller, O. F. (1786). *Animalcula infusoria fluviatilia et marina, que detexit, systematice descripsit et ad vivum deliineari curavit Otto Fridericus Müller, regi daniae quondam a consiliis conferentaie, plurium que academiarum et societatum scientarum sodalis, sistit opus hoc posthuman quod cum tabulis aeneis L. in lucem tradit vidua ejus nobillissima, cur Othonis Fabricii*. Nicolae Molleri, Hauniae.

Murgia, M., Milanesi, C., and Cresti, M. (1989). Origin of sperm cell association in the 'male germ unit' of *Brassica* pollen. *Protoplasma* **149**, 1–4.

Nawaschin, S. G. (1898). Resultate einer Revision der Befruchtungsvorgänge bei *Lilium Martagon* und *Fritillaria tenella*. *Bulletin de l'Académie Impérial des Sciences de Saint-Pétersbourg* **11**, 377–82.

Nawaschin, S. G. (1909). Über das selbständige Bewegungungsvermögen des Spermakerns bei einigen Angiospermen. *Östereichische botanische Zeitschrift* **59**, 457–67.

Nees von Esenbeck, C. D. (1822). I. Correspondenz. *Botanische Zeitung (Regensburg) 1822* **1**, 33–6.

Pennell, R. I., Geltz, N. R., Koren, E., and Russell, S. D. (1988). Production and partial characterization of hybridoma antibodies elicited to the sperm of *Plumbago zeylanica*. *Botanical Gazette* **148**, 401–6.

Russell, S. D. (1985). Preferential fertilization in *Plumbago:* ultrastructural evidence for gamete level recognition in an angiosperm. *Proceedings of the National Academy of Sciences USA* **82**, 6129–32.

Russell, S.D. (1986). A method for the isolation of sperm cells in *Plumbago zeylanica. Plant Physiology* **81**, 317–19.

Russell, S.D. and Cass, D.D. (1981). Ultrastructure of the sperms of *Plumbago zeylanica*. I. Cytology and association with the vegetative nucleus. *Protoplasma* **107**, 85–107.

Schmidel, C.C. (1762). Icones plantarum, J.C. Keller, Nürnberg.

Schmitz, C.J.F. (1879). Untersuchungen über die Zellkerne der Thallophyten. *Niederrheinishce Gesellschaft für Natur- und Heilkunde zu Bonn, Sitzunsberichte* **1878**, 345–76.

Shivanna, K.R., Xu. H., Taylor, P., and Knox, R.B. (1988). Isolation of sperms from the pollen tubes of flowering plants during fertilization. *Plant Physiology* **87**, 647–50.

Southworth, D. and Knox, R.B. (1989). Cell biology and isolation of sperm cells of *Gerbera jamesonii. Plant Science* **60**, 273–7.

Strasburger, E. (1892). *Histologische Beiträge*. Heft IV. Über das Verhalten des Pollens und die Befruchtungsgänge bei den Gymnospermen. Gustav Fischer, Jena.

Taylor, P., Kenrick, J., Li, Y., Kaul, V., Gunning, B.E.S., and Knox, R.B. (1989). The male germ unit of *Rhododendron*: quantitative cytology, three-dimensional reconstruction, isolation and detection using fluorescent probes. *Sexual Plant Reproduction* **2**, 254–64.

Theerakulpisut, P., Singh, M.B., Strother, S., and Knox, R.B. (1989). Isolation of cDNA clones specifically expressed in the pollen grains of *Brassica campestris*. In *Pollination '88*, (ed. R.B. Knox, M.B. Singh, and L. Troiani), pp. 110–14. School of Botany, University of Melbourne.

Thuret, G. (1840). Note sur l'anthére du *Chara* et les animalcules qu'elle renferme. *Annales des Sciences Naturelles Botanique, Sér. 2* **14**, 65–72.

Thuret, G. (1843). Recherches sur les organes locomoteurs des spores des algues. *Annales des Science Naturelles, Botanique, sér. 2* **19**, 266–77.

Thuret, G. (1850). Recherches sur les zoospores des algues et les anthèridiés des cryptogames. *Annales des Sciences Naturelles, Botanique, Sér. 3* **14**, 214–60.

Unger, F. (1839). Nouvelles observations sur les anthéres des mousses et sur les animalcules spermatiques. *Annales des Sciences Naturelles, Botanique, Sér. 2* **11**, 257–71.

Unger, F. (1843). *Die Pflanze im Moment der Thierwerdung*. Fr. Beck, Wien.

Wilms, H.J. and Aelst, A.C. van (1983). Ultrastructure of spinach sperm cells in mature pollen. In *Fertilisation and embryogenesis in ovulated plants* (ed. O. Erdelska), pp. 105–11. Veda, Bratislava.

Yuasa, A. (1954). Studies in the cytology of Pteridophyta; XXXIII. Cytomorphological study on the plant-spermatozoid, with special reference to its phylogenetic meaning. *Botanical Magazine, Tokyo* **67**, 6–14.

20. Underwater pollination, three-dimensional search, and pollen morphology: predictions from a supercomputer analysis

PAUL ALAN COX
Department of Botany and Range Science, Brigham Young University, Provo, Utah, USA
SCOTT CROMAR*
TYLER JARVIS
Department of Mathematics, Brigham Young University, Provo, Utah, USA

Abstract

Unlike terrestrial plants, which are pollinated by animals or the wind, many marine and freshwater angiosperms are pollinated by water. Yet little has been known of the mechanics of underwater pollination systems. Two-dimensional random search theory, which has been of use in studying two-dimensional hydrophilous pollination on the water surface, has yet to be developed for three dimensions. Numerical experiments of three-dimensional collisions and target encounter rates would be extremely useful in understanding the pollination strategies of underwater-pollinated plants, but the tremendous computational time requires the speed of a supercomputer for calculation purposes.

In an attempt to better understand underwater pollination systems, a number of Markov simulations were performed on the Palo Alto IBM 3090 supercomputer. In these numerical experiments, a series of spheres of equal volume were deformed by increasing their elliptical eccentricity. The efficiency of these spheroids in hitting arbitrarily placed targets

*Present address: *Department of Mathematics, Rutgers University, New Brunswick, New Jersey, USA*

Pollen and Spores (ed. S. Blackmore and S. H. Barnes), Systematics Association Special Volume No. 44, pp. 363–375, Clarendon Press, Oxford, 1991.

while they traced random trajectories was calculated. Both the number of hits per 10 000 trials and the length of search paths for successful trials were calculated for each spheroid.

The predictions from the supercomputer trials were compared to field observations of the pollination biologies of two marine and two freshwater angiosperm species: the seagrasses *Thalassia testudinum* in St. Croix (US Virgin Islands) and *Phyllospadix scouleri* in California; and the freshwater aquatics *Lepilaena bilocularis* in New Zealand, and *Zannichellia palustris* in Utah. Both the predictions from supercomputer experiments and the field observations agreed well: spheroids with high degrees of elliptical eccentricity were found in the supercomputer trials to be far more efficient in hitting targets during three-dimensional random searches, while in the field studies, underwater-pollinated plants were found to disperse their pollen in long filamentous strands. It therefore appears that there has been an evolutionary convergence towards efficient search strategies in underwater plants.

Introduction

The pollen of terrestrial plants is commonly carried from flower to flower by animals or the wind. However, flowering plants growing in freshwater or marine environments face unique problems in their reproductive biology. Although aquatic plants that float or grow in shallow water can be pollinated by insects or the wind (Cook 1988) if their flowers are raised above the water surface, 31 genera in 11 different families contain hydrophilous species, i.e. species whose pollen is transported by water (Cox 1988).

Hydrophilous plants are ecologically diverse. They occur throughout temperate and tropical areas, ranging in latitude from Alaska to Southern Argentina and can be found in every major continental area save Antarctica. They range in altitude from 30 m beneath the surface of the sea to 4800 m in the Andes where *Elodea potamogeton* is an impediment to navigation on Lake Titicaca. Water-pollinated plants are found in ecosystems as diverse as the rain forests of the Amazon and the seasonal pools of the high deserts of North America. Hydrophilous plants are also economically important, serving as food resources for fish and wildfowl in freshwater systems, and for invertebrates, fish, turtles, and manatees in marine ecosystems. The fruits and rhizomes of some seagrasses are also harvested by some indigenous peoples (Cox 1988).

Ecologically, hydrophilous plants can be divided into three general categories, depending on whether their pollen is transported to the female stigmas above, on, or under the water surface (Ernst-Schwarzenbach 1944; Cox 1988). Plants such as *Enhalus acoroides*

(Hydrocharitaceae) in which the pollen is carried above the water surface by floating male flowers (Troll 1931), or other conveyances, characterize Category 1 hydrophilous systems. Plants whose pollen is dispersed directly upon the surface of the water, such as *Halodule pinifolia* (Cymodoceaceae) (Cox and Knox 1989), are said to have Category 2 hydrophilous systems. Both Category 1 and Category 2 hydrophilous species have two-dimensional pollination in that their pollen is transported along a plane, i.e. the water surface. In Category 3 hydrophilous species, the pollen is transferred below the surface of the water, and hence moves in three-dimensions, such as in *Thalassia testudinum* (Hydrocharitaceae) (Cox and Tomlinson 1988). Although this later category includes plants with submarine pollination systems (Ducker and Knox 1976; Pettitt *et al.* 1981), some freshwater aquatic plants also disperse their pollen underwater.

In the first two categories of hydrophilous pollination, effective shapes of pollen and pollen assemblages can be predicted from two-dimensional search theory (Koopman 1956, 1980). Even though a pollen grain obviously cannot be described as 'searching' in the sense that it has volition or directedness, the likelihood of its encountering a 'target', in this sense the female stigma of another plant, can be predicted from mathematical search theory, which was originally developed for military purposes in the Second World War (Koopman 1980; Cox 1983). For example, in two dimensions, an object which moves with a random component to its trajectory can greatly increase its likelihood of encountering a target if it can increase its diameter or the diameter of its conveyance. The derivation of this principle is relatively simple (Cox 1983) and requires neither that the search path be entirely random nor that the search vehicle move with its long axis orthogonal to the direction of the flow; anything that increases the width of the search path will increase the probability of target encounter.

Detailed field observations of a variety of hydrophilous species with two-dimensional pollination systems indicate that their pollen is generally dispersed either in floating rafts or in flowers of large diameter (Cox 1988; Cox and Knox 1989); in some cases (i.e. *Halodule pinifolia* and *Halophila ovalis*) the search vehicles are assembled from filiform pollen or pollen embedded in filiform mucilaginous tubes (Cox and Knox, 1989), but in these cases the pollen is dispersed in snowflake-like rafts rather than as single grains. In these cases, there appears to have been evolutionary convergence by unrelated hydrophilous taxa to dispersal of pollen in conveyances or assemblages of large diameter, as predicted by search theory (Cox 1988).

The possible applications of search theory to three-dimensional hydrophilous systems has, however, remained unexplored for several reasons. First, little has been known about the pollination ecology of underwater-pollinated plants despite rather sophisticated studies of ultrastructure, physiology, and development of pollen of plants believed to be submarine pollinated (Pettitt and Jermy 1975; Ducker and Knox 1976; Yamashita 1976; Ducker *et al.* 1978; Pettitt *et al.* 1980, 1981; Pettitt 1981, 1984; McConchie *et al.* 1983). Few field observations of underwater pollination have been made, and as a result some species with surface pollination (e.g. *Amphibolis antarctica* and *Thalassodendron ciliatum*) have been incorrectly assumed to be obligately pollinated underwater (Ducker *et al.* 1978; Pettitt *et al.* 1983; Pettitt 1984; Cox and Knox 1988; Cox 1991), while important details of the pollination ecology of other species (e.g. mixed surface and submarine modes of pollination in *Phyllospadix scouleri* and *Zostera marina*, with pollen dispersed in aggregations rather than singly) have gone unreported (Ackerman 1986; Cox *et al.* 1991*a*,*b*). Secondly, three-dimensional search theory has yet to be developed due to the tremendous increase in complexity of the mathematics.

Theoretical extrapolations from two-dimensions to three in such cases is unlikely to be useful since there are significant qualitative differences of the nature of random motion in two and three dimensions. For example, Brownian walks are recurrent (i.e. they will eventually hit their starting point) in two dimensions but are not in three dimensions (Hersch and Griego 1969), and the differences between fractal paths in two and three dimensions are profound.

Since analytical three-dimensional search theory has yet to be developed, we designed a series of numerical experiments to explore the nature of three-dimensional random searches of the type that characterize underwater pollination systems. We particularly wished to discover whether the probability of successful pollination (pollen hitting a receptive stigma) in three dimensions is significantly influenced by the geometry of the pollen grain or pollen raft.

Our previous numerical experiments on two-dimensional random search required massive amounts of CPU time, as do Markov simulations in general. Since the predicted CPU time needed for the three-dimensional numerical experiments we planned far exceeded the capabilities of normal mainframe computing facilities, we requested and received significant time on an IBM 3090 supercomputer at the IBM Palo Alto Scientific Computing Center.

Since our study was designed to examine the efficiency of different pollen dispersal strategies in plants with underwater pollination mechanisms, we concurrently initiated field observations on four species of plants which we believed to have underwater pollination mechanisms. The theoretical predictions drawn from the numerical experiments were then compared to the observed dispersal strategies of underwater-pollinated plants.

Materials and methods

1. *Numerical experiments*

All of the numerical experiments were written in Fortran and run on an IBM 3090 supercomputer at the IBM Scientific Computing Center in Palo Alto, California. We were assisted by the IBM staff in code optimization to make use of the parallel processing and vectorization capabilities of the 3090; such capabilities were important since we sought to execute three-dimensional random searches in an unbounded domain.

In the numerical experiments we sought to determine if the shape of the pollen grain or pollen raft (hereafter called 'search vehicle') affected the likelihood of it encountering a point target representing the stigma. For ease of computation, we decided to fix search vehicles of different shape but constant volume at the origin in three-space and allow them to encounter a stigma with a random trajectory; this is equivalent to allowing a fixed stigma to be hit by search vehicles of different shapes but equal volumes and trajectories. If search vehicle shape has no effect on target encounter rates, we would expect equal numbers of collisions between the stigma and all search vehicles regardless of their shape. If search vehicle shape, however, affects encounter rates, search vehicles of different shapes should encounter different numbers of stigmas.

In the numerical experiments, we began with an unlimited domain and a set of 23 different pollen search vehicles, represented by spheroids of volume $8\pi/3$, centred on the origin. The spheroids had eccentricity $\alpha = 32, 30, 28, 26, \ldots 6, 4, 2, \sqrt[3]{2}, 1, 1/2, 1/8, 1/16$, and $1/32$. The spheroids were given by the equation

$$X^2/\alpha^2 + Y^2/\beta^2 + Z^2/\beta^2 = 1$$

where

$$\beta^2 = 2/\alpha.$$

Such spheroids are perfect spheres where $\alpha = \sqrt[3]{2}$. When the α values are large, the spheroids can be visualized as having long, string-like shapes, while α values below 1 result in shapes resembling Frisbees.

Within this domain, a point stigma was randomly distributed in a uniform distribution over the surface of a sphere with radius 75 units. The stigma was then moved in one unit steps in a direction chosen randomly within the spherical domain. At the end of the step, each stigma was checked to see which, if any, of the spheroids had collided with it. If the stigma encountered a spheroid, the number of steps required for the hit (pollination) was recorded. Since seagrass stigmas have a limited period of receptivity, in our numerical experiments each stigma was given a lifetime of 1 million moves, after which it was deemed to be unreceptive to pollen, and another trial was initiated. The experiment was run 10 000 times.

2. Pollination observations

The underwater pollination mechanisms of four different hydrophilous species, two marine and two freshwater, were studied. The pollination ecology of the seagrass *Thalassia testudinum* (Hydrocharitaceae) was studied, using snorkel gear, at Salt River Bay, St. Croix, in the US Virgin islands. The pollination ecology of the seagrass *Phyllospadix scouleri* (Zosteraceae) was studied at Monterrey Bay, California. The freshwater aquatic *Lepilaena bilocularis* (Zannichelliaceae) was studied at Lake Ellismere, New Zealand, while *Zannichellia palustris* (Zannichelliaceae) was studied in a canal in Provo, Utah. The shapes and mode of dispersal for the pollen search vehicles were observed and recorded for each of these species.

Results

1. Numerical experiments

The number of encounters between pollen search vehicles (spheroids) and stigmas (point targets) per 10 000 trials are indicated in Table 20.1, as are also the mean number of moves (= search path length) for trials that resulted in successful pollination events. The corresponding success rates are shown for spheroids of differing elliptical eccentricity, with perfect spheres having $\alpha = \beta = \sqrt[3]{2}$.

The ratio α/β is plotted against the average number of hits in Fig. 20.1. As can be seen, perfect spheres have the least number of mean hits, while pollen search vehicles with large elliptical eccentricity have a far greater likelihood of target encounter. This suggests that pollen grains or pollen rafts for underwater-pollinated plants should not be perfect spheres, but should have a high degree of elliptical eccentricity.

2. Pollination observations

In St. Croix, the male flowers of the dioecious seagrass *Thalassia testudinum* (Hydrocharitaceae) were found to open under water at night and to disperse spherical pollen grains embedded in long strands of mucilage. Each strand had many pollen grains embedded in it, with the

Table 20.1. Numerical experiments of Target encounter for spheroids of differing elliptical eccentricity in a three-dimensional random search

β	α	No. of encounters per 10 000 trials	Mean no. of moves per successful encounter
8	1/32	245	98243
$4\sqrt{2}$	1/16	216	92213
4	1/8	182	93664
$2\sqrt{2}$	1/4	163	96555
2	1/2	144	98799
$\sqrt{2}$	1	129	124687
$\sqrt[3]{2}$	$\sqrt[3]{2}$	126	115251
1	2	132	109872
$1/\sqrt{2}$	4	163	105017
$1/\sqrt{3}$	6	184	109116
1/2	8	199	118833
$1/\sqrt{5}$	10	220	101992
$1/\sqrt{6}$	12	235	109231
$1/\sqrt{7}$	14	244	110276
$1/2\sqrt{2}$	16	263	107151
1/3	18	271	100746
$1/\sqrt{10}$	20	280	100865
$1/\sqrt{11}$	22	285	96415
$1/2\sqrt{3}$	24	285	93691
$1/\sqrt{13}$	26	295	93656
$1/\sqrt{14}$	28	300	95062
$1/\sqrt{15}$	30	302	90912
1/4	32	307	93225

entire search vehicle resembling pearls on a string. The mucilage strands were slightly negatively buoyant, and after release were buffeted about by orbital wave motion in the open sea. Pollination occurs by collision of the mucilage strands with the densely papillate stigmas. Complete details of this study appear elsewhere (Cox and Tomlinson 1988).

The pollination ecology of the dioecious seagrass *Phyllospadix scouleri* (Zosteraceae) at Monterrey Bay, California proved to be more complex. The species was found to have both two- and three-dimensional pollen dispersal strategies, as previously suggested by Dudley nearly a century ago (1893, 1894). In two-dimensional strategy pollen dispersal, the long filiform pollen grains coalesce, forming floating search vehicles that resemble large snowflakes. However, in three-dimensional

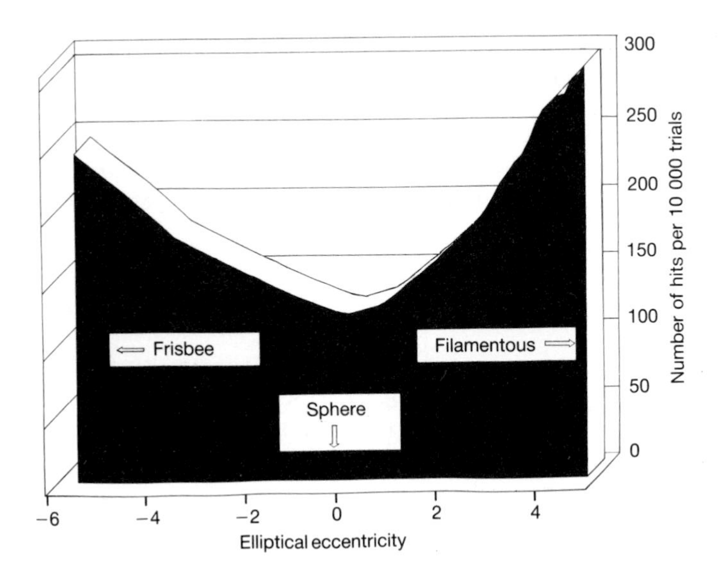

Fig. 20.1 Plot of IBM 3090 supercomputer results for a three-dimensional random search. The vertical axis represents the number of pollinations per 10 000 trials; the horizontal axis is log (α/β). Note that perfect spheres have the lowest success rate, while spheroids with high elliptical eccentricity, resembling Frisbees or filamentous strings, respectively, have high success rates. This suggests that underwater pollen should be dispersed in filamentous strings; see text for explanation.

pollination dispersal the pollen is dispersed under water and transported as individual filiform strands or as long groups of several strands. Submarine pollination occurs through collision of these mucilage strands with the filamentous stigmas. Complete details of the pollination ecology of this species also appear elsewhere (Cox *et al.* 1991*a*).

The pollen of the freshwater aquatic *Lepilaena bilocularis* (Zannichelliaceae), like that of *Thalassia testudinum*, is spherical. The male anthers containing the pollen grains abscise underwater and float to the surface. There they burst, releasing the pollen grains. Each pollen grain slowly sinks, trailing a long thread of mucilage from the anther. The pollen grain and mucilage thread are buffeted by the water currents and collide with the cup-like stigma under water.

The freshwater aquatic *Zannichellia palustris* (Zannichelliaceae) is monoecious, with male and female flowers borne at the same node. The spherical pollen is released in mucilaginous masses, which float if they encounter the water surface. As the mucilage quickly dissolves, however, the pollen grains are released and sink, trailing slender threads of mucilage. Similar release of pollen grains in a mucilaginous matrix has been described for *Z. palustris* plants in the southeastern United States (Haynes and Holm-Nielsen 1987; Haynes 1988). Pollination is by underwater collision of a pollen grain with the cup-like stigma. Given the close proximity of the male and female flowers produced by the same plant, self-pollination is likely, although some cross-pollinations do occur. Further details can be found in You-Hao *et al.* 1990).

Discussion

In three dimensions, increasing the elliptical eccentricity of spheroids of equal volume can result in two different types of shape. One series of deformations results in a long string-like morphology, while the alternative series of deformations results in a flat Frisbee-like shape. In the numerical experiments, spheroids with a high degree of elliptical eccentricity, i.e. those resembling either strings or Frisbees, were found to be far more likely to encounter point targets in three-dimensional searches, and to have far shorter searches in successful encounters. (Table 20.1). Either shape of pollen conveyance, therefore, would encounter far more stigmas in three dimensions than would spherical pollen conveyances.

Yet the production of rigid structures capable of holding together a large Frisbee-shaped mass of pollen and the evolution developmental processes necessary to assemble such structures from male flowers seems difficult in aquatic plants. However, production of filiform, string-like pollen or pollen agglutinations seems easier in a biological sense. Based on the numerical experiments, we therefore predicted that underwater pollen should be dispersed in long, string-like structures.

This prediction compares favourably with the observed pollen dispersal strategies of the four species of underwater-pollinated plants. The pollination biologies of these plants are remarkably similar, in that all have either filamentous pollen or have the pollen transported in filamentous strands of mucilage (Table 20.2). However, in *Zannichellia palustris* the oval pollen grains are sometimes transported singly, particularly after the mucilage dissolves. Such dispersal does not match the prediction made from the numerical experiments. Yet of the four species studies, only *Z. palustris* is monoecious, having male and female flowers produced on the same plant in proximity to each other. This apparent counter-example may not disprove the general conclusion that underwater plants should disperse their pollen in filamentous conveyances since *Z. palustris* is largely self-pollinated, particularly in still water, since anther dehiscence at some nodes occurs after filament elongation above the open stigma (You-Hao *et al.* 1990).

It appears, therefore, that there has been convergent evolution in

Table 20.2. Predicted and observed pollen search vehicle morphologies for underwater-pollinated plants

Plant	Habitat	Breeding system	Search vehicle	Morphology	Supercomputer prediction
Thalassia testudinum	Marine	Dioecious	Mucilage strand	Filamentous	Filamentous
Phyllospadix scouleri	Marine	Dioecious	Pollen/pollen clump	Filamentous	Filamentous
Lepilaena bilocularis	Freshwater	Dioecious	Pollen with mucilage	Filamentous	Filamentous
Zannichellia palustris	Freshwater	Monoecious	Pollen/mucilage	Oval/filamentous	Filamentous

underwater plants towards pollen dispersal strategies. Both the numerical experiments run on the supercomputer and the actual observations of underwater pollination seem to indicate that the most effective strategy for three-dimensional random search is to disperse pollen in long filamentous search vehicles. The precise nature of this convergence merits further investigation, however. For example, a recent phylogenetic analysis (Cox and Humphries 1991) indicates that filiform pollen is a synapomorphy uniting the clade composed of Cymodoceaceae, Posidoniaceae, and Zosteraceae. Hence the occurrence of filamentous pollen in different submarine-pollinated species of these families can in no sense be considered as convergent since their shared ancestor had filiform pollen. Yet it appears that this ancestor may have developed filiform pollen in association with submarine pollination, possibly through selection for search efficiency as indicated by our supercomputer analysis. However, this phylogenetic analysis strengthens the belief that dispersal of pollen in filamentous threads in unrelated underwater-pollinated species such as *Thalassia testudinum* and *Lepilaena bilocularis* is an example of evolutionary convergence. Although such convergent evolution is strongly indicated by both the numerical experiments and pollination observations, we hope to supplement this prediction with an analytical proof in the future.

Acknowledgements

We thank the IBM corporation for an Academic Computing Grant which allowed us supercomputer time and field expenses. We also thank the staff of the IBM Scientific Computing Center in Palo Alto for assistance with code optimization, vectorization, and machine orientation. Partial support was provided by National Science Foundation Grant Presidential Young Investigator Award BSR-8452090.

References

Ackerman, J. D. (1986). Mechanistic implications for pollination in the marine angiosperm *Zostera marina*. *Aquatic Botany* **24**, 343–53.

Cook, C. D. K. (1988). Wind pollination in aquatic angiosperms. *Annals of the Missouri Botanical Garden* **75**, 768–77.

Cox, P. A. (1983). Search theory, random motion, and the convergent evolution of pollen and spore morphologies in aquatic plants. *American Naturalist* **121**, 9–31.

Cox, P. A. (1988). Hydrophilous pollination. *Annual Reviews of Ecology and Systematics* **19**, 261–80.

Cox, P.A. (1991). Hydrophilous pollination of a dioecious seagrass, *Thalassodendron ciliatum* (Cymodoceaecae), in Kenya. *Biotropica*, **23**, 159–65.

Cox, P.A. and Humphries, C.J. (1991). Hydrophilous pollination and breeding system evolution in seagrasses: a phylogenetic approach to the evolutionary ecology of the Cymodoceaceae. Unpublished manuscript.

Cox, P.A. and Knox, R.B. (1988). Pollination postulates and two dimensional pollination in hydrophilous monocotyledons. Annals of the Missouri Botanical Garden **75**, 811–18.

Cox, P.A. and Knox, R.B. (1989). Two-dimensional pollination in hydrophilous plants: convergent evolution in the genera *Halodule* (Cymodoceaceae), *Halophila* (Hydrocharitaceae), *Ruppia* (Ruppiaceae), and *Lepilaena* (Zannichelliaceae). *American Journal of Botany* **76**, 164–75.

Cox, P.A. and Tomlinson, P.B. (1988). The pollination ecology of a Caribbean seagrass, *Thalassia testudinum* (Hydrocharitaceae). *American Journal of Botany* **75**, 958–65.

Cox, P.A. Tomlinson, P.B., and Nienanski, K. (1991*a*). Submarine and surface pollination in *Phyllospadix scouleri* (Zosteraceae), *Plant Evolution and Systematics*, in press.

Cox, P.A. Laushman, R., and Rucklehaus, M. (1991*b*). Surface and submarine pollination in the seagrass *Zostera marina L.*, *Botanical Journal of the Linnean Society*, in press.

Ducker, S.C. and Knox, R.B. (1976). Submarine pollination in seagrasses. *Nature* **263**, 705–6.

Ducker, S.C., Pettitt, J.M., and Knox, R.B. (1978). Biology of Australian seagrasses: pollen development and submarine pollination in *Amphibolis antarctica* and *Thalassodendron ciliatum* (Cymodoceaceae). *Australian Journal of Botany*, **26**, 265–85.

Dudley, W.R. (1893). The genus *Phyllospadix*. In *The Wilder Quarter-Century Book*, pp. 403–18. Comstock, Ithaca, New York.

Dudley, W.R. (1894). *Phyllospadix*, its systematic characters and distribution. *Zoe* **4**, 381–5.

Ernst-Schwarzenbach, M. (1944). Zur Blütenbiologie einiger Hydrocharitaceen. *Berichte Schweiz Botanical Geschichtliche* **54**, 33–69.

Guo, Y.H., Sperry, R., Cook C.D.K., and Cox, P.A. (1991). The pollination ecology of *Zannichellia palustris* (Zannichelliaceae). *Aquatic Botany* **38**, 341–56.

Haynes, R.R. (1988). Reproductive biology of selected aquatic plants. *Annals of the Missouri Botanical Garden* **75**, 805–10.

Haynes, R.R. and Holm-Nielsen, L.B. (1987). The Zannichelliaceae in the Southeastern United States. *Journal of the Arnold Arboretum* **68**, 259–68.

Hersch, R. and Griego, R.J. (1969). Brownian motion and potential theory. *Scientific American* **220**, 66–74.

Koopman, B.O. (1956). The theory of search. II: target detection. *Operations Research* **4**, 503–31.

Koopman, B.O. (1980). *Search and Screening*. Pergamon Press, New York.

McConchie, C.A., Knox, R.B., and Ducker, S.C. (1983). The unique pollen of the seagrass *Posidonia australis*. In *Pollen: biology and implications for plant breeding*, (ed. D.L. Mulcahy and E. Ottaviano), p. 426. Elsevier Biomedical, New York.

Pettitt, J.M. (1981). Reproduction in seagrasses: Pollen development in *Thalassia* hemprichii, *Halophila stipulacea* and *Thalassodendron ciliatum*. *Annals of Botany*. **48**, 609–22.

Pettitt, J.M. (1984). Aspects of flowering and pollination in marine angiosperms. *Oceanography and Marine Biology an Annual Review* **22**, 315–42.

Pettitt, J.M. and Jermy, A.C. (1975). Pollen in hydrophilous angiosperms. *Micron* **5**, 377–405.

Pettitt, J.M., McConchie, C.A., Ducker, S.C., and Knox, R.B. (1980). Unique adaptations for submarine pollination in seagrasses. *Nature* **286**, 487–9.

Pettitt, J.M., McConchie, C.A., Ducker, S.C., and Knox, R.B. (1981). Submarine pollination. *Scientific American* **244**, 134–43.

Pettitt, J.M., McConchie, C.A., Ducker, S.C., and Knox, R.B. (1983). Reproduction in seagrasses: pollination in *Amphibolis antarctica*. *Proceedings of the Royal Society of London* **B219**, 119–35.

Troll, W. (1931). Botanische Mitteilungen aus den Tropen II. Zur Morphologie und Biologie von *Enhalus acoroides* (Linn.f.) Rich. *Flora, Jena* **125**, 427–56.

Yamashita, V.T. (1976). Über die Pollenbildung bei *Halodule pinifolia* und *Halodule uninervis*. *Beitrage Zur Biologie Pflanzen*.

Index

Systematics Association Publications

Published by the Association (out of print)

1. Bibliography of key works for the identification of the British fauna and flora *3rd edition* (1967)
   Edited by G. J. Kerrich, R. D. Meikle and N. Tebble
2. Function and taxonomic importance (1959)
   Edited by A. J. Cain
3. The species concept in palaeontology (1956)
   Edited by P. C. Sylvester-Bradley
4. Taxonomy and geography (1962)
   Edited by D. Nichols
5. Speciation in the sea (1963)
   Edited by J. P. Harding and N. Tebble
6. Phenetic and phylogenetic classification (1964)
   Edited by V. H. Heywood and J. McNeill
7. Aspects of Tethyan biogeography (1967)
   Edited by C. G. Adams and D. V. Ager
8. The soil ecosystem (1969)
   Edited by H. Sheals
9. Organisms and continents through time (1973)†
   Edited by N. F. Hughes

Systematics Association Special Volumes

1. The new systematics (1940)
   Edited by Julian Huxley (Reprinted 1971)
2. Chematoxonomy and serotaxonomy (1968)*
   Edited by J. G. Hawkes
3. Data processing in biology and geology (1971)*
   Edited by J. L. Cutbill
   Out of print
4. Scanning electron microscopy (1971)*
   Edited by V. H. Heywood
   Out of print
5. Taxonomy and ecology (1973)*
   Edited by V. H. Heywood

6. The changing flora and fauna of Britain (1974)*
   Edited by D. L. Hawksworth
   Out of print
7. Biological identification with computers (1975)*
   Edited by R. J. Pankhurst
8. Lichenology: progress and problems (1976)*
   Edited by D. H. Brown, D. L. Hawksworth, and R. H. Bailey
9. Key works to the fauna and flora of the British Isles and northwestern Europe, *4th edition* (1978)*
   Edited by G. J. Kerrich, D. L. Hawksworth, and R. W. Sims
10. Modern approaches to the taxonomy of red and brown algae (1978)*
    Edited by D. E. G. Irvine and J. H. Price
11. Biology and systematics of colonial organisms (1979)*
    Edited by G. Larwood and B. R. Rosen
12. The origin of major invertebrate groups (1979)*
    Edited by M. R. House
13. Advances in bryozoology (1979)*
    Edited by G. P. Larwood and M. B. Abbot
14. Bryophyte systematics (1979)*
    Edited by G. C. S. Clarke and J. G. Duckett
15. The terrestrial environment and the origin of land vertebrates (1980)*
    Edited by A. L. Panchen
16. Chemosystematics: principles and practice (1980)*
    Edited by F. A. Bisby, J. G. Vaughan, and C. A. Wright
17. The shore environment: methods and ecosystems (2 Volumes) (1980)*
    Edited by J. H. Price, D. E. G. Irvine and W. F. Farnham
18. The Ammonoidea (1981)*
    Edited by M. R. House and J. R. Senior
19. Biosystematics of social insects (1981)*
    Edited by P. E. House and J.-L. Clément
20. Genome evolution (1982)*
    Edited by G. A. Dover and R. B. Flavell
21. Problems of phylogenetic reconstruction (1982)*
    Edited by K. A. Joysey and A. E. Friday
22. Concepts in nematode systematics (1983)*
    Edited by A. R. Stone, H. M. Platt, and L. F. Khalil
23. Evolution, time and space: the emergence of the biosphere (1983)*
    Edited by R. W. Sims, J. H. Price, and P. E. S. Whalley
24. Protein polymorphism: adaptive and taxonomic significance (1983)*
    Edited by G. S. Oxford and D. Rollinson
25. Current concepts in plant taxonomy (1983)*
    Edited by V. H. Heywood and D. M. Moore
26. Databases in systematics (1984)*
    Edited by R. Allkin and F. A. Bisby
27. Systematics of the green algae (1984)*
    Edited by D. E. G. Irvine and D. M. John
28. The origins and relationships of lower invertebrates (1985)‡
    Edited by S. Conway Morris, J. D. George, R. Gibson, and H. M. Platt

* Published by Academic Press for the Systematics Association
† Published by the Palaeontological Association in conjunction with the Systematics Association
‡ Published by the Oxford University Press for the Systematics Association.